International Journal on Advances in Life Sciences

2016 vol. 8 nr. 3&4

Printed from e-media with permission by:

Curran Associates, Inc.
57 Morehouse Lane
Red Hook, NY 12571

Some format issues inherent in the e-media version may also appear in this print version.

Printed by Curran Associates, Inc. (2018)

International Academy, Research, and Industry Association (IARIA)
412 Derby Way
Wilmington, DE 19810

Phone: (408) 893-6407
Fax: (408) 527-6351

petre@iaria.org

Additional copies of this publication are available from:

Curran Associates, Inc.
57 Morehouse Lane
Red Hook, NY 12571 USA
Phone: 845-758-0400
Fax: 845-758-2633
Email: curran@proceedings.com
Web: www.proceedings.com

The *International Journal on Advances in Life Sciences* is published by IARIA.
ISSN: 1942-2660
journals site: http://www.iariajournals.org
contact: petre@iaria.org

Sponsored by IARIA
www.iaria.org

International Journal on Advances in Life Sciences
Volume 8, Number 3 & 4, 2016

International Journal on Advances in Life Sciences
Volume 8, Numbers 3 & 4, 2016

CONTENTS

A Virtual Presence System Design with Indoor Navigation Capabilities for Patients with Locked-In Syndrome

Jens Garstka*, Simone Eidam†, and Gabriele Peters*

Human-Computer Interaction
Faculty of Mathematics and Computer Science
FernUniversität in Hagen – University of Hagen
D-58084 Hagen, Germany

Email: {jens.garstka, gabriele.peters}@fernuni-hagen.de*, eidam.simone@gmail.com†

***Abstract*—In this article, we present a prototype of a virtual presence system combined with an eye-tracking based communication interface and an indoor navigation component to support patients with locked-in syndrome. The common locked-in syndrome is a state of paralysis of all four limbs while a patient retains full consciousness. Furthermore, also the vocal tract and the respiration system are paralyzed. Thus, the virtually only possibility to communicate consists in the utilization of eye movements for system control. Our prototype allows the patient to control movements of the virtual presence system by eye gestures while observing a live view of the scene that is displayed on a screen via an on-board camera. The system comprises an object classification module to provide the patient with different interaction and communication options depending on the object he or she has chosen via an eye gesture. In addition, our system has an indoor navigation component, which can be used to prevent the patient from navigating the virtual presence systems to critical areas and to allow for an autonomous return to the base station using the shortest path. The proposed prototype may open up new possibilities for locked-in syndrome patients to regain a little more mobility and interaction capabilities within their familiar environment.**

***Index Terms*—biomedical communication; human computer interaction; eye tracking; indoor navigation; virtual presence.**

I. INTRODUCTION

This article describes an extension of the previous work of Eidam et al. [1]. Undoubtedly, it is a major challenge for locked-in syndrome (LIS) patients to communicate with their environment and to express their needs. Patients with LIS have, for example, to face severe limitations in their daily life. LIS is mostly the result of a stroke of the ventral pons in the brain stem [2]. The incurred impairments of the pons cause paralysis, but the person keeps his or her clear consciousness. The grade of paralysis determines the type of LIS and has been classified in classic, total and incomplete LIS. Incomplete LIS means that some parts of the body are motile. Total LIS patients are like classic LIS patients completely paralyzed. However, the latter ones still can perform eyelid movements and vertical eye movements that can be used for communication. Therefore, several communication systems for classic LIS patients have been designed in the past.

This article introduces an eye-gesture based communication interface for controlling movements of a virtual presence system (VPS) and for selecting objects of the environment with the aim to interact with them.

In the presented prototype, the patients will see exemplary scenes of the local environment instead of the typically used on-screen keyboard. These scenes contain everyday objects, e.g., a book the impaired person wants to get read, which can be selected using a special eye gesture. After selection, the patient can choose one of various actions, e.g., "I want to get read a book" or "please, turn the page over". A selection can either lead to a direct action (light on/off) or to a notification of a caregiver via text-to-speech. Moreover, the prototype allows the LIS patient to control a VPS. For this purpose, different eye gestures controlling the VPS are presented and discussed in this article. We also show an effective but cheap implementation of an indoor navigation component to enable the VPS to maneuver itself back to the base station taking the shortest way possible.

In a long-term perspective, the aim is to build a system where the object selection screen mentioned above shows the live view of the environment captured by the on-board camera of the VPS. This requires the implementation of an object classification approach for the most common objects. Each of the recognizable object classes provides an adjustable set of particular interactions/instructions. By this means, a VPS enables the LIS patient to interact with an environment in a very direct way.

The article is organized as follows: Section II gives a short introduction to eye tracking and describes different existing communication systems for LIS patients using eye tracking approaches. Furthermore, the section provides a brief overview of indoor navigation approaches. In Section III the concept and implementation details of our object-based interaction are presented. In the subsequent Section IV we introduce the models of the eye tracking interface controlling the VPS and the indoor navigation used to enable the VPS to autonomously move to the base station on the shortest path. Finally, the evaluation results will be presented in Section V and discussed in Section VI. The article concludes with a description of future work in Section VII.

II. RELATED WORK

This section starts with a brief overview on eye tracking techniques and already existing systems that support LIS patients with their communication. Finally, a short sub-section gives an overview on methods for indoor navigation with focus on impaired persons.

A. Eye Tracking

Many existing eye tracking systems use the one or other kind of light reflection on eyes to determine the direction of view. The human eye reflects incident light at several layers. The eye tracking device used for controlling the prototype employs the so-called method of dark-pupil tracking. Dark-pupil-tracking belongs to the video-based eye tracking methods. Further examples are bright-pupil- and dual-Purkinje-tracking [3].

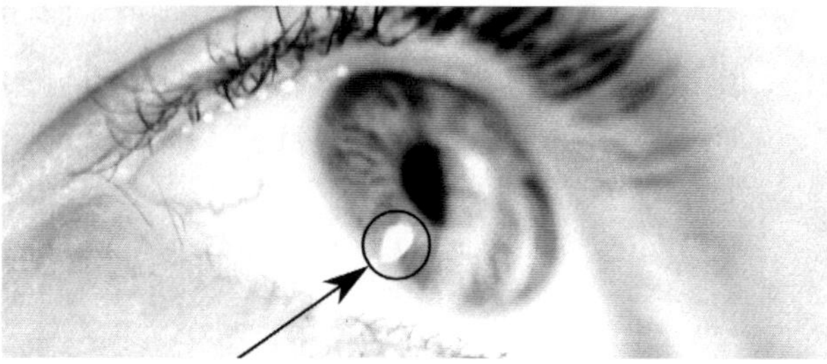

Fig. 1. The light reflection used by many eye trackers is called the glint.

For video-based systems, a light source (typically infrared light) is set up in a given angle to the eye. The pupils are tracked with a camera and the recorded positions of pupil and reflections are analyzed. Based on the pupil and reflection information, the point of regard (POR) can be calculated [3]. In Figure 1, the white spot just below the pupil shows a reflection of an infrared light on the cornea. This reflection is called the glint. In case of dark-pupil tracking, it is important to detect both, the pupil center and the glint. The position of the pupil center provides the main information about the eye gaze direction while the glint position is used as reference. Since every person has individually shaped pupils, a onetime calibration is needed. In case of a stationary eye tracker, also the distance between the eyes is determined to calculate the position of the head relative to the eye tracker.

B. Communication Systems for LIS Patients

There are many prototypes that have been developed in order to support LIS patients with their communication. Many of them are video-based eye tracking systems. One of the first systems was the communication project ERICA developed in 1989 [4]. With the help of the system users were enabled to control menus with eyes. They were able to play computer games, to hear digitized music, to use educational programs and to use a small library of books and other texts. Additionally, ERICA offered the possibility to synthesize speech and control nearby devices. Currently available and commercial communication systems for LIS patients are basically based on ERICA. These systems include the Eyegaze Edge Talker from LC Technologies and the Tobii Dynavox PCEye Go Series. The Tobii solution provides another interaction possibility called "Gaze Selection" in addition to an eye controlled mouse emulation. It allows a two stage selection, whereas starring at the task bar on the right side of the screen enables a selection of mouse options like right/left button click or the icon to display a keyboard. Subsequently, starring on a regular GUI-element triggers the final event (such as "open document"). Two-stage means that the gaze on the target task triggers a zoom-in event. It is said, that this interaction solution is more accurate, faster and reduces unwanted clicks in comparison to a single stage interaction.

Furthermore, current studies present alternative eye based communication systems for LIS patients. For example, the prototype developed by Arai and Mardiyanto, which controls the application surface using an eye gaze controlled mouse cursor with the eyelids to trigger the respective events [5]. This prototype offers the possibility to phone, to visit websites, to read e-books, or to watch TV. An infrared sensor/emitter-based eye tracking prototype was developed from Liu et al., which represents a low-cost alternative to the usual expensive video-based systems [6]. With this eye tracking principle, only up/down/right/left eye gaze moves can be detected as well as staying in the center using the eyelids to trigger an event. By using the eye movement, the user can move a cursor in a 3×3 grid from field to field. And by using the eyelids, the user can finally select the target field. Barea, Boquete, Mazo, and Lpez developed another prototype that is based on electrooculography [7]. This prototype allows by means of eye movements to control a wheelchair allowing an LIS patient to freely move through the room.

All prototypes that have been discussed so far are based on an interaction with static contents on screen, for example a virtual keyboard. However, the prototype presented in this contribution shows a way to select objects in images of typical household scenes by a simulated object classification. This allows an evaluation of the system without the need of a full classification engine. The latter will lead to a selection of real objects in the patient's proximity.

C. Indoor Navigation

The use of GPS for indoor navigation is often not possible as ceilings and walls almost completely absorb the weak GPS signal. However, there are numerous alternatives including ultrasonic, infrared, magnetic, and radio sensors. Unfortunately, in many cases the position is not determined by the mobile device. Instead, it is determined from the outside. This requires a permanent electronic infrastructure, which often can not be retrofitted without major effort.

The following publications provide a brief overview of indoor navigation solutions. The survey by Mautz and Tilch [8] contains a good overview of optical indoor positioning systems. Nuaimi and Kamel [9] explore various indoor positioning systems and evaluate some of the proposed solutions. Moreover, Karimi [10] provides a wide overview of general approaches to indoor navigation in his book.

Considering that QR codes are used for positioning in our approach, an overview of recent publications focusing on QR codes follows.

The indoor navigation described by Mulloni et al. is an inexpensive, building-wide orientation guide that relies solely on mobile phones with cameras [11]. The approach uses bar codes, such as QR codes, to determine the current position with a mobile phone. This method was primarily used at conferences. Information boards containing appropriate QR codes were used to determine the current location of visitors.

The work of Li et al. is focused on robot navigation and the question of how QR codes can be identified and read even under bad lightning conditions [12]. For this purpose, they combine and optimize various image filters for the mentioned use case.

Gionata et al. use a combination of an IMU (rotational and translational sensors) and QR codes for an automated indoor navigation of wheelchairs [13]. The QR codes are used as initial landmarks and to correct the estimated position of a wheelchair after driving a certain period. The movement of the wheelchair between two QR codes is approximated with an IMU.

A somewhat different intended use of the QR codes is shown in the paper of Lee et al. [14]. They use QR codes to transfer navigational instructions to a mobile robot along a predefined route. These instructions hint the robot where it needs to turn, for example, left or right.

Zhang et al. use QR codes as landmarks to provide global pose references [15]. The QR codes are placed on the ceilings and contain navigational information. The pose of the robot is estimated according to the positional relationship between QR codes and the robot.

In brief, it has been found that that QR codes or similar markers represent an effective and proven means for indoor navigation. In context of our work presented in this article, we will combine the work of Zhang et al. with a simple floor plan [15]. This will be discussed more in detail in Section IV.

III. INTERACTION

This section describes the concepts and the implementation of our interface for object-based interaction and communication using an eye tracker.

A. Concept

The following section provides an overview of the basic concept of this work. As already mentioned, the impaired person will see an image of a scene with typical everyday objects. This image is representative for a real scene, which is to be captured by a camera and analyzed by an object classification framework in future work. Figure 2 shows an image of one possible scene. The plant can be used by a LIS patient to let a caregiver know, that one would like to be in the garden or park, the TV can be used to express the desire to watch TV, while the remote control directly relates to the function of the room light. The red circle shown at the center of the TV illustrates the POR calculated by the eye tracker. The visual feedback by the circle can be activated or deactivated, depending on individual preferences.

Fig. 2. An example scene used with this prototype.

An object is selected by starring a predetermined time on the object, what we call a “fixation”. With a successful fixation a set of options will be displayed on the screen. A closing of the eyelids is used to choose one of these options. Depending on the selected object, a direct action (e.g., light on/off) or an audio synthesis of a corresponding text is triggered (e.g., “I want you to read me a book.”).

Furthermore, other eye gestures have been implemented to control the prototype. By means of a horizontal eye movement, the object image is changed. However, the latter is only an aid during the test phase without an implementation of a real object classification to avoid the use of a keyboard or mouse. By means of a vertical eye movement, the object-based interaction and communication mode is switched to the robot controlling mode and vice versa.

B. Implementation

The eye tracking hardware used is a stationary unit with the name RED manufactured by SensoMotoric Instruments (SMI). RED comes with an eye tracking workstation (a notebook) running a software, which is named iView X. The latter provides a network component to allow an easy communication between the hardware and any software through a well-defined network protocol.

Figure 3 gives a brief overview of all components of our prototype. Area 1 shows the patient’s components to display test scenes with different objects. The stationary eye tracking unit is shown in area 2. Area 3 shows the eye tracking

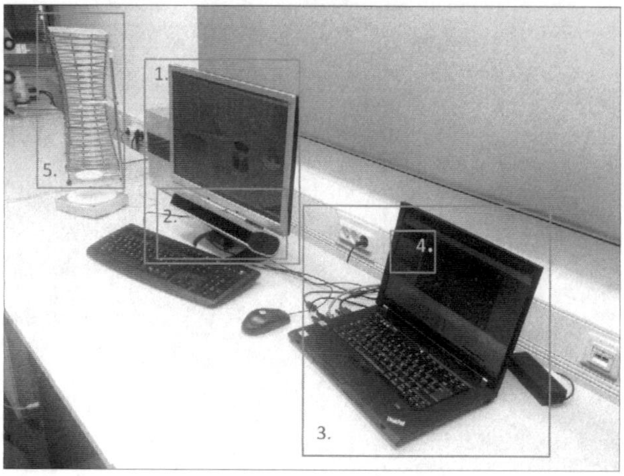

Fig. 3. The eye tracking components of the prototype.

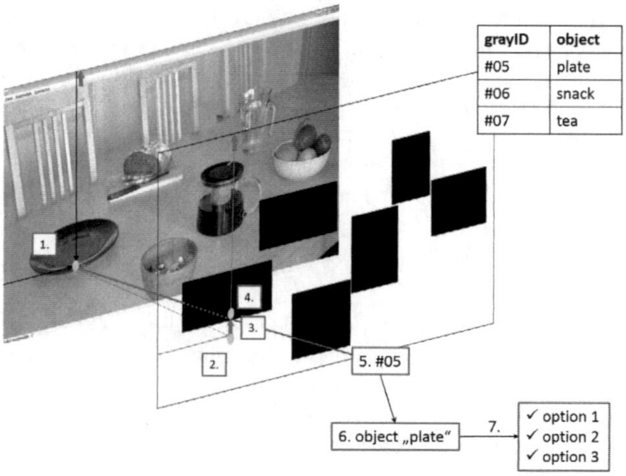

Fig. 4. Elements used to simulate the object classification.

workstation with the eye tracking control software in area 4. Finally, area 5 contains a desk lamp, which can be turned on and off directly with a fixation of the remote control shown in Figure 2.

C. Eye Gesture Recognition

Eye gesture recognition is based on the following principle: the received POR-coordinates from the eye tracker are stored in circular buffer. At each coordinate insertion the buffer is analyzed for eye gestures. These eye gestures are a fixation, a closing of the eyelids, and a horizontal/vertical eye movement. The following values can be used to detect these eye gestures:

- the maximum x- and y-value: $x_{\max}$, $y_{\max}$
- the minimum x- and y-value: $x_{\min}$, $y_{\min}$
- the number of subsequent zero values: c

The detection of the fixation is performed as follows:

$$|x_{\max} - x_{\min}| + |y_{\max} - y_{\min}| \leq d_{\max}, \tag{1}$$

where $d_{\max}$ is the maximum dispersion while the eye movements are still recognized as fixation. The value of $d_{\max}$ is individually adjustable.

The detection of a closing of the eyelids is realized by counting the amount c of subsequent coordinate pairs with zero values for x and y. Zeros are transmitted by the eye tracker, when the eyes couldn't be recognized. This occurs on the one hand when the eyelids are closed, but on the other hand when the user turns the head or disappears from the field of view of the eye tracker. Therefore, this event should only be detected if the number of zeros corresponds to a given time interval:

$$(c > c_{\min}) \wedge (c < c_{\max}) \tag{2}$$

All variables $c_{\min}$ and $c_{\max}$ can be customized by the impaired person or the caregiver, respectively.

The combination of these two different approaches is a benefit, because object selection is realized through the fixation while option selection is done by closing the eyelids. The latter allows the LIS patient to rest the eyes while the option panel is open. Hence, the patient can calmly look over the offered options in order to get an overview.

For the horizontal eye gesture detection, a given range of x-values must be exceeded while the y-values remain in a small range, and vice versa for the vertical eye gesture. As already mentioned, the horizontal eye movement is used to switch between different images. But this functionality is not a part of a later system and is merely a simple additional operation to present a variety of objects while using this prototype. The vertical eye movement (vertical eye gesture) is used to switch between the object-based interaction and communication mode and the robot controlling mode.

D. Simulated Object Classification

Figure 4 shows schematically the principle of the simulated object classification. It is based on a gray-scale image that serves as a mask for the scene image. On this mask the available objects from the scene image are filled with a certain gray value. Thus, each object can be identified by a unique gray value ($grayID$). The rear plane illustrates the screen. The coordinates that correspond to a fixation of an object (1.) refer to the screen and not to a potentially smaller image. Thus, these raw coordinates require a correction by an offset (2. & 3.). The corrected values correspond to a pixel (4.) of the gray-scale image whose value (5.) may belong to one of the objects shown. In case of the example illustrated in Figure 4 this pixel has a gray value of 5 and corresponds to the object "plate" (6.). Finally, either all available options will be displayed (7.) or nothing will happen in the case the coordinates do not refer to a known object.

IV. NAVIGATION

Control and navigation of a VPS should primarily take place through eye gestures of an impaired person. But the system should autonomously return to the base station in times when the VPS is not in use. If the latter shall be achieved without boring random movements, as it can be observed frequently on robotic vacuum cleaners, the system must have knowledge of the local environment. For the tasks outlined in this article, QR codes are an effective means, mainly because they are very inexpensive and they are easy to install. However, the location by itself as described by Alessandro Mulloni et al. is not enough [11]. Even the approach of Zhang et al. putting some navigational information in the QR codes is not sufficient for some application scenarios [15].

Ideally, the robot knows a complete map of the local indoor environment.

A. Maps

One possibility to achieve a map of the local environment can be a commonly used method with the acronym SLAM ("Simultaneous Localization and Mapping"). Jeong and Lee describe a SLAM approach where they only use ceilings captured with a camera pointing upwards to create a map of the indoor environment [16]. Using this method, it is possible to identify QR codes that are placed on the ceiling (see Zhang et al. [15]) and put them into the map.

Alternatively, one can use a manually created floor plan. The latter would have the advantage that the floor plan is complete and can contain various extra information. These additional information may include:

- The exact position of the base station.
- The exact position and orientation of each QR code placed on the ceilings.
- The ceiling height.
  This is mainly of interest for a precise positioning or position correction of the robot based on the QR codes, which are placed on the ceiling. With the knowledge of the ceiling height, the opening angle of the camera, and the viewing direction upwards, the relative displacement of the robot with respect to the QR codes can easily be triangulated.
- Regions that should not be entered.
  Considering the fact that the target group of the approach presented in this article will have difficulties to control the VPS even with simplest eye gestures, it is useful to be able to mark certain regions that should be avoided. This could be, for example, a table with chairs where the robot can get stuck, or an area with sensitive objects like plants.

An exemplary floor plan is shown in Figure 5. It contains the positions and orientations of the QR codes. The QR codes themselves initially contain only an ID for the identification of each code. However, there is also the possibility to encode extra information in each QR code.

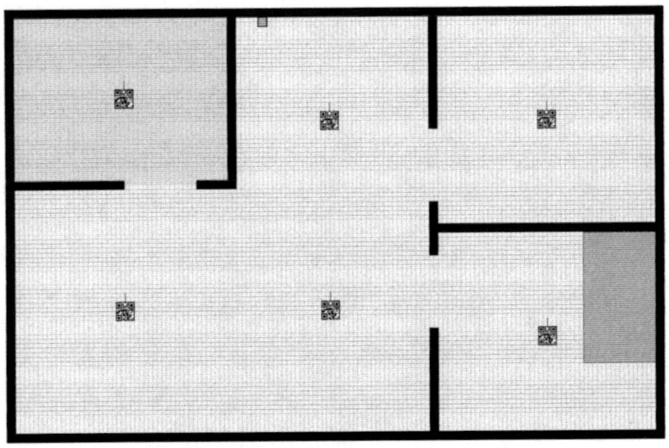

Fig. 5. An exemplary floor plan used for indoor navigation.

The floor plan can be implemented as a pixel image. In our case (see Figure 5) each pixel has an edge length of $5 \times 5\,cm$. The different yellow shades shown in Figure 5 indicate different ceiling heights. The area marked in red indicates a region that should be avoided.

B. Control

When controlling a robot with eye gestures several questions have to be answered:

- What eyes gestures can be used to activate or deactivate the control?
- What should happen if the eye tracker fails to detect the eyes?
- What eye gestures should be used to control the VPS?
- When should the robot return to the base station?
- Are there ways to define regions on the screen where the eyes can rest without triggering an eye gaze event?

To enable and disable the VPS control, we use an eye lid closure similar to Subsection III-C, i.e., the eye lid closure is within a given time interval $(c > c_{\min}) \wedge (c < c_{\max})$, where $c_{\min}$ and $c_{\max}$ can be customized. When switched off, an impaired person can switch between the object-based interaction and communication mode and the robot controlling mode by a vertical eye movement.

If the eye tracker fails to detect the eye gaze position for a period of $> c_{\max}$, it gets into a fail state. This results in an immediate stop of the VPS. To continue, a patient needs to reactivate the eye gaze control with a lid closure.

In general, a live view of the area in front of the VPS is always visible on the screen. This ensures that a patient can examine how and where the VPS is moving. Three different models of eye gestures to control the VPS are currently tested. The first model, shown in Figure 6, corresponds to the model of a joystick.

This means that an eye gaze pointing to the upper half of the screen accelerates the VPS in a forward motion. Pointing to the left and to the right causes a corresponding rotation. Since

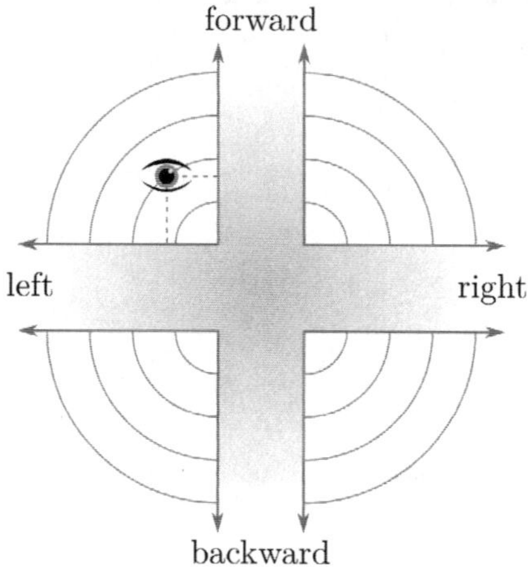

Fig. 6. Eye gaze control model I: the joystick mode.

an exact positioning of eye gazes can be very stressful, the area of a neutral position has been widened. This is visualized through the gray gradient shown in Figure 6.

There is also the possibility to drive backwards. However, according to the current prototype, the VPS has no rear camera. Thus, a reverse drive would be a blind drive. For this reason this ability has been removed in a second control model (see Figure 7).

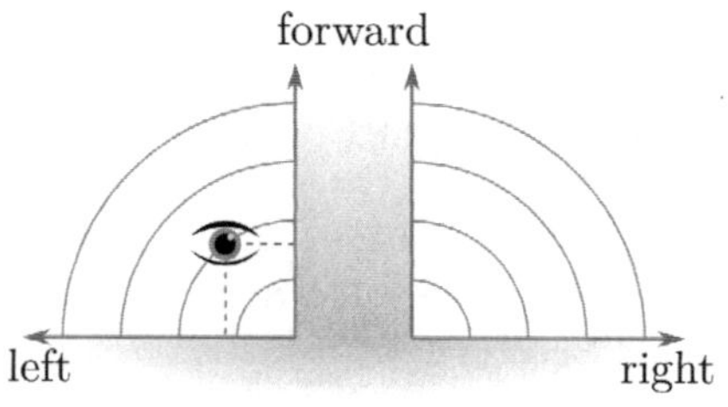

Fig. 7. Eye gaze control model II: the half joystick mode.

The latter model has another advantage: If the control sensitive area is located only on the upper half of the screen, the entire lower half of the screen can be used to rest the eyes.

The third model corresponds to a vertical slider. It can be used to do a turn-on-the-spot or to move straight forward by pointing to the upper or lower half of the screen. To switch between the two control states, we will use a fixation in a small area in the center of the screen (see Figure 8).

The horizontal region left and right of this central area (gray-shaded region in Figure 8) can be used to rest the eyes. Moreover, it will make no difference where the eye gaze position is located horizontally. Therefore, this model is suitable especially for the aforementioned LIS patients whose movements have been degraded to the extent that they are limited to vertical eye movements.

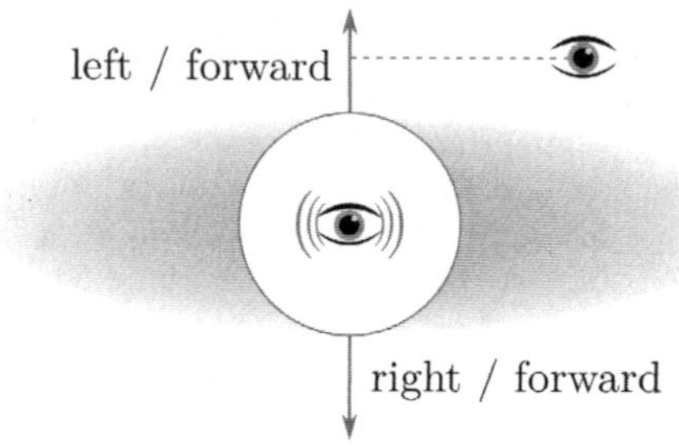

Fig. 8. Eye gaze control model III: slider mode.

C. Shortest Path to Base Station

An autonomous movement of a robot from a point A to a point B is a common and well solved problem in robotics. Path planning algorithms are measured by their computational complexity. The results depend on the accuracy of the map (floor plan), on the robot localization and on the number of obstacles.

If the underlying map is a raster map (e.g., a pixel image), one of the many variants of the A^* algorithm introduced by Hart et al. [17] is often used. Modern modifications and improvements like the work by Duchoň et al. [18] optimize the A^* algorithm for fast computation and optimal path planning in indoor environments.

In order to avoid contact with walls and doors and to pass the restricted areas in sufficient distance, the thickness of wall and blocked regions is enlarged by dilatation. Our robot has a radius of about $15\,cm$. In addition to the radius $10\,cm$ safety distance are used, to take account of inaccuracies in localization and movement of the robot. Accordingly, a dilation by $25\,cm$ or 5 pixels in the case of the presented map in Section IV-A is applied to the base map.

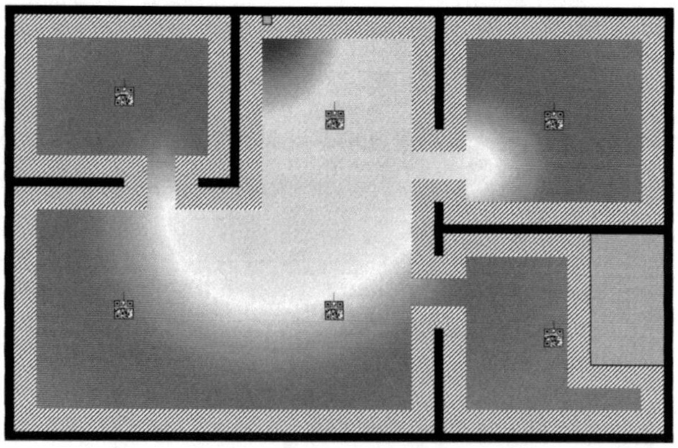

Fig. 9. Color gradient of shortest path to base station.

Figure 9 illustrates with a color gradient how a robot can find a direct path to the base station through gradient descent. The base station is depicted by the small green rectangle on the upper wall of the middle room. The colors indicate from purple

($> 8.5\,m$), over red ($\approx 6.8\,m$), yellow ($\approx 5.1\,m$), green ($\approx 3.4\,m$), and cyan ($\approx 1.7\,m$) to blue, the shortest path distance from an arbitrary point on the floor plan to the base station. The shaded areas show the $25\,cm$ wide safety distance along the walls.

D. Prototype

To build a prototype, an iRobot Roomba 620 vacuum cleaning robot is used as platform. It was extended by an access point and a USB to UART converter to send serial control command via network. In addition, two wireless cameras were mounted on top of the Roomba. One camera points forward, while the other camera points towards the ceiling. All devices get their electricity from the batteries of the Roomba. The prototype is shown in Figure 10.

Fig. 10. Prototype configuration based on a vacuum cleaning robot.

The Roomba has two separately controllable drive wheels. This enables the system to do a turn-on-the-spot and easily enables the implementation of the above-mentioned joystick mode.

Let x and y be the coordinates of the eye gaze position on the screen and c_x and c_y be the center coordinates of the screen. Further let s be a configurable speed factor. Then the speed values of the left and right wheel are:

$$v_l = ((x - x_c) + (y - y_c)) \cdot s \text{ and} \quad (3)$$
$$v_r = ((x - x_c) - (y - y_c)) \cdot s, \quad (4)$$

where v_l and v_r are the velocities of the left and right driving wheel. Figure 11 shows an exemplary view of the front camera.

Fig. 11. An exemplary view of the front camera.

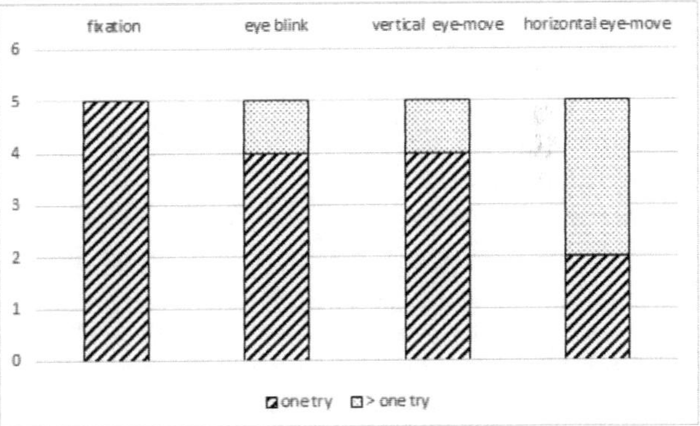

Fig. 12. Bar diagram of the eye gesture recognition.

V. RESULTS

The results can be divided into two parts. The first part deals with the object-based interaction, while the second part deals with the control of the robot.

A. Object-Based Interaction

The interface for object-based interaction has been tested by five persons to analyze its basic usability. Figure 12 briefly illustrates the results of the usability test. It shows whether a test person (subject) required one or more attempts to use a specific function successfully. During these tests, the subjects were able to validate the detected position of the eye tracker by means of the POR visualization. The diagram shows that none of the test persons had problems with the fixation. While the options were selected due to closing the eyelids, only one subject required several attempts. The same applies to the vertical eye movement. In a second pass, it turned out that precisely this subject requires other settings for a successful eye gesture recognition. Thus, more time for training and personal settings will help to achieve better results. However, it should be stated that this combination of object selection via fixation and option selection by closing the eyelids turned out to be a workable solution.

Figure 12 further shows that three of five test persons had difficulties to deal with the horizontal eye movement. Interviews with the subjects showed that it appears to be very difficult to control the horizontal eye movement to get a straight motion. Apart from that, it must be considered that in general LIS patients are not able to do horizontal eye movements.

In summary, it can be noted that the usability can be assessed as stable and accurate. With a well-calibrated eye tracker, the basic handling consisting of the combination of fixation and closing the eyelids is perceived as comfortable. Additionally, it is possible to adjust the eye gesture settings individually at any time. This enables an impaired person to achieve optimal eye gesture-recognition results and a reliable handling.

B. Controlling the Robot

The development of the controlling interface of the robot has nearly been completed. It stands to reason that the second model seems to be the the interface with the easiest control and the least symptoms of fatigue for the eyes. However, a detailed test is still pending.

VI. RESULTS AND DISCUSSION

Since this work is in progress, there are different parts of this work that need to be discussed, implemented and evaluated in the near future. We list the main points – even in parts – below:

- Currently, the LIS patient can only deactivate the eye tracking during the object-based interaction mode by switching to robot control mode. Thus, there should be a way to disable the fixation detection. Since eye gestures based eye movements have proved to be difficult, our idea is a combination of two consecutive fixations, e.g., in the upper left and lower right corners.
- Instead of the currently used static pictures displayed in object-based interaction mode, a live view of the VPS should be shown. But this requires a well functioning object classification.
- Thus, a major part of this work will be the classification of a useful set of everyday objects. Recently, deep convolutional neural networks trained from large datasets have considerably improved the performance of object classification (e.g., [19], [20]). At the moment, they represent our first choice.

In addition, there are many other minor issues to deal with. However, at this point these issues are not listed individually.

VII. CONCLUSION AND FUTURE WORK

The presented prototype demonstrates an interface to drive a VPS through a local environment and offers a novel communication and interaction model for LIS patients, where visible objects selected by eye gestures can be used to express the needs of the patients in a user-friendly way.

In contrast to the discussed state-of-art methods, which are based on an interaction with static content on screen, the direct interaction with the environment is a benefit in two ways. On the one hand, compared to the methods that use a virtual keyboard, our method is faster and less complex. And on the other hand, compared to the methods where pictograms are used, our method eliminates the search for the matching icon. Thus, the advantage of such a system is a larger flexibility and a greater interaction area, i.e., a direct connection to controllable things like the light, a TV, or a radio.

Our current work examines different models to control the movements of the prototype with eye gestures in a live view from the on-board camera of a VPS. Moreover, an autonomous navigation of a VPS using QR codes and a floor plan is currently tested to fit the particular situation of LIS patients.

Future work will include the ability to select objects individually from the local environment. This will enable the patients to use real objects for communication tasks with the help of an eye tracker. The interaction with the real environment via a live view will ensure a more intuitive interaction than the communication via static screen content and thus will provide LIS patients with even more freedom. In addition, in this scenario dynamic changes within the room (displacement or exchange of objects) will not affect the interaction range of a patient.

Independently of this, a LIS patient should alway have the ability to select a virtual keyboard to send individual messages as fall-back option.

REFERENCES

[1] S. Eidam, J. Garstka, and G. Peters, "Towards regaining mobility through virtual presence for patients with locked-in syndrome," in *Proceedings of the 8th International Conference on Advanced Cognitive Technologies and Applications*, Rome, Italy, 2016, pp. 120–123.

[2] E. Smith and M. Delargy, "Locked-in syndrome," *BMJ: British Medical Journal*, vol. 330, no. 7488, pp. 406–409, 2005.

[3] A. Duchowski, *Eye Tracking Methodology, Theory and Practice*. Springer-Verlag, 2007, ch. Eye Tracking Techniques, pp. 51–59.

[4] T. E. Hutchinson, K. P. White, W. N. Martin, K. C. Reichert, and L. A. Frey, "Human-computer interaction using eye-gaze input," *IEEE Systems, Man, and Cybernetics*, vol. 19, no. 6, pp. 1527–1534, November 1989.

[5] K. Arai and R. Mardiyanto, "Eye-based human computer interaction allowing phoning, reading e-book/e-comic/e-learning, internet browsing, and tv information extraction," *IJACSA: International Journal of Advanced Computer Science and Applications*, vol. 2, no. 12, pp. 26–32, 2011.

[6] S. S. Liu, A. Rawicz, S. Rezaei, T. Ma, C. Zhang, K. Lin, and E. Wu, "An eye-gaze tracking and human computer interface system for people with als and other locked-in diseases," *JMBE: Journal of Medical and Biological Engineering*, vol. 32, no. 2, pp. 111–116, 2011.

[7] R. Barea, L. Boquete, M. Mazo, and E. Lpez, "System for assisted mobility using eye movements based on electrooculography," *IEEE Transactions on Neural Systems and Rehabilitation Engineering*, vol. 10, no. 4, pp. 209–218, 2002.

[8] R. Mautz and S. Tilch, "Survey of optical indoor positioning systems." in *IPIN*, 2011, pp. 1–7.

[9] K. A. Nuaimi and H. Kamel, "A survey of indoor positioning systems and algorithms," in *Innovations in Information Technology (IIT), 2011 International Conference on*. IEEE, 2011, pp. 185–190.

[10] H. A. Karimi, *Indoor wayfinding and navigation*. CRC Press, 2015.

[11] A. Mulloni, D. Wagner, I. Barakonyi, and D. Schmalstieg, "Indoor positioning and navigation with camera phones," *IEEE Pervasive Computing*, vol. 8, no. 2, pp. 22–31, 2009.

[12] W. Li, F. Duan, B. Chen, J. Yuan, J. T. C. Tan, and B. Xu, "Mobile robot action based on qr code identification," in *Robotics and Biomimetics (ROBIO), 2012 IEEE International Conference on*. IEEE, 2012, pp. 860–865.

[13] C. Gionata, F. Francesco, F. Alessandro, I. Sabrina, and M. Andrea, "An inertial and qr code landmarks-based navigation system for impaired wheelchair users," in *Ambient Assisted Living*. Springer, 2014, pp. 205–214.

[14] S. J. Lee, J. Lim, G. Tewolde, and J. Kwon, "Autonomous tour guide robot by using ultrasonic range sensors and qr code recognition in indoor environment," in *Electro/Information Technology (EIT), 2014 IEEE International Conference on*. IEEE, 2014, pp. 410–415.

[15] H. Zhang, C. Zhang, W. Yang, and C. Chen, "Localization and navigation using QR code for mobile robot in indoor environment," in *2015 IEEE International Conference on Robotics and Biomimetics, ROBIO 2015, Zhuhai, China, December 6-9, 2015*, 2015, pp. 2501–2506.

[16] W. Y. Jeong and K. M. Lee, "CV-SLAM: a new ceiling vision-based SLAM technique," in *2005 IEEE/RSJ International Conference on Intelligent Robots and Systems, Edmonton, Alberta, Canada, August 2-6, 2005*, 2005, pp. 3195–3200.

[17] P. E. Hart, N. J. Nilsson, and B. Raphael, "A formal basis for the heuristic determination of minimum cost paths," *IEEE Transactions on Systems Science and Cybernetics*, vol. 4, no. 2, pp. 100–107, July 1968.

[18] F. Duchoň, A. Babinec, M. Kajan, P. Beňo, M. Florek, T. Fico, and L. Jurišica, "Modelling of Mechanical and Mechatronic Systems Path Planning with Modified a Star Algorithm for a Mobile Robot," *Procedia Engineering*, vol. 96, pp. 59–69, 2014.

[19] A. Krizhevsky, I. Sutskever, and G. E. Hinton, "Imagenet classification with deep convolutional neural networks," in *Advances in Neural Information Processing Systems 25: 26th Annual Conference on Neural Information Processing Systems 2012. Proceedings of a meeting held December 3-6, 2012, Lake Tahoe, Nevada, United States.*, 2012, pp. 1106–1114.

[20] K. Simonyan and A. Zisserman, "Very deep convolutional networks for large-scale image recognition," *CoRR*, vol. abs/1409.1556, 2014.

The Natural-Constructive Approach to Representation of Emotions and a Sense of Humor in an Artificial Cognitive System

Olga Chernavskaya
Laboratory of elementary particles
Lebedev Physical Institute (LPI)
Moscow, Russia
e-mail: olgadmitcher@gmail.com

Yaroslav Rozhylo
BICA Labs
Kyiv, Ukraine
e-mail: yarikas@gmail.com

Abstract— **The Natural-Constructive Approach is proposed to describe and simulate the emotions and a sense of humor in an artificial cognitive system. The approach relates to the neuromorphic models and is based on the original concept of dynamical formal neuron. The main design feature of the cognitive architecture consists in decoupling the cognitive system into two linked subsystems: one responsible for the generation of information (with the required presence of random component usually called "noise"), the other one – for processing the well-known information. The whole system is represented by complex multi-level hierarchical composition of neural processors of two types that evolves according to certain principle of self-organization. Various levels are shown to correspond to the functional areas of the human-brain cortex. Human emotions are treated as a trigger for switching the subsystem activity that could be imitated and mathematically expressed as variation of the noise amplitude. Typical patterns of the noise-amplitude variation in the process of problem solving are presented. The sense of humor is treated as an ability of quick adaptation to unexpected information (incorrect and/or incomplete forecast, surprise) with getting positive emotions. Specific human humor response (the *laughter*) is displayed as an abrupt "spike" in the noise amplitude. Thus, it is shown that human emotional manifestations could be imitated by specific behavior of the noise amplitude.**

Keywords- noise; emotions; explanatory gap; spike; surprise.

I. INTRODUCTION

Recently, the paper concerning the interpretation of emotions and the sense of humor in an artificial cognitive system was published and presented at the conference COGNITIVE 2016 [1]. This paper represents an invited extended version.

The problem of modeling and imitation of the cognitive process is actual and very popular now, especially in the context of Artificial Intelligence (AI) creation. Among the most popular approaches, there are Active Agent paradigm (e.g., the SOAR architecture, see [2], [3]), Deep Learning paradigm [4]–[6], Brain Re-Engineering [7], [8], Robotics [9], Resonance theory [10], etc. The majority of imitation models proposed are aimed to construct the artificial cognitive systems for solving certain (even broad) set of problems *better* than human beings do. Hence, those systems have to be *efficient, reliable,* and *fast-acting*.

In our works [11], [12], so called Natural-Constructive Approach (NCA) has been elaborated, which is focused on modeling just the human-like cognitive systems. Therefore, the priority is given to the features inherent to the *human* cognition, such as *individuality, intuitive* and *logical* thinking, *emotional impact* on cognitive process, etc. This approach is based on the Dynamical Theory of Information [13]–[15], data from Neurophysiology [16]–[18], and Neuropsychology [19], and Neural Computing [20]–[22] (with the latter being used in a modified form). Note that NCA could be related to the Human-Level Artificial Intelligence (so called *HLAI* track, see, e.g., [23]) and is close to some extent to the Deep Learning paradigm [4]–[6], but possesses certain important and original peculiarities presented below.

This paper is focused on modeling the manifestation of emotions in the cognitive process. The version of the human-like cognitive architecture elaborated under NCA is presented schematically. The main constructive feature of this architecture consists in *decoupling* the cognitive system into two linked subsystems: one responsible for generation of information (with required presence of random component, i.e., "noise"), the other one — for reception and processing the well-known information. The activity of these subsystems is proposed to be controlled by the *emotional* mechanism. Switching the subsystem activity is associated with the noise amplitude variation, which could be related to the change in neurotransmitter composition. This paradigm is applied to simulate the human reactions under stress conditions (including "smooth" stress, i.e., surprises). A particular case of the noise-amplitude behavior, — namely, the abrupt up-and-down change ("spike"), — is treated as an analogue to human *laughter*.

The paper is organized as follows. Section II presents a brief overview of modern approaches to representation of emotions in AI. Section III describes basic components of NCA. Section IV describes the main constructive blocks of cognitive architecture designed under NCA. In Section V, we discuss the role and place of emotions in the proposed architecture and present the example of application of the proposed model to describe the effects of stress/shock. In Section VI, typical manifestations of emotions in course of solving different problems are considered; special attention is paid to representation of the sense of humor in AI. Perspectives on practical validation of the results obtained are discussed in Section VII. In Conclusion, main results are summarized and future perspectives are discussed.

II. MODERN STATE OF THE EMOTION REPREENTATION PROBLEM

Simulation of the human-like cognitive process implies inherently the integration of rational reasoning and emotions into one cognitive system. This problem represents one of the main challenges as for *AI*, as well as for any human-level cognitive architecture (*HLAI*, [23]). The main problem here is connected with the so-called "explanatory gap" [24], i.e., the gap between the "Brain" (cortical and sub-cortical structures) and the "Mind" (consciousness). This means that there is a lot of information from the B*rain* side (neurophysiology) on the structure and functions of single neuron, and even on the neuron ensemble (e.g., [14]). On the other ("Mind") side, there is a lot of information from philosophy and psychology (including personal experience) on the consciousness manifestations (e.g., [25], [26]). However, there is a lack of ideas on how the first could provide the second.

A particular consequence of this fact is surprisingly poor and vague definitions of such concepts as *emotions, intuition, logical thinking*, *subconscious,* etc., which are presented in such respective Dictionaries as Miriam-Webster [27]. However, definitions from the Wikipedia [28] seem more meaningful, modern, and reasonable in our view.

The same "gap" concerns as well the representation of emotions. On the "Mind" side, emotions represent, according to definition "*...subjective self-appraisal of the ...current/future state*" [28]. On the other side, from the "Brain" viewpoint (see, e.g., [7], [16]), emotions are treated as a *composition of neurotransmitters* produced by certain sub-cortical structures. This value is objective and experimentally measurable. But where is the "bridge" between the neurotransmitter composition and personal feeling of satisfaction, disappointment, etc. — that is the question.

This problem actually attracts attention and evokes a lot of studies (see, e.g., [29] –[39]). However, the variety of approaches to the problem of emotion representation indicates itself that the problem is not solved yet, so that, "*...emotions still remain an elusive phenomenon*" [40].

Below, we try to collect the interpretations and main features of emotions provided by different approaches and propose our view on accounting for emotional component in the artificial cognitive system.

The approaches from the "Brain" viewpoint refer mainly by the Brain Re-Engineering paradigm (e.g., [7], [8], [29], [30], [31]). It is based on the analysis of complementary role of cerebral cortex and certain sub-cortical structures — *thalamus*, *basal ganglia*, *amygdale*, etc., — directly related to the control of the emotions in cognitive process. This way looks very close to the goal, but the consideration actually seems mostly verbal: the mathematical apparatus used seems rather poor. Moreover, the role of emotions is attributed mainly to the *reinforcement learning* process, while it is important but far not the only act of cognition. Besides, these studies focused on the motor (acting) training, leaving aside the cognitive process itself.

Another, somewhat more abstract "Brain-inspired" approach is presented by the works of Lovheim and followers (see [32], [33]). Here, the three-component model was proposed that involved three systems of monoamine neurotransmitters (namely, *serotonin*, *dopamine*, and *nor-adrenaline*), which provide cubic representation of various emotional states. This model is popular and provides good results for describing several medical problems (deceases), but seems not so well in modeling regular cognitive process.

From the "Mind" viewpoint, the majority of researches refer to the *active agent* concept ([2], [34], [35]). Here, the agents are supposed to have the ability of self-appraisal from the very beginning, and the question is: how this appraisal does influence their reasoning. There were suggested various principles of organization of the "emotional space" that affect the cognitive process. However, the main problem from our viewpoint is to understand the very mechanism that could provide the self-appraisal ability.

A similar way is to introduce several discrete emotional states that would affect (with certain weight coefficients) the model calculations for *AI*. Their number may vary — from two (positive and negative ones) up to 27 in [34]. However, clear mechanisms of emotion emergence are not revealed in any of these cases.

The other approach ([36], [37]) involves two sets of dynamical variables, emotional and rational ones, so that their (nonlinear!) interaction results in various states of the system providing certain nontrivial regimes of transition between those states. However, the neurophysiology interpretation of the emotional, as well as rational, variables under this approach remains somewhat dissatisfactory.

An interesting (but somewhat shocking) idea was put forward by Schmidhuber [38]: the ultimate goal of living activity that provides the most positive emotions is connected with the *compression* of information. Being seemingly not the most actual goal for a human being (as compared with, e.g., *survival*), it could be reformulated in terms of "image-to-symbol conversion" (see below). Then, this idea surprisingly meets our final inferences.

The last but not least, let us turn to the concept suggested by Huron [39] that emotions are evoked by *anticipations*. In spite of this hypothesis is formulated rather verbally than mathematically, it seems the most promising and could serve as a basis for mathematical modeling.

Note that common modern trend consists in associating emotions not with *particular state*, but with certain *transitions* between different states (see [35], [39]). This trend seems to be the most promising since it does not fix or limit the number of mechanisms (as well as neurotransmitters) that provide emotional manifestations, but is focused on the *variability* of the cognitive process.

This study represents an attempt to merge the "Brain" and "Mind" paradigms under NCA by revealing (or introducing) proper variables and coupling them into unified dynamical system (i.e., "emotional block", see below).

III. Basic Components of NCA

The approach NCA is aimed to understand and reproduce in mathematical model the human-like cognitive features like *spontaneity*, *paradoxicality* (the ability to formulate and solve paradoxes), *individuality*, *intuitive* and *logical* reasoning, integration of *emotions* and rational reasoning. Therefore, certain paradigms typical just for the living objects are required. NCA involves one of such paradigms provided by the Dynamical Theory of Information.

Being biologically inspired, the approach belongs to so called neuromorphic models, which implies that the neuron is the basic element (in some sense, the "*active agent*") of the whole cognitive architecture. Hence, both neurophysiology and neuropsychology data should be taken into account.

Neural computer paradigm is used for computation and numerical simulations. Under NCA, somewhat modified representation of the neuron that was called the "*dynamical formal neuron*" model is employed.

Thus, NCA combines actually three areas of expertise.

A. Dynamical Theory of Information

The Dynamical Theory of Information (DTI) is relatively new theory elaborated in the post-middle of XX*th* century, almost at the same time as the well-known theory of communication of Shannon (see [41], [42]). However, Shannon's theory was focused on the process of information transmission, while DTI analyses the process of its origin and evolution. This theory, being the subfield of Synergetics (see [13], [43]), was elaborated in the works of Haken [13] and Chernavskii [14], [15]. It is based on the idea that the information is a specific kind of object that possesses simultaneously as solid (material), as well as virtual features. The information appears as a result of evolution and interaction within certain community of living subjects. Let us stress that the brain, being an ensemble of neurons, represents a specific case of such community.

The most constructive and explicit definition of Information belongs to Quastler [44]: "*The Information is the memorized choice of one version of N possible (and similar) ones*". This definition provides immediately the possibility to reveal different types of information:

- *Objective* Information — the choice done by the Nature as a result of its evolution, i.e., physical (objective) laws reflecting real structure of the surrounding world.
- *Conventional* (*Subjective)* Information — the choice done by a group of subjects as a result of their interaction, communication, fight, agreement, convention, etc., that is individual for a given community.

In the first (Nature) case, the choice appears to be done according to the principle of minimum energy expenses. In the second (people) case, the particular choice should not be *the best* one, but should be done and stored. The most widely-known examples of conventional information are the following: language, alphabet, traffic signs, symbols, etc. A particular language could be neither better nor worse than other, but it reflects the *mentality* (individuality) of a given society (see, e.g., [45]).

Moreover, that definition provides the idea of *how* the information could emerge. There are two mechanisms:

- *Perception* — superimposed (externally forced) choice associated with the Supervisor learning.
- *Generation* — free (random) choice that should be done without external control (internally).

It was shown in [13]–[15], that the information generating process requires mandatory the participation of chaotic element (so called "mixing layer") that is commonly called the *noise*.

The main inference of DTI is that these two mechanisms are *dual* (or *complementary*), and hence, *two subsystems are required to perform both these functions.* In analogy with two cerebral hemispheres of human brain, let us call these subsystem Left Hemi-system (**LH**) and the Right Hemi-system (**RH**), respectively.

From the positions of DTI, the *cognition* is considered as a process *of processing the information.* Therefore, the cognitive process could be defined as "*the self-organizing process of recording (perception), memorizing (storage), coding, processing, generation and propagation of the personal conventional information"* [11]. Note that this definition does presume the *subjective* (individual) character of human thinking.

B. Neurophysiology and Neuropsychology Data

Let us stress that both, the "Brain" and the "Mind" evidences should be taken into account. "Brain" data concern the neuron structure and mechanisms of their interactions.

1) Neuron Representation: NCA refers to so called "neuromorphic" models. This implies that the basic element for any structure is the neuron. In neurophysiology (see, e.g., [46]), the neuron model presented by Hodgkin-Huxley [47], as well as its somewhat reduced version suggested by FitzHugh-Nagumo [48], [49], are considered still as the most relevant ones. Starting from the Fitz-Hugh model, we have elaborated the *dynamical formal neuron* concept (see [11]) that represents a particular case of this model. Accordingly, nonlinear differential equations were used to describe the single-neuron behavior and the neuron interactions. This enables us to trace the dynamics and reasons for symbol formation.

2) Neuron Interaction Representation: Experimental data on interaction in the neuron ensemble show:

a) Numerous experiments indicate that the perception of new information is accompanied by amplification of the connections between neurons involved in this process. This is called the "Hebbian rule" [17].

b) Modern experimental data on the neuron structure [18] show very intriguing fact: those neurons that participate in acquiring certain experience ("skill") appear to be modified as compared to free (unemployed) neurons. This inference is based on the experimentally observed distribution for the expression of so called c-FoS gen responsible for changing the neuron structure. Thus, the proper model representing a neuron should involve the possibility of a certain mutation for engaged (trained) neurons.

3) Neuropsychology Evidence: Another challenge for any relevant model of a human-level cognitive system is the question: why there are just two cerebral hemispheres in the human brain – the right (**RH**) and the left (**LH**) ones. From psychological viewpoint, we take into account the wide-spread opinion that **RH** is associated with non-verbal, imaginary, *parallel* thinking and intuition (see, e.g., [26], [50]). Correspondingly, **LH** is associated with *sequential verbalized* thinking and the logical reasoning. However, while there is no clear explanation of intuition and logic, these statements seem ambiguous.

Another, more constructive from our viewpoint, idea had been put forward by E. Goldberg (practicing psychologist) [19]. He infered that **RH** is responsible for processing *new* information, i.e., learning, while **LH** has to process the *well-known* information. Note that this concept entirely coincides with the main inference of DTI, that any cognitive system should contain two subsystems, one for generation of new information, the other one for reception and processing the existing information.

C. Neurocomputing

A cognitive system could be presented as a composition of neural processors, i.e., the plates populated with model neurons. It should be stressed that, in contrast to common neural computing (see, e.g., [51]) based on the simple formal neural paradigm suggested by McCulloch and Pitts [52], NCA is based on the concept of *dynamical* formal neuron presented in [11].

Two types of neural computers are employed:

1) Distributed memory:

This concept refers to the Hopfield-type processor (***H***) with *cooperative* intra-plate ("horizontal") interaction [20]. Any real object is represented as a "chain" of activated (excited) neurons, which is called the *"image"* of this object. The main advantage of such type of representation is connected with the fact that the damage of few neurons of this chain does not lead to the damage of the image as a whole. The integrity of the image is secured by trained connections between the neurons involved into the image formation.

Note that real objects having similar fragments are to be written by the *overlapping* chains of neurons, which provide *associative* connections between these objects.

The model of the ***H***-type processor with dynamical formal neurons could be written in the form:

$$\frac{dH_i(t)}{dt} = \frac{1}{\tau^H}[\{H_i - \beta_i \cdot (H_i^2 - 1) - H_i^3\} + \sum_{i \neq j}^{n} \Omega_{ij} \cdot H_j] \equiv \frac{1}{\tau^H}[\Im_H\{H_i, \beta_i\} + \sum_{i \neq j}^{n} \Omega_{ij} \cdot H_j], \quad (1)$$

where $H_i(t)$ is variable describing the state of i-th dynamical formal model neuron, τ_i^H— activation characteristic time, β_i — parameters that characterize the neuron excitation threshold. The functional $\Im_H\{H_i,\beta_i\}$ describes the internal dynamics of a single H-type neuron, the second term refers to interaction with neighbors, with Ω_{ij} being the matrix of connections between neurons, $i, j = 1.....n$. Stationary states are: $H_i = +1$ (active) and $H_i = -1$ (passive), that provides the effect of neuron switching on/off under its neighbor's impact. Note that the parameters β referring to the excitation threshold could be modified as the result of learning process.

It should be stressed that the functions performed by the H-type plates depend essentially on the principle of the connection training. Under NCA, two types of training rules are used. The first one that is required for recording corresponds to well-known Hebb's rule [17] of connection amplification, which implies that the strength of connections between excited neurons *increases* as

$$\Omega_{ij}^{Hebb}(t) = \frac{\Omega_0}{4 \cdot \tau_\Omega} \cdot \int_0^t [H_i(t')+1] \cdot [H_j(t')+1] dt' \cdot \zeta(t'), \quad (2)$$

where Ω_0, τ_Ω — training parameters, $\zeta(t)$ is monotonic integrable function to provide the saturation effect.

Another version of the connection-training principle had been proposed in original work of J. Hopfield [20] as a tool for *recognition* of the already learned (stored) images. This version reads:

$$\Omega_{ij}^{Hopf}(t) = \Omega_0\{1 - \frac{1}{2\tau_0} \int_0^t [1 - H_i(t')H_j(t')] \cdot \zeta(t')dt'\}, \quad (3)$$

that corresponds to the "redundant cut-off" principle. This means that the "informative" connections between excited neurons are initially strong and do not change in the training process, while irrelevant (waste) connections should die out. This principle corresponds actually not to the *choice* of recording, but rather to the *selection* of trained connections.

It should be stressed that such way of training leads to the fact that this processor could perceive *any* (even new) image as one of the already learned (stored). This results in two effects:

- refinement of the damaged (noisy) image: due to the hard influence of neighbors, the irrelevant neurons would die, while missing ones would be excited;
- there are problems with re-learning of this processor to incorporate new images.

Thus, the necessity and reasons for exploring two versions of the H-type processor are apparent.

2) Symbolic memory:

This concept involves the coding (localization) procedure combined with possibility of further cooperative (Hebbian) interaction. These two functions could be realized by means of the Grossberg-type (***G***) processor [22] with competitive intra-plate (horizontal) interaction, which works at the first stage for choosing one neuron to be the *symbol* (representer of the certain group of neurons, i.e., the *image*). At the next stage, competitive interaction should be altered to cooperative. The model of processor possessing all these abilities could be written in the form:

$$\frac{dG_k(t)}{dt} = \frac{1}{\tau_G}[\{-(\alpha_k - 1)\cdot G_k + \alpha_k \cdot G_k^2 - G_k^3\} - \theta(\Psi_0 - \Psi)\cdot \sum_{l\neq k}^{n} \Gamma_{kl}\cdot G_k \cdot G_l + \vartheta(\Psi - \Psi_0)\cdot \sum_{l\neq k}^{n} \Omega_{kl}\cdot G_l] + Z(t)\cdot \xi(t)$$
$$\equiv \frac{1}{\tau_G}[\Im_G\{G_k,\alpha_k\} - \theta(\Psi_0 - \Psi)\cdot \sum_{l\neq k}^{n} \Gamma_{kl}\cdot G_k \cdot G_l + + \theta(\Psi - \Psi_0)\cdot \sum_{l\neq k}^{n} \Omega_{kl}\cdot G_l] + Z(t) \quad (4)$$

where the variable G_k refers to the state of k-th G-type neuron, τ^G is activation characteristic time, with its internal dynamics being described by the functional $\Im_G\{G_k, \alpha_k\}$. The term $Z(t)\cdot\xi(t)$ stays for the random component, with $Z(t)$ being the noise amplitude, $0<\xi(t)<1$ is random function. Two step-wise *theta* functions $\vartheta(\Psi\text{-}\Psi_0)$, $\vartheta(-\Psi+\Psi_0)$ stop the competitive process and start the cooperation (depending on the argument's sign).

Note that this representation differs from given for the H-type neuron since the stable state here are equal to: G=1 (active) and G=0 (passive). Such choice of representation enables us to account for both, competitive and cooperative interactions depending on the state of inter-plate (so called "vertical") Ψ connections.

The *competitive* connections Γ provide the symbol-choosing procedure that requires mandatory participation of random component (noise), see [11]. The dynamics of connection training is determined by the equation:

$$\frac{d\Gamma_{kl}(t)}{dt} = -\frac{\Gamma_0}{\tau^{\Gamma}}\{G_k \cdot G_l (G_k - G_l)\}, \quad (5)$$

where Γ_0 and τ^{Γ} are training parameters. This training rule provides so-called "localization" reaction, when only one neuron from the activated chain wins the round. So this type of neuroprocessor serves to convert the chain corresponding to the real object (i.e., the "*image*") into single neuron referred further as the "*symbol*" (in other terminology, the "*name*") of this object.

After the choosing procedure was finished, the inter-plate (vertical) Ψ connections should formed to link the chosen symbol with its image neurons at the previous hierarchy level:

$$\frac{d\Psi^{R,\sigma}_{km}(t)}{dt} = \frac{\Psi_0}{\tau^{\Psi}}\cdot G_k^{R,\sigma}\cdot G_m^{R,(\sigma-1)}, \quad (6)$$

where Ψ_0 and τ^{Ψ} are characteristic parameters of training. Such connections secure the *semantic* content of the chosen symbol; therefore, they are called the "*semantic*" connections. These very connections do realize the Kohonen paradigm "Winner Takes All" (WTA) [38], providing a possibility to decompose the symbol into distributed image.

Note that this processor differs from the standard versions of ACT procedure (see, e.g., [53], [54]) by at least two factors:

- there is no fixed rule for conversion process, it proceeds due to *competitive interaction* between neurons only;
- symbol-formation procedure in the given processor is *unstable,* thus providing uncertainty and "individuality" of the position of chosen symbol.

Let us stress that this very mechanism of the winner-choosing procedure is derived not from the neurocomputing, but from the analysis of choices done within given society, and is known in DTI as the "conventional information struggle" (see [14], [15]). It is typical not only for the human society, but for all living objects as well. This very choice should not be "*the best*" (i.e., the most efficient, or fast, or reliable, as it is typical for neural computing), but should be individual for the given system. Thus, the symbol formation procedure under NCA represents an example of creating the *conventional* information.

After the semantic connections between the chosen symbol and its image were formed up to sufficiently ("black") Ψ_0 value the competitive interaction stops due the presence of step-wise function in (4). Then, the cooperative interaction with neighboring symbols could start that correspond to the last term in (4). The cooperative connections are trained according to the Hebbian principle:

$$\frac{d\Omega^{\sigma}{}_{kl}(t)}{dt} = \frac{\Omega_0}{\tau^{\Omega}}\{G^{R,\sigma}{}_k \cdot G^{R,\sigma}{}_l\}. \quad (7)$$

These connections provide the possibility to form the *generalized* image, i.e., "image-of-symbols", which could get its symbol at the next hierarchy level. Note that this process may be reproduced at each step of the system's evolution. Thus, this processor actually possesses the properties of distributed memory as well.

Note that in our previous works [11], [12] this effect was secured by the mechanism of *parametric modification* of the neuron-symbol, which takes it out from the competitive interaction, simultaneously providing the possibility of cooperative interactions with neighbors. It has been proposed that after the given G-neuron got a status of symbol and had formed the inter-plate connections Ψ with his image neurons, it should leave a competitive struggle for the right to be a symbol of another image. This effect could be provided by parametric modification of the neuron-symbol: $\alpha_k \to \alpha_k(\{\Psi_{ik}\})$. Actually, both mechanisms, dynamical and parametric ones, could work together.

In any case at the time scale $t >> \tau^{\Gamma}$, the neuron-symbol stops its competitive interaction with neighbors, but acquires a possibility to participate in the *cooperative* interactions with the other neuron-symbols by the same Hebbian mechanism as *H*-type neurons do. Note that "free" ***G***-neurons (that were failed to become a symbol of any image) could compete only.

Another very important point should be stressed. Encoding (i.e., symbol formation) means as well the *comprehension* of the image information received from outside. The very fact of symbol formation implies that the system had apprehended the given chain of *M* active neurons at the plate ***H*** as a representation of a single real object and had awarded a proper symbol ("name") to it. That is why the inter-plate (vertical) connections between the symbol and its progenitor image neurons are called *semantic* ones.

Let us stress ones more that, the instability of the conversion procedure under NCA results in just *random (free) choice* of the symbol among possible "nominant" neurons. This means that this procedure represents a particular case of generation of *conventional* information – this choice should not be the best (the most efficient), it should be individual. Thus, this process does secure the *individuality* of any (even artificial) cognitive system.

IV. Architecture of Cognitive System

The architecture of cognitive system has been designed under NCA in the works [11], [12] of Chernavskaya et al. Let us recall briefly main peculiar features.

A. *Basic Elements of NCA Architecture*

The schematic representation of NCA cognitive architecture is plotted in Fig. 1. This system represents a composition of several neural processors of Hopfield (***H***) and Grossberg (***G***) types, which are composed into hierarchical structure, with σ being the number of hierarchical level. Each processor is represented as a plate populated with *n dynamical* formal neurons described in Section III. The total number of levels (symbolic plates) is neither fixed nor limited since they appear "as required" in course of the system evolution as a response to the operational complexity of the perceptible world.

Each symbol G^{σ} is linked by *semantic* connections $\Psi^{(\sigma-1)}$ and $\Psi^{(\sigma+1)}$ defined in (6) with its "parent" image at the previous level and the "descendant" symbol at the next level σ+1, respectively. Besides, it is linked with its *neighbors* by cooperative connections Ω^{σ} (defined in (7)), which create new (independent) image. Using imagination, one may say that each symbol has its "*legs*" (to rely to the ground) and "*hands*" (to reach the ceiling). Such "pyramid" is replicated at every level of hierarchy, thus forming the fractal-type multi-level structure.

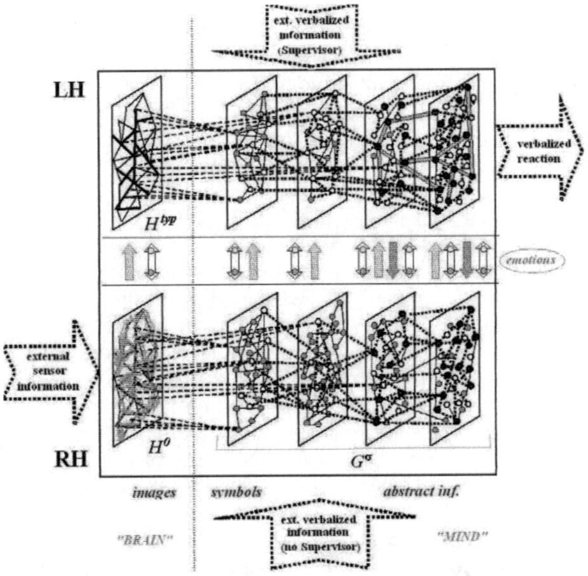

Figure 1. Schematic representation of NCA cognitive architecture.

According to DTI principles, the system is divided into two coupled subsystems, the right hemi-system (**RH**) and the left hemi-system (**LH**). These terms were chosen to correlate these subsystems with cerebral hemispheres, with the cross-subsystem connections $\Lambda(t)$ being an analogue to the *corpus callosum.* These connections should provide the interaction ("dialog") between the subsystems ("up-down" arrows in Fig. 1). One subsystem (**RH**) is responsible for *learning* and processing *new* information; the other one (**LH**) is dealing with the *well-known* information. This functional specialization coincides completely with that proposed (from the "mind" viewpoint) by Goldberg [19], that represents a pleasant surprise as well as an indirect validation of our approach. Under NCA we can also *reveal its mechanism* from the "brain" viewpoint. It is secured by three factors:

- the presence of random component (noise) in **RH** provides the conditions for generation of information, i.e., *free choice* of the version of recording new information;
- different connection-training principles in the different subsystems: the Hebb's principle of active connection amplification [17] in **RH**, and the Hopfield's principle of the "redundant cut-off" [20] in **LH**;
- the "connection-blackening" principle of self-organization, which implies that strong enough ("black") images in **RH** are replicated in **LH**. Hence, **RH** acts as a Supervisor for **LH**.

Let us consider the connection-blackening principle in more details by analyzing the elementary act of system's evolution.

B. Elementary Learning Act: "Connection-Blackening" Principle

The elementary act of cognitive process realization (in particular, learning) should involve implementation of the functions of recording, storing and coding the image of new object.

The functions of recording and storing "raw" images could be implemented by means of two *H*-type cross-linked processors (see Fig. 2a), with the connection-training rules being *different* on those plates (Fig. 2b). One of them (called H^0) should be trained by Hebbian mechanism, while the other one (called H^{typ}) — according to the original Hopfield principle "redundant cut-off". They are correlated by the value of well-trained connections Ω_0 (see Fig. 2b).

Primary ("raw") images are recorded at the plate H^0 by Hebbian-trained connections, with their strength being vary from weak ("grey") to strong ("black") state. When the strength of trained connections achieves the "black" value Ω_0, the "black" image should be transferred by direct (one-to-one) inter-plate connections and replicated at the *typical image* plate H^{typ} for storing. This procedure corresponds to the implementation of so called "connection blackening" principle.

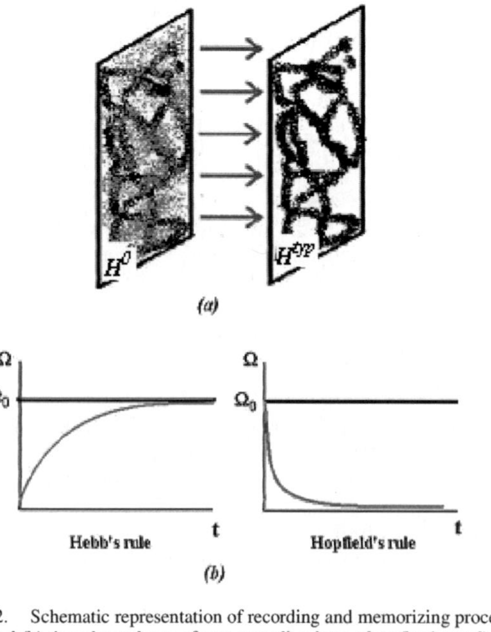

Figure 2. Schematic representation of recording and memorizing process (a) and (b) time dependence of corresponding intra-plate (horizontal) connection strength $\Omega(t)$.

The combination of this process with the encoding procedure provides the "*elementary act*" of the system's formation and is presented in Fig. 3. This process again corresponds to the self-organization principle of "connection blackening" and proceeds in three steps:

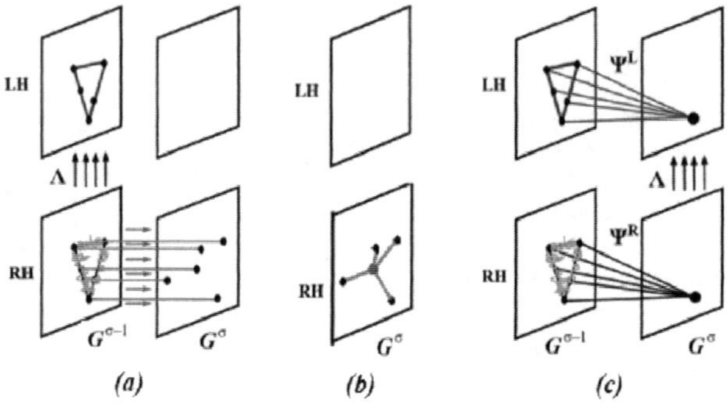

Figure 3. The elementary act of learning.

a) The First Step: an image formed at the previous-level ($\sigma-1$) in **RH**, after its cooperative connections Ω^R become strong ("black") enough, is delivered by the direct (one-to-one) inter-plate ("vertical") connections to the next-level plate $\mathbf{G}^\sigma$ and, simultaneously, by the inter-subsystem connections Λ to the same level plate $\mathbf{G}^{\sigma-1}$ in **LH** (see Fig. 3a); **LH** level is free.

b) The Second step: NCA conversion procedure image-into-symbol occurs at the next-level plate $\mathbf{G}^\sigma$ in **RH** (Fig. 3b); **LH** level is free.

c) The Third (Final) Step: New symbol is formed together with its semantic (one-to-many) inter-plate connections Ψ^R and is replicated at the same level in **LH**, where vertical connections Ψ^L are forming according to Hopfield-type rule. Here again, the "connection blackening" principle for Ψ^R connections controls the symbol-formation process (Fig. 3c).

This process could be repeated at each level of hierarchy thus generating a multi-fractal structure.

It is important to stress that the raw images in **RH** with relatively weak ("grey") connections (those that didn't achieve the level typical for **LH**) are neither transferred to the next level in **RH**, nor replicated in **LH**. They remain only at the given level and not acquire their symbol at the next level. Thus, they represent latent (hidden) information, which is "auxiliary" for the given system.

C. Specialization of Various Hierarchical Levels

Let us discuss the roles of different hierarchy levels and their correspondence to the cerebral functional areas.

1) Hierarchy-Level Specialization: The whole system represents complex multi-level block-hierarchical construction that does evolve by itself (in Fig. 1 —from the left to the right) due to the self-organization principle of "connection blackening". This implies that at each level, the elementary act presented in Fig. 3 is repeated. New levels (symbol layers) appear "as required", i.e., after a new image was formed at the previous level. In physics, there is special term "scaling" for such principle of organization and the whole structure is called a fractal.

The lowest level $\sigma = 0$ is represented by the *H*-type plates containing the *image* information. The plate H^0 in

RH carries the *whole* image information received by the given system by means of the "sense organs", i.e., from the receptors. The intra-plate (horizontal) connections vary from weak ("grey") up to strong ("black") ones. Note that the images recorded by "grey" (rather weak) connections, according to the connection blackening principle described above, are neither delivered to the next level, nor replicated in **LH.** They are stored at H^0 only, thus representing some vague (fuzzy) information. That is why the plate H^0 hereinafter is referred to as the "fuzzy set". This plate is responsible for recording new sensor images.

The plate H^{typ} in **LH** contains the information selected for storing (memorization). This plate is "filling up" in course of learning (with the role of Supervisor being played by the plate H^0) with those images that are recorded by sufficiently "black" connections ("up" green arrow in Fig. 1). These images are referred to as *typical* ones. This plate does play the main role in recognition of already learned objects; in some sense, it is a classifier.

The next level $\sigma = 1$ is occupied by the *symbols of typical images,* which are formed in **RH.** These symbols do carry a semantic content, that is, a *comprehension* of the fact that the given chain of active neurons represents one real object. Semantic content (sense) of such symbol consists in its *decomposition* (by means of semantic inter-plate connections Ψ) into its image corresponding to this very real object. Only after formation of sufficiently "black" connections Ψ^R, this symbol could be replicated in **LH**.

At the same very level, the process of *primary verbalization* starts. This implies that there occur the *internal words* as the *names* of already learned objects. These names occur in **RH**, i.e., they are chosen *arbitrary* and individually thus are understandable for a given system only. If simultaneously **LH** receive an external information (from external Supervisor, see top external arrow in Fig. 1) on conventional name for this object, the 'internal" name would be replaced (after certain conflict) by the conventional one (by means of inverse training **LH→RH**, see "down" purple arrow in the middle part in Fig. 1). Such process, that is similar to the process of children speech trials, was considered and discussed in [14], [15].

At the same level in **RH**, the symbols could cooperate and create the *generalized images* (image-of-symbols), which acquire their own symbols at the next level $\sigma+1$. These images are rather primitive, since they correspond to concrete real objects. However, even at this level, a Poet could create, using primitive words, a pronounced pattern ("*night, street, lamp, drugstore...*" as in a famous Alexander Block's poetry).

At the next levels $\sigma>1$, this process is repeated with increasing degree of "abstraction" of created images. This implies that new generalized images could hardly be related to any real object and explained at the image level.

At the higher levels of hierarchy $\sigma>>1$, the *abstract information* emerges, that is, the infrastructure of symbols and their connections, which are not mediated by "raw" images, i.e., the neuron-progenitors of *H*-type plates. Here, the *concept symbols* arise, that could not be related to any concrete pattern (e.g., *conscience, infinity, beauty, consciousness, love,* etc.). This information appears in the already well-trained system as a result of interactions of all the plates (not "perceptible", but "deduced" knowledge). This very information could be completely *verbalized*, i.e., expressed in the symbolic form (with relevant grammar and syntax) by means of *conventional language* of a given society. These very higher levels provide a possibility of communication with similar systems. This implies a possibility to propagate personal conventional information ("to explain by words") and understand semantic content of external symbolic (verbal) information. Besides, at such level **LH** obtains a possibility to receive new information not only from **RH**, but also from outside, in symbolic form, from external Supervisor. In psychology, such knowledge is called "semantic", in distinguish to "episodic" one that the system (**RH**) obtains in process of acquiring its individual experience. This knowledge could appear to be active only after incorporation into the existing architecture due to **LH→ RH** connections ("down" purple arrow at the right part of Fig. 1). Note that **RH** itself can get the symbolic verbalized information from outside, without Supervisor (bottom external arrow in Fig. 1), and this information is processing just as internal one, i.e., by forming the Hebbian connections between different external symbolic images.

Thus, the system as a whole does *grow up* from the lower *image* information levels, over *semantic* information (understandable for a given individual system only), to the higher levels of *abstract* information, which could be verbalized and *propagated* (understood) within the given society. At every stage of new level formation, the same process is repeated. New connections are forming in **RH** up to the "black" state, and after that, the new-formed symbol is transferred to **LH**. In this process, certain part of information (*inessential* details recorded by "grey" connections) appears to be lost. Speaking more exactly, it is not delivered to the next level, but is stored at the previous one as *auxiliary* or *latent* information specific for a given individual system.

Note that the label "emotions" in Fig. 1 refers neither to **RH** nor to **LH.** Below, it will be shown that emotions are directly related to switching the cross-subsystem connections Λ ("up-down" arrows in Fig. 1) providing the "dialog" between subsystems. The color of arrows reflects emotional "valence" (green for positive and rose for negative ones).

2) Corresponendence with Cerebral Cortex Areas: Let us point out that the geometry of the NCA architecture corresponds to the functional areas of the human cerebral neocortex (see Fig. 4). The neocortex could be (conventionally) divided into areas ("lobes"), which are responsible for the vision (occipital lobe), motor activity (parietal lobe), auditory activity (temporal lobe), abstract thinking (frontal lobes), etc. Temporal lobes embraces Wernicke's and Broca's areas that are responsible, respectively, for language hearing (word perception) and reproducing (word production), but not for the speech itself.

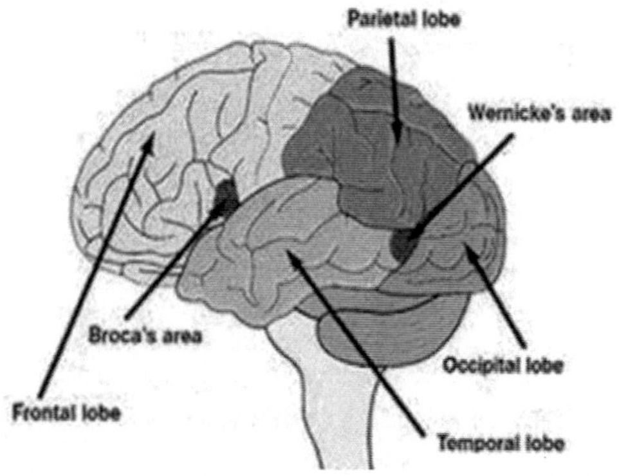

Figure 4. Map of the functional areas of human cerebral cortex (extracted from [55])

The speech function, i.e., coherent and sensible transmission of information, relates to the frontal lobe that is associated with abstract thinking.

Note that similar allocation of functional levels is realized in the NCA scheme: low levels (σ=0) provides *images*, i.e., visual patterns; middle levels (σ>1) contain symbol-words, that is, elements of *language*.

The correspondence between the *abstract information* in the NCA architecture (σ>>1) and the "abstract thinking" typical for the frontal lobes, is obvious. Thus, the map in Fig. 4 actually corresponds to the mirror reflection of the scheme in Fig. 1.

D. Interpreting the Concepts of Intuition, Sub-consciousness, Consciousness, and Logic

Now, let us turn to interpretation and revealing the mechanisms of specific human features of cognitive process, — namely, intuition, logic, sub-consciousness, etc.

If the *intuition* is treated as occasional, spontaneous, unreasoned solution, or, following Immanuel Kant [56], "the direct discretion of the truth" without any reasons and proofs – then, it apparently emerges from **RH** (more exactly, from the noise in **RH**). Typical feature of intuition consists in unconscious way of getting the result.

Treating the *logic* as all the cause-and-effect unbroken chains (causal relationships), one could infer that all the processes in **LH** are related. In this sense, the inference of our early paper [57] (where there was no symbolic structure) remains still valid. At this level, the inference of [50] seems valid also.

However, these concepts could be considered in more detail. Thus, the *logical thinking*, according to [28], is defined as "*correct provable reasoning*". This definition immediately leads to the inference that only verbalized reasoning (thereby, conclusive and commonly understandable) is related. At that, the term "correct" implies that these reasons should be based on the *conventional axioms.* Then, between the "pure logic" and "pure intuition" there should be a place for some other, intermediate, thinking algorithms.

Similar reasons concern the concepts of *consciousness* and *sub-consciousness.* Defying the consciousness as "the state of being aware of and responsive to one's surroundings" [28], we infer that it could emerge after verbalization only.

The sub-consciousness is defined as "...aggregate of processes lacking the subjective control" [28]. This implies that it should be based on the randomly stored information, that had not acquired any symbol and thus, could not be activated from outside by means of symbols (i.e., words).

Keeping in mind previous reasons, we can interpret the notions of intuition, logic, and (sub-)consciousness under NCA.

The architecture described above has large number (N>>1) of levels. The lower levels contain *auxiliary* or *hidden individual* information for a given system, the "thing in itself". Only verbalized information that occurs at higher levels of hierarchy could be comprehended in a common sense (not individually). Then, we can try to answer the question "How the brain makes a thought?" Since a speech represents a *consecutive set of symbols*, this is the very tool to form (separate) a pattern called a "thought" from all the variety of the brain-activity patterns. There exists a picturesque formula "the language is a means for our brain to speak with us". Thereby the *consciousness* could be defined as the system's ability to draw up the cognitive activity into consecutive content set by means of a speech. Here, the main role is played by **LH**.

As it was shown above, a part of information appears to be lost at any transition from previous level to the next one. More exactly, it converts into form of "*latent*" (auxiliary), or "*hidden*" information for a given system. Let us consider this in more details.

The innermost level of latent information is represented by weak ("grey") connections at the *fuzzy* set, i.e., the image plate $\boldsymbol{H^0}$. Its role consists in storing the "occasional" (i.e., "randomly collected") information that could appear to be important some time later. This information is transferred neither to **LH** nor to the level $\boldsymbol{G^1}$, thus, could not be associated with any symbol. This means that it remains *not comprehended* and not *controlled* by the system, i.e., just what has been defined as the "sub-consciousness". Such ("grey") chains could be activated only due to the noise, by chance ("to see suddenly by internal view"), that could be interpreted as the "aha *moment*" (see, e.g., [26], [58]).

At the transition from semantic information to verbalized one, there remain a lot of symbols that are not associated with any standard word. This implies certain "pictures" that could be described only by means of decomposition, i.e., one internal symbol can be described by several standard words. Verbalization of this information requires not an insight, but assortment of necessary words. This is always possible, but not always simple. Using the terms of recognition theory, this process could be called "formalizing the expert knowledge".

Thus, the latent information is disposed at various levels of depth, and this fact does control the efforts for extracting it up to the consciousness level. It seems natural to interpret the inferences based on the latent information, as intuitive thinking (*insight*). It is worth noting that in the proposed scheme, the majority of latent information is actually concentrated inside **RH**.

Logical thinking could be specified as "...unbroken sequential thoughts" [27], as well as "...operating by verbalized (abstract) concepts and their connections" [28]. This process is typical for **LH** at higher hierarchy levels.

An abstract information as itself has its own levels and infrastructure, which emerges gradually, in course of system's evolution (for human beings, this implies "with years"). This developed infrastructure that combines higher levels of **RH** and **LH** could be associated with the *wisdom*. This implies that the wisdom is broader than logic.

Specific features of the "latent" elements become rather pronounced in the process of solving the problems related to fixing the similarity/difference of the objects. These problems are solved automatically, at the image levels. The similarity is emphasized by shared neurons, while the difference is specified by diverse ones, and the system *does know* it. However, this knowledge could not be *comprehended* until those common/diverse neurons were not associated with combinations of internal symbols. Then, the *auxiliary-image* knowledge ("feeling") could be converted into *semantic* one. Further verbalization of this knowledge implies ascertainment of the connections between internal symbols and the words. The obtained result is valuable for a given system (individual), but could appear to be fault objectively, since the mode of recording the image information is individual as well. The obtained solution is *intuitive*, since it is based on the recorded experience, i.e., the individual "worldview pattern". This solution should not be proved (the system itself does not need any proof since it just knows that it is so). However, being verbalized, this solution could be explained to others and argued. If the arguments fit the conventional axioms, it would be a proof of its truth. Actually, the method of "converting the intuitive expert knowledge into logic one" is presented aforesaid.

E. Master Equations: Mathematics & Phylosophy

The mathematical foundations for the architecture presented in Fig. 1 were discussed in details in [11]. Let us recall the key points and present the mathematical basis in generalized form:

$$\frac{dH_i^0(t)}{dt} = \frac{1}{\tau_i^H}[\Im_H\{H,\beta_i(G^R{}_{\{i\}})\} + \sum_{i\neq j}^{n}\Omega_{ij}^{Hebb}H_j^0 + \sum_k \Psi_{ik}G_k{}^{R,1} - \Lambda(t)\cdot H_i^{typ}] + Z(t)\xi_i(t) \quad (8)$$

$$\frac{dH_i^{typ}(t)}{dt} = \frac{1}{\tau_i^H}[\Im_H\{H,\beta_i(G^L{}_{\{i\}})\} + \sum_{i\neq j}^{n}\Omega_{ij}^{Hopf}\cdot H_j^{typ} + \sum_k \Psi_{ik}\cdot G_k{}^{L,1} + \Lambda(t)\cdot H_i^0] \quad (9)$$

....

$$\frac{dG_k^{R,\sigma}}{dt} = \frac{1}{\tau_G}[\Im_G\{G_k,\alpha^\sigma{}_k(\{\Psi_{ik}^{R,(\sigma-1)}\},G^{\sigma+\nu})\} + \hat{Y}\{G_k^{R,\sigma},G_l^{R,(\sigma+\nu)}\} - \Lambda(t)\cdot G_k^{L,\sigma}] + Z(t)\cdot\xi(t) \quad (10)$$

$$\frac{dG_k^{L,\sigma}}{dt} = \frac{1}{\tau_G}[\Im_G\{G_k,\alpha^\sigma{}_k(\{\Psi_{ik}^{L,(\sigma-1)}\},G^{L,(\sigma+\nu)})\} + \hat{Y}\{G_k^{L,\sigma},G_l^{L,(\sigma+\nu)}\} + \Lambda(t)\cdot G_k^{R,\sigma}] \quad (11)$$

Here, variables H_i and G_k refer to purely "rational" components that are associated with neocortex, various 'τ' parameters stay for characteristic times. The term $Z(t)\xi(t)$ corresponds to the random (stochastic) component (noise) which is presented in the subsystem **RH** only; $Z(t)$ is the noise amplitude. The functionals $\Im_H\{H,\beta\}$ and $\Im_G\{G,\alpha\}$ describe the internal dynamics of corresponding neurons; the functionals $Y^R\{G^\sigma,G^{\sigma+\nu}\}$ and $Y^L\{G^\sigma,G^{\sigma+\nu}\}$ in the equations for symbolic plates describe the horizontal and vertical interactions of symbols (see [11] for details); $\Lambda(t)$ specifies the cross-subsystem connections.

Let us present several remarks on the meaning of certain terms.

1) "Brain vs. Mind" Border:

First two equations relate to the lowest (zero) level of hierarchy, while the others (***G***) variables describe σ=1,...N symbolic levels. Note that the *dotted line* after two first equations indicates the analogy with the dotted line in Fig. 1. This line symbolizes the *virtual border* between the Brain and the Mind. Indeed, the ***H***-plates (zero-level of the hierarchy) containing only the "raw" images, serve to represent the sensible information received from the organs of sense. This information is (roughly speaking) objective, so this level belongs (roughly speaking) to the Brain.

The level σ=1, that is, the level of the typical-image symbols, already belongs to the Mind, since any symbol represents not objective, but *conventional,* i.e., *subjective* and *individual* (for a given system) information. The same is true even more for all other hierarchy levels, up to the highest level associated with the abstract information. Thus, we come to

Philosophical Inference #1: The "bridge" between the "Brain" and the "Mind" is made of semantic connections between symbols and their images, i.e., by conventional (individual) information generated by the neuron ensemble itself.

2) Reflection of a single-neuron history:

The functionals $\Im_H\{H,\beta\}$ and $\Im_G\{G,\alpha\}$ defined by (1) and (4), respectively, describe the internal dynamics of the corresponding *dynamical formal neurons*. This very representation provides the possibility to describe the *parametric mutations* of the "trained" neurons, — i.e., those neurons that actually participated in creation of images and symbols forming the architecture as a whole. This effect corresponds to experimental evidences from [18].

One of the parametric-modification mechanisms consists in the influence of high-level symbols on the corresponding image neurons: $\beta_i \rightarrow \beta_i\{G^\sigma_{\{i\}}\}$. First of all, this refers to so called *symbol of class*, that is, the symbol, which was induced not by the image of certain object, but by a set of *common attributes* of certain class of objects. Excitation of such symbol could not excite all "referring" images, but switches them into the "standby mode" by lowering the activation threshold of common image neurons β_i. Thus, these images acquire the *right of priority* for activation, i.e., an *attention.*

All these arguments refer as well to the parameters α^σ_k of symbolic neurons. The *k*-th neuron at the plate $\boldsymbol{G^\sigma}$, being a member of new "generalized" image, plays the role of the *image* neuron for all the higher-level symbols $G^{\sigma+v}_{\{k\}}$ that it is related, thereby its parameter should be modified as: $\alpha^\sigma_k \rightarrow \alpha^\sigma_k\{G^{\sigma+v}_{\{k\}}\}$. Besides, as it was considered above, the neuron-symbol should be modified parametrically after its semantic content (i.e., the inter-plate connections $\Psi^{(\sigma-1)}_{ik}$ with its image) was formed: $\alpha^\sigma_k \rightarrow \alpha^\sigma_k(\{\Psi^{(\sigma-1)}_{ik}\})$. This modification takes the neuron out from the *competitive* interactions and turns on the *cooperative* ones. This factor secures complex multi-level interactions of the neuron-symbols and leaves "off screen" those *G*-neurons that failed to become a symbol.

Thereby, complete modification of a *G*-neuron reflecting the "history" of his relations with other neurons (his "skill") could be presented in the form: $\alpha^\sigma_k \rightarrow \alpha^\sigma_k(\{\Psi^{(\sigma-1)}_{ik}\}, G^{\sigma+v}_{\{k\}})$.

Thus, the model of dynamical formal neuron enables us not only to reproduce the fact of mutation of the "trained" neurons observed in [18], but also to specify and distinguish concrete modifications associated with different "skills".

Philosophical inference #2: The account for the neuron internal structure enables us to reproduce the effect of mutation of the neurons participated in certain "skill" acquirement. This provides the interpretation for the effect of "neuron memory" concentrated not in the inter-neuron connections, but inside the neurons themselves.

3) What is the tool for switching the subsystem activity?

The variable $\Lambda(t)$ controls the dialog between two subsystems. This is the only variable presenting in each equation, thus 'sewing' all the components together. Therefore, it deserves special discussion. These connections should not be trained, but should provide *switching* the subsystem activity in course of the problem solving. Here, the connections $\Lambda^{R\rightarrow L}$ activating **LH** are treated as positive $\Lambda^{R\rightarrow L} = +\Lambda_0$, and vice versa, connections $\Lambda^{L\rightarrow R}$ activating **RH** are treated as negative ones $\Lambda^{L\rightarrow R} = -\Lambda_0$. All the processes requiring the generation of new information, namely — forming either new image, or new symbol — are to proceed in **RH** with necessary noise participation. Then, the result of this process should be transferred to **LH** by direct cross-subsystem connections: $+\Lambda_0$. The reverse connections $-\Lambda_0$ are switching on in the already trained system, when an incoming external information appears to be unknown, i.e., *new*. Then, the system should pass over the re-training stage by means of **RH.** Let us stress that the mechanism of the $\Lambda(t)$ switching is not specified in (8) – (11) yet; it will be considered in the next Section.

Note that this system of equations is not complete in mathematical sense (as it was also in [11]), since not all the variables are determined via their mutual interactions. Namely, Z(t) was considered as a model parameter, and the mechanism of $\Lambda(t)$ switching is not clear. Since the considered cognitive architecture is in a good agreement with functional areas of *neocortex* (not subcortical structures), we come to

Philosophical Inference #3: Proper system of equations that describes the whole cognitive process could be completed only after taking into account the participation of *emotions.*

V. The Role and Place of Emotions

The incorporation of emotions and rational thinking into cognitive system represents really the challenge, since we need to ride over the explanatory gap between "Brain" and "Mind". Under NCA, this implies that two different "tools" are required, the one relating to the "Brain" structures, and the other one expressed in the "Mind" terms. Then, mutual influence of these "tools" could provide *integral* representation of emotions in the cognitive process.

From the evolutionary point of view (see, e.g., [30]), emotions represent far more ancient mechanism of the analysis of environment, than rational reasoning. Therefore, the sources of emotional bursts relate to so called "old cerebellum", — i.e., certain sub-cortical structures like *thalamus, basal ganglia, amygdale, substance negro,* etc. (see [7], [30]). Then, the production of these very structures could be considered as the required "Brain tool" for emotion representation.

From the other hand, the rational reasoning as rather "young" (evolutionary) ability relates to cerebral *neocortex.* Thus, the required "Mind tool" should relate also to this very structure.

Emotions provide a *synthetic* (integral) reaction that appears before the analysis of concrete reasons and motives. For humans, the specification of "*emotio*" and "*ratio*" becomes meaningful after formation of the common *language* (that is, the developed system of conventional symbols) within a certain community (see, e.g., [45]). Let us point out that any language-delivered information (speech) represents a *successive time set* of symbols. Hence, the reasoning, or rational thinking, represents a *consecutive* method of information processing. Therefore, it seems reasonable to assume that not-rational or emotional reactions correspond to the *parallel* information processing. Recalling that these functions are attributed to the left and right hemispheres, respectively (see [25], [50]), one may come to a big *temptation* to infer that rational and not-rational (emotional) thinking correspond to **LH** and **RH**, respectively. Below, it is shown that that all these arguments actually are related to the problem, but realization of this program calls for more accurate consideration.

A. *The Problem of Emotion Formalization*

In order to formalize the above arguments, let us consider the approaches to emotion classification.

In psychology, the self-appraisal (emotion) is ordinarily associated with achieving a certain *goal.* Commonly, emotions are divided into positive and negative ones, with increasing probability of the goal attainment leads to positive emotions, and vice-versa. Furthermore, it is known that any *new* (*unexpected*) thing/situation calls for *negative* emotions (see, e.g., [19]), since it requires additional efforts to hit the new goal (in the given case, to adapt to unexpected situation). Hence, to the first approximation emotions could be divided into positive and negative ones.

From the neurophysiology viewpoint, emotions are controlled by concentration and composition of the neurotransmitters inside the organism [7], [25]. All the exciting variety of known neurotransmitters (more than 4000 known species) can be sorted into two groups: the *stimulants* (like *adrenalin*, *caffeine*, etc.) and the *inhibitors* (*opiates, endorphins*, etc.). Note that this fact indicates indirectly that the binary emotion classification — positive vs. negative ones — seems bearable despite its primitiveness. However, there is no direct correspondence between positive self-appraisal and the excess of inhibitors or stimulants, the problem is more intriguing.

Anyway, the simplest "Brain tool" to represent the emotions is rather apparent: it is the *effective* (aggregated) *composition* of neurotransmitters $\mu(t)$ representing the *difference between the stimulants and inhibitors.*

According to DTI, emotions could be divided into two types: *impulsive* (impelling the generation of information) and *fixing* (effective for reception). Since the generating process requires the noise, it seems natural to associate impulsive emotions (*anxiety, nervousness*) with the *growth of noise amplitude $Z(t)$.* Vice-versa, fixing emotions could be associated with *decreasing* noise amplitude (*relief, delight*). By defining the goal of the living organism as the maintenance of *homeostasis*, (i.e., calm, undisturbed, stable state), one may infer that, speaking very roughly, this classification could correlate with negative and positive emotions, respectively.

Thus, we may infer that it is the noise amplitude $Z(t)$ (relating actually to the *neocortex*) that could be treated as the required "Mind tool" for accounting emotions.

B. *Main Hypotheses on Emotion Representation in AI*

We propose the following hypothesis on the nature of emotions: *The random component (noise) in artificial systems does correspond to the emotional background of living systems, as well as free (random) choice imitates the human emotional choice.*

This concept gives immediately *three* tools directly connected with emotions, and all of them are individual for any given artificial system:

Z_0 — stationary-state background, i.e., the value that characterizes the state "at rest";

$\Delta Z(t) = Z(t) - Z_0$ is the excess of the noise level over the background, which reflects the *measure* of cognitive activity;

dZ/dt — is the time derivative of the noise amplitude, which apparently is the most promising candidate to the analogue to emotional reaction of human being. The absolute value of derivative dZ/dt corresponds to the *degree* of emotional manifestation: drastic change of noise amplitude imitates either *panic* ($dZ/dt>0$), or *euphoria* ($dZ/dt<0$), and so on.

Various combinations of these values reveal a wide field for speculations and interpretations. For example, the calibrated value Z_0 could serve as the indicator of *individual temperament.* The states with $Z(t) < Z_0$ could be interpreted as *depression*, etc. These parameters could be applied to construct artificial cognitive systems (*robots*) of various "psychology" types.

The influence of the "Brain" component should be accounted by linking the value of dZ/dt with an *aggregated* variable $\mu(t)$ that represents the *effective* composition of neural transmitters. In an artificial cognitive system and AI, an additional (artificial) variable $\mu(t)$ should be introduced as an external factor to control the "emotional" state of the system.

Thus, the Main Hypothesis results in the following set of basic hypotheses:

- ***Hypothesis #1***: The impact of neurotransmitters should be described by the system of equations linking the noise amplitude $Z(t)$ with the aggregated variable $\mu(t)$ that corresponds to the effective composition of neural transmitters (the difference between *stimulants* and *inhibitors*).
- ***Hypothesis #2***: The *apprehended* emotional reaction of human beings could be described as the *time derivative* of the noise amplitude $dZ(t)/dt$.

Note, that this value could be either positive or negative that could be (very roughly) related to negative and positive emotions, respectively. The absolute value of derivative corresponds to the *degree* of emotional manifestation and can take *any* values to describe various emotional shades.

- ***Hypothesis #3***: The same derivative should control the "dialog" between subsystems: increasing $Z(t)$ (*negative* emotions) corresponds to activation of **RH**, while decreasing $Z(t)$ (*positive* emotions) switches on the **LH** activity.

Basing on these hypotheses, we can write the system of equations describing mutual interaction of the variables $\mu(t)$ and $Z(t)$ in course of cognitive process in the form:

$$\frac{dZ(t)}{dt} = \frac{1}{\tau^Z} \cdot [a_{Z\mu} \cdot \mu + a_{ZZ} \cdot (Z - Z_0) + F_Z(\mu, Z) + \\ X\{\mu, G_k^{R,o}\} + \{\chi \cdot (D - \omega \cdot dD/dt) - \eta \cdot \delta(t - t_{D=0})\}] \quad , (12)$$

$$\frac{d\mu}{dt} = \frac{1}{\tau^\mu} \cdot [a_{\mu\mu} \cdot \mu + a_{\mu Z} \cdot (Z - Z_0) + F_\mu(\mu, Z)], \quad (13)$$

$$\Lambda(t) = -\Lambda_0 \cdot th\left(\gamma \cdot \frac{dZ}{dt}\right), \quad (14)$$

where a, χ, η, τ, ω, and γ are model parameters, the functional $X\{\mu, G_k^{R,\sigma}\}$ refers to the process of new symbol formation (which decreases $Z(t)$ value, see details in [12]). The linear in Z and μ part in (12), (13) provides the system's homeostasis: stationary stable state corresponds to $\{Z=Z_0, \mu=0\}$. The functions $F_Z(\mu,Z)$ in (12) and $F_\mu(\mu,Z)$ in (13) are written to account for possible nonlinear effects, which may arise from mutual influence of "emotional" (neurophysiology) and "rational" (referring to the neocortex ensemble) variables (see below).

The last term in (12) refers to processing the incoming information. The term D stays for the *discrepancy* between the *incoming* and *internal* (stored) information that provokes *Z(t) increasing.* This very situation refers to the "effect of surprise", which evokes human's negative emotions. Vise versa, finding the solution to the problem (D=0) results in momentary decrease of *Z(t),* that corresponds to positive emotional splash. Thus, the model seems quite reasonable.

Finally, the hypothesis #3 results in (14), where Λ_0 being the characteristic value of the cross-subsystem connections; γ is the model parameter, which specifies the Λ dynamics. Note that *hyperbolic tangent* function in (14) provides the step-wise behavior at γ>>1. This implies that $\Lambda = \Lambda_0 = \Lambda^{R\to L}$ at $dZ(t)/dt<0$ and $\Lambda = -\Lambda_0 = \Lambda^{L\to R}$ at $dZ(t)/dt>0$, with Λ being zero at $dZ(t)/dt=0$. Small/moderate variations of dZ/dt around zero provide corresponding oscillations of Λ(t) that represent permanent (normal) "dialog" between subsystems. Besides, the solution to standard problems can be found in **LH** only and commonly does not provide any emotional reaction: $\Lambda \sim dZ/dt = 0$ (any inter-subsystem connections are not activated). Hence, this equation fits completely our previous psychological considerations.

Thus, the system of equations (8) – (14) appears to be fully complete since all the variables are defined via their mutual interactions. Let us stress that linking the cross-subsystem connections Λ(t) with the emotional variable dZ(t)/dt gives quite original and necessary mechanism to control the subsystem activity and provides desired tool for realization of an artificial two-subsystem schemes (robots).

C. Application of the Model to the Stress/Shock Effect

Let us consider an example of applying this model to reproduce certain observable effect. The effect of "stress and shock", that occurs when people find themselves in a stressful situation, was investigated for several years by the group of neurophysiologists [59]. Two specific characteristics of electrocardiogram were measured, one of them being an appraisal of vegetative imbalance, another one being the measure of heart-rate variability. It was observed that under small or moderate external impact, people gradually calm down after several oscillations of measured characteristics. But in the case of strong impact, initial excitation changes for *depression* and only after sufficiently long time the person can return to ordinary (regular) reactions. This type of behavior was identified as "*stress"*. Moreover, there was detected the regime called a "*shock*": the probationer, after too strong initial excitation, falls down to *deep depression* (*stupor* or *coma),* and cannot relax independently, without medical assistance. In the latter case, the vegetative balance is controlled by the *opiates* only (pronounced inhibitors), with the variability index comes to zero. It is worse noting that the levels of initial excitation resulting in "irregular" regimes of behavior were detected to be individual. All these regimes could be reproduced in the proposed model by choosing an appropriate parameter set.

The first attempt to describe these effects was done in [12], where *two different* sets of parameters had been used to reproduce the "normal\stress" and "shock" regimes respectively. This means that, the transition between the stress and shock states was treated as *parametric* modification of the system. Alternative version of this model (different choice of parameters) is presented in this article. It enables us to reproduce *all the regimes* within *single* combination of parameters, by varying the initial conditions. Besides, modern description of the stress-to-shock transition seems to be more interesting and relevant (see below).

In Fig. 5, the phase portrait for the model (12) – (13) is presented, where the parameters are chosen to provide the *N*-shape isoclinic curve dZ/dt = 0 with just *two stable* statio-

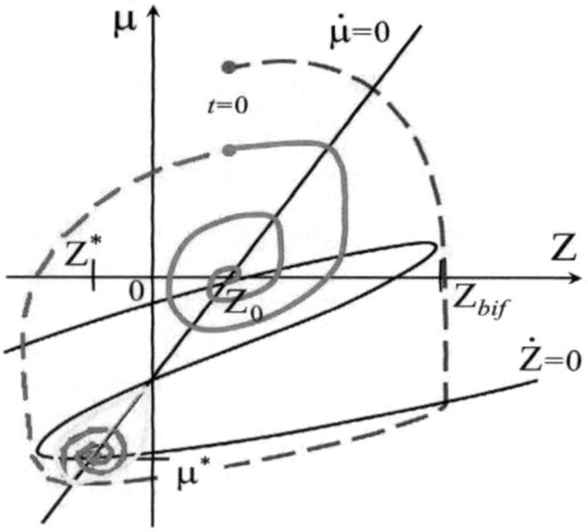

Figure 5. Model phase portrait in terms of "noise amplitude *Z(t)* vs. an aggregated neurotransmitter composition *μ(t)*".

nary states. The normal stationary state $\{Z=Z_0, \mu=0\}$ corresponds to homeostasis. The second one $\{Z=Z^*, \mu=\mu^*\}$ corresponds to *abnormal* state (pathology), where the noise is deeply suppressed ($Z^*<0$), and the neurotransmitter imbalance is shifted to deep inhibitor region ($\mu^*<<0$). This state just corresponds to that of the "shock" — this implies deep depression with possible transition into a *coma*. Both stationary states represent *stable focuses*.

Normally, the dynamical regime represents *damping oscillations* around the homeostasis point $\{Z_0,0\}$. Initial excitation $\mu(t=0)$ (that imitates an external impact) provokes growth of Z supplied by the following decrease of μ down to negative values, which then changes for decreasing Z with μ growth, and so on. Thus, the values of $Z(t)$ and $\mu(t)$ gradually (over several cycles) trend to their stable points (solid green curve). But if the trajectory, starting from somewhat larger initial value $\mu(0)$, would pass beyond some *bifurcation* value Z_{bif}, the dynamical regime changes (dashed red curve). The trajectory falls down to negative μ (inhibitor) zone where spends a long time. Then it slowly, over the *depression* zone $Z<0$, returns to regular (oscillatory) mode. This regime qualitatively corresponds to the "*stress*" behavior.

The yellow curve in Fig. 5 represents the *separatrix* between the attraction zone of abnormal stationary state $\{Z^*,\mu^*\}$ and other behavioral modes. Since this state is also a stable focus, the *affix*, ones getting inside the attraction zone, will be "sucking" up (over several damping oscillation around) to the abnormal stationary state, and cannot leave this zone independently, without serious external impact. It should be stressed that normally, the trajectory cannot cross the separatrix from outside; this could occur only occasionally (due to some small excitation when the affix is near the separatrix). This implies that commonly, the *stress* regime returns to a normal mode and should not result in the shock state. But, since at certain stage of the process the trajectory comes very close to the separatrix, the least excitation could result in hitting the shock zone and fall down to the coma state. Thus, this model enables us to infer that the stress regime is *dangerous* for human beings, since this process includes the stage (just before the stress mode turns to increasing μ values, i.e., to rather normal behavior) when the weak external excitation could provoke momentary stress-to-shock transition. This is *novel* model prediction, which could be tested experimentally. Note that certain evidences in favor of this effect were already detected [59].

This model could be applied to analyze possible results of use of different medical impacts, such as adding certain *stimulants* at different stages of the stress process. Such research could lead to pronounced applied results.

The described effects are in good qualitative agreement with the experimentally observed ones [59]. Quantitative correspondence is intricate, since the characteristics that are measured experimentally are close *per se* to $Z(t)$ as a measure of irregularity, and $\mu(t)$ as a measure of mediator imbalance. However, there is no direct correspondence between theoretical and experimentally measured variables.

VI. Emotional Manifestations in Course of Information Processing

Let us discuss the role of emotions in solving the problems of *recognition* and *prediction,* which could be accompanied by certain dynamical variation of the noise amplitude $Z(t)$. Typical patterns of $Z(t)$ behavior will be presented below.

A. Recognition

Note that he extended set of images with distinguished "borders" is needed for good quality of recognition (classification). Usually such classifier is built in course of training the recognition system. Under NCA, **RH** plays the role of Supervisor for **LH,** and that is trained **LH** that implements the function of the object recognition (classification).

The problem of object/phenomenon recognition is solving in already trained system (with at least two trained lower levels $\sigma=0, 1$) by means of image plates.

The incoming information is perceived by both subsystems. If this information is well known, the problem is solved in the subsystem **LH** by means of Hopfield-type mechanism of *refinement*: all the images are treated as already known ones — by fitting them to coincide with already stored patterns. In the case of insufficient recognition (when the fitting procedure fails), the participation of **RH** becomes necessary. An unrecognized image is treated as a new one and undergoes the common procedure of new symbol formation.

The problem setting consists in excitation of certain group of neurons ("*examinee* object") in the *fuzzy set* $\boldsymbol{H}^0$ in **RH.** Here, this "object" is processing "as it is", i.e., by *blackening* connections between all the examinee neurons. This neuron group could contain several "skilled" neurons (that belong to certain already known image), with already black connections between them. This means that the examinee object is (to some extent) similar to some familiar (already learned) one.

Then, this image is transferred (by direct cross-subsystem connections Λ) to the typical-image plate $\boldsymbol{H}^{typ}$ in **LH.** Further procedure is controlled by the value of the discrepancy D, which could be defined as

$$D(t) \equiv \sum_{i}^{M} \left\| H_i^0 - H_i^{typ} \right\|, \tag{15}$$

where summation is performed over M excited examinee neurons.

There are several possible cases.

1) Familiar object: If the examinee object is well-known to the system, i.e., its image completely coincides with one of typical images in **LH,** so that $D(0)=0$, it would be straight away (quickly!) associated with corresponding symbol, with all the following consequences concerning its position in the hierarchy. In this case, **RH** does not participate further in the process. Accordingly, $dZ(t)/dt=0$, so this (practically automatic) procedure do nor call for any emotions.

2) Examinee object is close to familiar one: The examinee object can be sufficiently similar to one of the known typical images (fits its "attraction area"), i.e., $D(0)\neq 0<D_{cr}$, where D_{cr} represents certain critical value of discrepancy. Then, it is treated as familiar one together with its symbol at the next level $G^{L,1}$. However, in this case the recognition propriety requires *verification*. For this purpose, the *symbol* should be transferred to **RH** for decomposition, and the result should be compared with the examinee image. Thus, there arises the *loop*, i.e., iterative process presented in Fig. 6:

$$\begin{array}{ccc} H^{typ}(LH) & \xrightarrow{\Psi^L} & G^{L,1}(LH) \\ \uparrow\Lambda^{R\to L} & & \downarrow\Lambda^{L\to R} \\ H^{0}(RH) & \xleftarrow[\Psi^R]{} & G^{R,1}(RH) \end{array}$$

Figure 6. Schematic representation of the iterative recognition procedure

If the result of comparison was satisfactory (it could be estimated by running D value), the examinee object would be associated with an existing symbol. If not, the discrepancy provokes repeating, and the procedure should pass over several iterations. At that time, the image in the fuzzy set $\boldsymbol{H^0}$ gradually blackens.

3) Examinee object is far from familiar one: If $D(0)>D_{cr}$, at some moment the connections recording the object in **RH** turn to be sufficiently "black" ($\Omega^0_{ij}\geq\Omega_0$), but the typical-image plate do not recognize the object, — then, it turns out to be the *new typical image* and should take its place at the plate $\boldsymbol{H^{typ}}$, so that $D(t_{D=0})$ = 0. Then, the common procedure of new symbol formation should provide its own symbol, which should be linked to high-level symbols, and so on. The moment $t_{D=0}$ is accompanied by $Z(t)$ decrease — the system had solved the recognition problem and could relax. Typical pattern of $Z(t)$ dynamics in course of recognition procedure is presented in Fig. 7a.

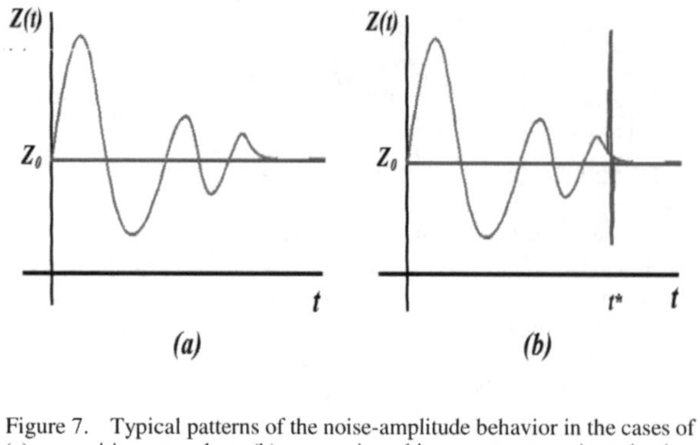

Figure 7. Typical patterns of the noise-amplitude behavior in the cases of (a) recognition procedure; (b) prognosis and incorrect prognosis at the time moment t* (illustration for the sense of humor).

Thus, we can infer that the given system is capable to process and recognize even *new* objects, yet only with participation of the fuzzy set $\boldsymbol{H^0}$.

B. Prognosis

The prognosis (forecast) can be treated as a "recognition of a time-depending process". It proceeds in **LH** after the symbol of the given *process* is formed. This generalized symbol collects all the information about the "process pattern" (image of symbols) in a compressed form. Then, the information on some middle stage of the given process activates it's symbol, providing the activation of the entire chain of symbols enclosed in this process.

Therefore, emotional manifestations, as well as the pattern of noise-amplitude $Z(t)$ behavior here is similar to that in the case of recognition (Fig. 7b). Note that this statement is true up to the moment when the prediction is failed. This means that the information coming at some moment t^* appears to be *unexpected*. This case refers to the problem of the sense of humor (see below).

C. Interpretation of the Sence of Humor

Under the presented concept, the sense of humor could be interpreted as an "*ability to adapt quickly to unexpected information with getting positive emotions*". This process is illustrated in Fig. 8.

Let the incoming data represent a time sequence of symbols that is perceived *consequently* by **LH**, as it is shown in Fig. 8. At the initial stages, the information perceived is usually not concrete enough to correspond to one *symbol of process* at $\boldsymbol{G^2}$, thus the system makes no predictions. A prognosis could be done when accumulated information enables the subsystem to choose one symbol among the others (in Fig.8, "black" symbol at $\boldsymbol{G^2}$ plate, which has more strong connections than the "green" one, i.e., it corresponds to more "common" process). Then, the system *waits* for further details of the predicted process (this means activation of the "black"-symbol chain at $\boldsymbol{G^1}$ plate).

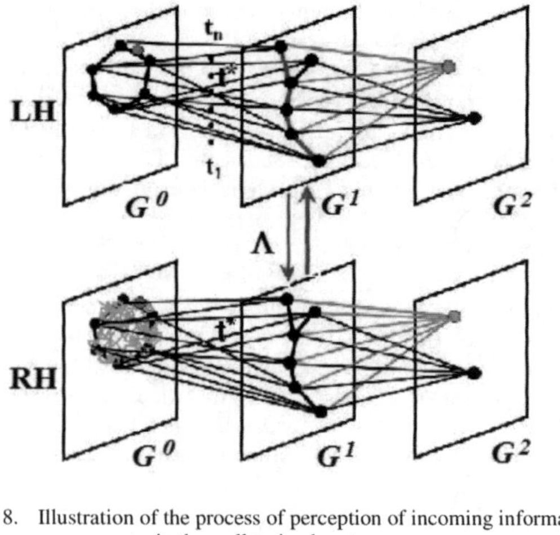

Figure 8. Illustration of the process of perception of incoming information in the well-trained system.

Up to certain moment t^*, the incoming information ("violet" chain in Fig. 8) fits these expectations. At the moment t^*, the prognosis on further information could appear to be *incorrect*, — the next symbol at G^1 plate belonging to "violet" chain, actually is not involved into the "black"-symbol chain, and thus unexpected. Then the system has to appeal to **RH** (down Λ-arrow in Fig. 8); in this process, the emotions are negative: $dZ/dt > 0$. However, the system may rapidly find a new solution — this implies that there *already exists* the symbol of another process that matches completely both, former and current information ("green" symbol at G^2 plate in Fig. 8). This leads to positive emotions ("aha" moment) and hence switches the connections $\Lambda^{R\to L}$ (up arrow in Fig. 8).

According to this concept, a good anecdote should be a story that, up to certain moment t*, permits a well-known interpretation. The next information block should *not deny* the previous version, but suggest another (alternative) also well-known interpretation. In this case, the system has to return to the turning point at t^* and then choose the "right" chain of symbols fitting all the incoming information. The very process of returning and jumping to the "right" trajectory requires definite specific efforts — so again it leads to the spike of the noise amplitude that simulates laughter.

Let us stress that all this is possible, if the system is reach enough with symbols of processes, i.e., has large enough "repertoire" of various symbols and images. Then this process is rapid, both trends appear to be superimposed: the value $Z(t)$ undergoes abrupt increase-and-decrease ("spike") that could be interpreted as an analogy to human *laughter* (abrupt involuntary reaction). Thus, we infer that a sense of humor could be inherent to the well-trained *(erudite)* system only, just as it is for human beings.

D. Interpretation of Aesthetic Emotions (annons)

NCA could be applied to the problem of analysis of the nature of so called *aesthetic* emotions. Emotions of this type are not connected with any rational (pragmatic) reasons, but are evoked by pure Nature phenomena (rainbow, fire, etc.), pieces of Art, etc. Under NCA, one may suppose that these emotions are associated with the *recognition paradox*: these phenomena seem familiar and surprising simultaneously. In this case, $Z(t)$ should display small variations (*vibration*) around the normal level Z_0, that correspond to the human feeling called the "*goosbumps*". In many aspects, the mechanism of aesthetic emotion production is similar to the incorrect\undone prognosis, therefore to the sense of humor. That is why pronounced emotions are often accompanied by the laughter (or tears). However, there is important difference: the unexpectedness in the case of aesthetic emotions could not be "resolved" by switching to another, already known symbol. Nevertheless, this problem deserves further study [60].

VII. Perspectives on Practical Validation

This theoretical study has *per se* fundamental character and could be related to the human-level AI (HLAI) trend. However, its experimental verification represents the most interesting problem.

The comparison of our model predictions with the experimental results on Electro Cardiogram (ECG) analysis under the stress/shock conditions [59] has shown good qualitative agreement. Note that these experiments were based on the analysis of ECG, with *model-dependent* interpretation of the correspondence between ECG pattern and the activity of certain brain areas.

However, the model variable corresponding to the noise amplitude Z(t) has no direct analogues within the experimental technique used. This requires special efforts to extract this information from experimental data on the neocortex activity.

The model predictions concern certain peculiarities in the brain activity, including the cerebral cortex and subcortical structures. The sub-cortical production could be estimated indirectly, by analysis of certain vegetative indices, as it was done, e.g., in [59]. However, meaningful experiments should involve combined study using ECG, Electro Encephalogram (EEG), and functional Magnetic Resonance Imaging (fMRI). In this process, the main attention should be paid to the dynamic variations in the brain-structures activity, thus good enough time resolution of the experimental devices is required.

These techniques are actually available now [61]. We plan to perform such experiments in collaboration with the group of V. L. Ushakov in Kurchatov Research Center, Moscow, in the nearest future. In particular, we plan to perform combined analysis of ECG, EEG, and fMRI data for people under the "light stress" experimental conditions (e.g., time trouble in solving specific cognitive problems).

VIII. Conclusions and Future Work

In summary, the main inference of the paper is that NCA architecture inherently contains the possibility and even necessity to incorporate the emotions into the cognitive process.

The main constructive feature of this architecture is representation of the whole system as a combination of two linked subsystem, **RH** and **LH**, with the presence of random element (noise) in **RH** only. These subsystems could be associated with cerebral hemispheres, with the connections Λ(t) between them representing *corpus callosum*. It was shown that **RH** is responsible for processing the new (therefore, unexpected) information, while **LH** stores and processes the well-known one. This functional specialization is in entire agreement with the practical inferences of Goldberg [19]. The coincidence of theoretical (DTI-based) and practical (practicing psychologist E. Goldberg) inferences represents a pleasant surprise and indirect verification of NCA.

However, this design requires a specific mechanism to control the subsystem activity. It is quite natural to associate this mechanism with the emotional response to incoming information.

It is shown that emotional self-appraisal in an artificial cognitive system could be associated with the variation of the noise amplitude. In order to reproduce human-level emotional process, which is regulated by the neural transmitters, it should be linked to certain additional variable reflecting aggregated composition of neurotransmitters. Their mutual dynamical interaction in course of cognitive process provides the tool for regulating the activity of subsystems. Thus, the problem of "Explanatory Gap" between the "Brain" and "Mind" approaches to representation of emotions is solved.

Returning to the wide-spread and somewhat "vulgar" idea that **RH** is a "container" for emotions while **LH** provides rational reasoning, we may infer that emotions actually lie *deeper* (in all senses). They belong neither to **RH** nor to **LH**, but actually control their activity. That is why in Fig. 1, "emotions" are virtually displayed beyond both subsystems (associated with *neocortex).*

The emotional response is described by the *derivative* dZ/dt of the variable that indicates the level of the noise $Z(t)$. Negative emotions imitated by the noise increasing ($dZ/dt>0$) correspond to unexpected incoming information (incorrect and/or undone prognosis, surprise); in this process, **RH** should be activated. Vice-versa, solving any problem results in positive emotions and, correspondingly, decrease of the noise amplitude ($dZ/dt <0$) — then, only **LH** remains active, while **RH** gets an opportunity to be "at rest". Specific case of an abrupt up-and-down jump ("spike") of the function $Z(t)$ could be associated with specific human manifestation of emotions (the *laughter*).

Realization of this program in AI could be accompanied by certain sound effects, such as artificial "*laughter*" in the case of abrupt spike of $Z(t)$. In addition, variation of the noise amplitude during the process of problem solving could be accompanied by the display of visual "symbols", such as cheery or sorrowful "faces", etc.

This approach opens a wide field for imitation and model analysis of various human peculiar features. This implies, e.g., that various types of temperament could be associated with certain values of the rest-state noise amplitude Z_0 and thus classified. Furthermore, the model described the stress/shock effect could be employed for working up new medical-treatment techniques for specific (neural) diseases. All these tasks require further study.

It should be stressed that all these possibilities emerge from the human-like cognitive architecture proposed under NCA. Let us accentuate several *key points* of NCA that distinguish it from other neuromorphic approaches and could be applied successfully to artificial cognitive systems (particularly, in Robotics):

- *Continual* representations of neural processors involving nonlinear differential equations.

This representation enables us to interpret and reproduce the experimentally observed effect of mutation of the "*skilled*" neurons (participated in acquisition of certain experience) by the *parametric* modification.

- The whole system represents a combination of *two linked subsystems* (**RH** and **LH**) – for generation and reception of information, respectively.
- *Different training* principles in **RH** and **LH** secure the hemisphere specialization.

New information processing requires the amplification of the new connections (Hebbian principle), while the processing of well-known information (recognition) requires the selection principle "redundant cut-off" (Hopfield's rule).

- Account for *random component* ("noise") presented in **RH** only.

This fact immediately specifies the role of **RH** in the response to unknown/unexpected conditions and leads to:

- Interpretation of emotions as a tool for controlling the subsystem activity, that could be realized via the noise-amplitude derivative dZ/dt.
- *Instability* of the image-to-symbol conversion process that leads to unpredictable patterns.

This very factor could secure the *individuality* of an artificial cognitive system.

- The "connection-blackening" principle of self-organization, which provides the possibility for **RH** to acts as a Supervisor for **LH;** no external supervising is needed for permanent learning.

Thus, these design features make it possible to reproduces the peculiarities of *human* cognition — that is, unpredictable character, individuality, permanent learning, ability of logical and intuitive thinking, etc. Note that these problems are actually not considered in other approaches.

It should be stressed that under NCA, the noise (random element) is treated not as unavoidable obstacle (as it is in radio physics, information-delivery tasks, etc.), but as necessary *full member* of all the processes referring to generation of information. Note that the noise (concerning the living systems, this implies fortuitous, spontaneous, sudden act), represents the *survival mechanism* that prevents precise and speed acting (particular for robots) in common situations, but provides an ability to find *occasionally* quite sudden and unpredictable exit from a critical situation. This very factor could provide the human-like features in an artificial system.

Actually, modern AI systems correspond to **LH** under NCA, but this is the **RH** that secures the emergence and individuality of such intellect. Moreover, even in the well-trained cognitive system, the combination of **LH** and **RH** provides rather broad spectrum of abilities than **LH** only — without **RH**, the cognitive system appears to be poor. Thus, we can infer that the NCA architecture, in spite of its seeming complexity and awkwardness, has several advantages comparing with popular AI architectures. Some loss of materials for doubling the system could gain a profit in system's self-development.

It is worth noting that the idea of using two subsystems, with the noise being presented in the one, has already attracted an attention in Robotics [62]. However, this idea requires specific mechanism for switching the activity of certain subsystem depending on the process stage. Under NCA, this mechanism is actually proposed. According to our main hypothesis, it should be controlled by emotions displayed as the noise-amplitude variation.

Thus, it is shown that under NCA, emotional response to external information (including unexpected, i.e., surprising one) could be imitated by specific behavior of the noise amplitude.

These ideas deserve further research and experimental verification.

REFERENCES

[1] O. D. Chernavskaya and Ya. A. Rozhylo, "On the Possibility to Imitate the Emotions and "Sense of Humor" in an Artificial Cognitive System," Proc. of the Eighth International Conference on Advanced Cognitive Technologies and Applications (COGNITIVE 2016) IARIA, Mar. 2016, pp. 42–47; ISBN: 978-1-61208-462-6.

[2] J. E. Laird, The Soar cognitive architecture, MIT Press, 2012.

[3] A. Samsonovich, "Bringing consciousness to cognitive neuroscience: a computational perspective," Journal of Integrated Design and Process Science, vol. 1, pp. 19–30, 2007.

[4] Y. Bengio and Y. Le Cun, Scaling Learning Algorithms towards AI. In: Large Scale Kernel Mashines, L. Botton, O. Chapelle, D. DeCoste, and J. Weston (Eds.), MIT Press, 2007.

[5] H. Lee, R. Ekanadham, and A. Ng, "Sparse deep belief net model for visual area V2,"Advances in Neural Information Processing Systems (NIPS), vol.7, pp.873–880, 2007.

[6] Yann LeCun, Yoshua Bengio, and Geoffrey Hinton, "Deep Learning," Nature, vol. 521, pp. 436–444, 2015.

[7] L. F. Koziol and D. E. Budding. Subcortical Structures and Cognition. Implications for Neurophysiological Assessment. Springer, 2009 .

[8] K. Doya, "Complementary roles of basal ganglia and cerebellum in learning and motor control," Current Opinion in Neurobiology, vol. 10, pp.732–739, 2000.

[9] Springer Handbook of Robotics. Editors: Bruno Siciliano, Oussama Khatib , Springer, 2016. ISBN: 978-3-319-32552-1.

[10] G. A. Carpenter and S. Grossberg, Adaptive resonance theory. In: Encyclopedia of Machine Learning and Data Mining, C. Sammut and G. Webb (Eds.) Berlin: Springer-Verlag, 2016.

[11] O. D. Chernavskaya, D. S. Chernavskii, V. P. Karp, A. P. Nikitin, and D. S. Shchepetov, "An architecture of thinking system within the Dynamical Theory of Information," Biologically Inspired Cognitive Architectures (BICA), vol. 6, pp. 147–158, 2013.

[12] O. D. Chernavskaya, D. S. Chernavskii, V. P. Karp, A. P. Nikitin, D. S. Shchepetov, and Ya.A.Rozhylo, "An architecture of the cognitive system with account for emotional component," Biological Inspired Cogitive Architectures (BICA), vol.12, pp. 144–154, 2015.

[13] H. Haken, Information and Self-Organization: A macro-scopic approach to complex systems. Springer, 2000.

[14] D. S. Chernavskii,"The origin of life and thinking from the viewpoint of modern physics," Physics-Uspekhi, vol. 43, pp. 151–176, 2000.

[15] D. S. Chernavskii, Synergetics and Information. Dynamical Theory of Information. Moscow, URSS, 2004 (in Russian).

[16] J. G. Nicholls, A. R. Martin, and P. A. Fuchs, D. A. Brown, M. E. Diamond, and D. A. Weisblat, From Neuron to Brain, 5th ed. Sunderland, Mass: Sinauer Associates, Inc., 2012.

[17] D. O. Hebb, The organization of behavior. John Wiley & Sons, 1949.

[18] O. E. Svarnik, K. V. Anokhin, and Yu. I. Aleksandrov, "Experience of a First Whisker-Dependent Skill Affects the Induction of c-Fos Expression in Somatosensory Cortex Barrel Field Neurons in Rats on Training the Second Skill," Neuroscience and Behavioral Physiology, July 2015, DOI: 10.1007/s11055-015-0135-3

[19] E. Goldberg, The New Executive Brain. Oxford University Press, 2009.

[20] J. J. Hopfield, "Neural networks and physical systems with emergent collective computational abilities," PNAS, vol. 79, p. 2554, 1982.

[21] S. Grossberg, Studies of Mind and Brain. Boston, Riedel, 1982.

[22] T. Kohonen, Self-Organizing Maps. Springer, 2001.

[23] S. Grossberg, "Towards Solving the Hard Problem of Consciousness," Invited talk given at the international conference on Human-Level Artificial Intelligence (HLAI). Available from: http://agi-conf.org/hlai2016/wpcontent/uploads/2016/03/HLAI2016 Retrieved 2016.11.30

[24] J. Levin, "Materialism and Qualia: The Explanatory Gap," Pacific Philosophical Quarterly, vol. 64(4), pp. 354–361, 1983.

[25] J. Stirling and R. Eliott, Introducing Neuropsychology, 2nd ed. (Psychology Focus), Psychology Press, 2010.

[26] R. L. Solso, Cognitive psychology, 6th ed., Pearson, 2000.

[27] Merriam-Webster Dictionary. Available from: http://www.merriam-webster.com/definition Retrieved 2016.11.30

[28] Wikipedia. Available from: http://www.en.wikipedia.org/wiki/Main_Page Retrieved 2016.11.30

[29] E. M. Izhikevich and G.M. Edelman, "Large-scale model of mammalian thalamocortical systems". In: Proceedings of the national academy of sciences (PNAS), vol.105, pp. 9, 2008.

[30] J. Panksepp and L. Biven, The Archaeology of Mind: Neuroevolutionary Origins of Human Emotions. N.Y.: Norton, 2012.

[31] P. Vershure, "The Distributed Adaptive Control: A theory of the mind, brain, body nexus," BICA, vol. 1, pp. 55–72, 2012; "The Distributed Adaptive Control of Consciousness," Lecture given at FIRCES on BICA, April 20–24, Moscow, 2016.

[32] H. Lovehiem, "A new three-dimensional model for emotions and monoamine neurotransmitters," Med. Hypotheses, vol. 78, pp. 341– 348, 2012; DOI: 10.1016./j.mehy.2011.11.016.

[33] J. Valerdu, M. Talanov, S. Distefano, M. Mazzara, A. Tchitchigin, and I. Nurgaliev, "A cognitive architecture for the implementation of emotions in computing systems," Biologically Inspired Cognitive Architectures (BICA), vol. 15, pp.34–40, 2016.

[34] A. Samsonovich "Emotional biologically inspired cognitive architecture," Biologically Inspired Cognitive Architectures (BICA), vol. 6, pp. 109–125, 2013.

[35] M. Sellers, "Toward a comprehensive theory of emotion for biological and artificial agents," Biologically Inspired Cognitive Architectures (BICA), vol. 6, pp. 3–26, 2013.

[36] M. I. Rabinovich and M.K. Muezzinoglu, "Nonlinear dynamics of the brain: emotions and cognition" Physics-Uspekhi, vol. 53, pp. 357–372, 2010.

[37] J. Treur, "An integrative dynamical systems perspective on emotions," Biologically Inspired Cognitive Architectures (BICA), vol. 6, pp. 27–40, 2013.

[38] J. Schmidhuber, "Simple algorithmic theory of subjective beauty, novelty, surprise, interestingness, attention, curiosity, creativity, science, music, jokes," Journal of Science, vol. 48 (1), pp. 21–32, 2009.

[39] D. Huron, Sweet Anticipation: Music and Physiology of Expectation. MIT Press, 2006.

[40] E. Hudlycka, "Affective BICA: Challenges and open questions," Biologically Inspired Cognitive Architectures (BICA), vol. 7, pp. 98–125, 2014.

[41] W. Weaver and C. Shannon, The Mathematical Theory of Communication. Univ. of Illinois Press, 1963; ISBN 0-252-72548-4.

[42] S. Verdu, "Fifty years of Shannon theory". In: Sergio Verdu and Steven McLaughlin: Information Theory: 50 years of discovery, IEEE Press, pp.13–34, 2000.

[43] I. Prigogine, End of Certainty. The Free Press, 1997; ISBN 0684837056.

[44] H. Quastler, The Emergence of Biological Organization. New Haven: Yale University Press, 1964.

[45] T. W. Deacon, The Symbolic Species: The Co-Evolution of Language and the Brain. Norton &Co, NY, 1997.

[46] E. M. Izhikevich, Dynamical systems in neuroscience: the geometry of excitability and bursting. MIT Press, 2007.

[47] A. L. Hodgkin and A. F. Huxley, "A quantitative description of membrane current and its application to conduction and excitation in nerve," Journal of Physiology, vol. 117, pp. 500–544, 1963.

[48] R. FitzHugh, "Impulses and physiological states in theoretical models of nerve membrane," Biophys. J., vol. 1, p. 445, 1961.

[49] J. Nagumo, S. Arimoto, and S. Yashizawa, "An active pulse transmission line simulating nerve axon," Proc. IRE, vol. 50, p. 2062, 1962.

[50] V. L. Bianki, "Parallel and sequential information processing in animals as a function of different hemispheres," Neuroscience and Behavioral Physiology, vol. 14 (6), pp. 497–501, 1984.

[51] C. M. Bishop, Pattern Recognition and Machine Learning. Springer, 2007.

[52] W. S. McCulloch and W. Pitts, "A Logical Calculus of the Ideas Immanent in Nervous Activity," Bulletin of Mathematical Biophysics, vol. 5, p. 115, 1943.

[53] J. R. Anderson, How can the human mind occur in the physical universe? Oxford University Press, NY, 2007; ISBN 0-19-532425-0.

[54] C. L. Dancy "ACT-R<PHI>: A cognitive architecture with physiology and affect," Biologically Inspired Cognitive Architectures (BICA), vol. 6, pp. 40–45, 2013.

[55] The Cerabral Cortex Map. Available from: http://www.Antranik.org/functional-areas-of-the-cerebreal-cortex / Retrieved 2016.11.30

[56] I. Kant, Critick of Pure Reason, ed. William Pickering, London, 1838; original 3-d edition: Kritic der reinen Vernunft, ed. J.F. Hartknock, 1790 (in Deutch).

[57] O. D. Chernavskaya, D. S. Chernavskii, and A. P. Nikitin, "The concept of intuitive and logical in neurocomputing," Biophysics, vol. 54, pp. 727–735, 2009.

[58] G. A. Wiggins, "Learning and Creativity in the Global Workspace," Advances in Intelligent Systems and Computing, vol. 196, p. 57, 2013.

[59] S. B. Parin, A. V. Tsverlov, and V. G. Yakhno, "Models of Neurochemistry Mechanism of Stress and Shock Based on Neuron-like Network," Proceedings of Int. Simp. "Topical Problems of Biotonics," Aug. 2007, pp. 245–246.

[60] O. D. Chernavskaya, D. S. Chernavskii, Ya. A. Rozhylo, and D. S. Shchepetov, "Hypothesis on the nature of aesthetic emotions and the concept of Chef-d'eouvre" (unpublished).

[61] A. S. Sedov, D. A. Devetiarov, U. N. Semenova, V. V. Zavyalova, V. L. Ushakov, R. S. Medvednik, M. V. Ublinsky, T. A. Akhadov, and N. A. Semenova, "Dynamics of Brain Activity during Voluntary Movement: fMRIStudy," Zh. Vyssh. Nerv., vol. 65, pp. 436–445, 2015; DOI:10.7868/S0044467715040115

[62] K. Kushiro, Y. Harada, and J. Takeno, "Robot uses emotions to detect and learn the unknown," Biologically Inspired Cognitive Architectures (BICA), vol. 4, pp. 69–78, 2013.

Constraining the Connectivity of Sparse Neural Associative Memories

Philippe Tigréat, Vincent Gripon, Pierre-Henri Horrein
Electronics Department, Telecom Bretagne
Brest, France
Email: {philippe.tigreat, vincent.gripon, ph.horrein}@telecom-bretagne.eu

***Abstract*—Willshaw networks are a type of associative memories with a storing mechanism characterized by a strong redundancy. Namely, all the subparts of a message get connected to one another. We introduce an additional specificity, by imposing the constraint of a minimal space separating every two elements of a message. This approach results from biological observations, knowing that in some brain regions, a neuron receiving a stronger stimulation can inhibit its neighbors within a given radius. Theoretical arguments are derived to quantify the benefits of this method in terms of memory usage as well as pattern completion ability. We experiment with different values of the inhibition radius introduced, and we study its impact on the error rate in the retrieval of stored messages. We show that this added constraint can result in significatively better performance of the Willshaw network, either when reducing its set of connections, or when extending its set of neurons while maintaining the memory resource.**

Keywords—Willshaw Networks; Clique-Based Neural Networks; Content-Adressable Memory; Lateral Inhibition; Sparse neural networks.

I. INTRODUCTION

Associative memories are a type of computer memories that are part of the broader category of content-adressable memories. A new model of associative memory based on neural networks was formerly introduced in [1]. Where adressable memories associate an adress with a piece of data, associative memories have the characteristic of associating patterns to one another. Among this group, we distinguish between hetero-associative memories, and auto-associative memories. An hetero-associative memory will associate together patterns in pairs. For instance, if the pattern p_1 was associated with pattern p_2, the request p_1 will bring the response p_2. Auto-associative memories follow a different principle, as they will associate a pattern with itself. The main use case of these memories is pattern completion, where a request made of a subpart of a stored message will get as response the completed pattern. Associative memories can be found in several types of real-world applications, such as database engines [2], network routers [3], data compression devices [4], and computer vision systems [5].

Today, it is widely accepted that the working principle of the brain can often be likened to the operation of an associative memory. The focus is put here on the phenomenon observed in biological neural networks, called lateral inhibition [6]. It can also be referred to as surround suppression [7]. This translates in the inhibition exerted by some neurons on their close neighbors when these have an activity inferior to their own. Starting from the Willshaw model [8], we propose a neural associative memory that is improved in terms of plausibility, by the introduction of local inhibition that results in the prohibition of short-range connections. We show that this modification brings a performance improvement in the retrieval of stored messages.

Section II introduces three associative memory models with relevant relationship to this work. Section III gives a formal presentation of Willshaw networks, including the usual message retrieval algorithm, and biological considerations motivating the modifications we introduce. Section IV details modifications in our implementation as compared to the classic Willshaw model, including the constraint applied on the space between connected neurons. Section V provides theoretical arguments showing the better usage of memory brought by this constraint. Section VI presents the results we obtain in pattern completion, and gives some theoretical explanations.

II. RELATED WORK

A. Hopfield Networks

The prominent model for associative memories was introduced by John Hopfield [9], [10]. Hopfield networks are made of a set of N neurons that are fully interconnected. The training of these networks, given n binary vectors x^μ of length N, consists in modifying the weight matrix W according to the formula:

$$w_{ij} = \frac{1}{n}\sum_{\mu=1}^{n} x_i^\mu x_j^\mu, \tag{1}$$

where element w_{ij} at the crossing between line i and column j of W is the real-valued connection weight from neuron i to neuron j.

As connections are reciprocal and not oriented, we have:

$$w_{ij} = w_{ji} \quad \forall i, j \in [\![1, N]\!] \tag{2}$$

for any indices i and j in the list of neurons, which makes W symmetrical.

The binary values considered for the stored messages are usually -1 and 1, but can be adapted to work with other binary alphabets. The Hopfield model has a limited efficiency, in particular it does not allow a storage of more than $0.14N$ messages [11]. The limits of the model can be explained by the facts that each entry of the matrix is modified at every time step of the storing procedure, and that the changes are made in both directions and can, therefore, cancel each other out. This overfitted characteristics of associative memories is very different from that observed in learning applications. Indeed, an overfitted learning system recognizes only the training samples and fails at generalizing to novel inputs. To the contrary, an overfitted storing system recognizes everything and does not discriminate anymore between stored and nonstored data.

B. Willshaw Networks

Willshaw networks are another model of associative memories in which information is carried by the existence or absence of connections [8], [12]. Its material is made of a set of N neurons and N^2 potential connections between them. A message is then a fixed size subset of the N neurons, and can be represented by a sparse vector of length N with ones at these neurons' positions and zeros everywhere else. The connection weights are binary, and the active units in a message get fully interconnected as soon as it is memorized, thus forming a clique. Figure 1 gives an example of such a network. The performances of Willshaw networks are way superior to those of Hopfield memories, given that stored messages are sparse (i.e., they contain a small proportion of nonzero elements). Further theoretical and numerical comparison between Hopfield and Willshaw networks can be found in [13]–[16].

C. Clustered Cliques Networks

Recently, a novel type of associative memories was proposed by Gripon et al., called Gripon-Berrou Neural Networks (GBNNs) or clustered cliques networks (CCNs) [17], [18]. These associative memories make use of powerful yet simple error correcting codes. These networks consider input messages to be nonbinary, and more precisely to be words in a finite alphabet of size l. This specific structure allows the separation of nodes into different clusters, each being constituted of the same number l of nodes. Connections between nodes inside a given cluster are forbidden, only the connections between nodes in two different clusters are allowed. There again, this model brings a significantly improved performance as compared to the former state-of-the-art of associative memories, namely Willshaw networks [19]–[21]. For instance, it can be found experimentally that with 2,048 nodes and 10,000 stored messages of order 4 and 2-erasures queries, a Willshaw network will have an error rate close to 80%, while a clique-based neural network will only make 20% of wrong retrievals.

In both Hopfield and Willshaw models, the number of messages the network can store and retrieve successfully is linearly proportional to the number of nodes, with a greater proportionality constant for Willshaw networks [14]. In clique-based neural networks however, storage capacity grows quadratically as a function of the number of units.

One of the objectives of the present work is to explain the performance improvement brought by the separation of the network into clusters. Therefore, we study a network that can be considered as an intermediate between the Willshaw and Gripon-Berrou models. More precisely, our proposed model adds a locally exclusive rule for nodes to be active in the network.

III. WILLSHAW NETWORKS AND BIOLOGICAL CONSIDERATIONS

Willshaw networks are models of associative memories constituted of a given number of neurons. A stored message, or memory, is a combination of nodes taken in this set.

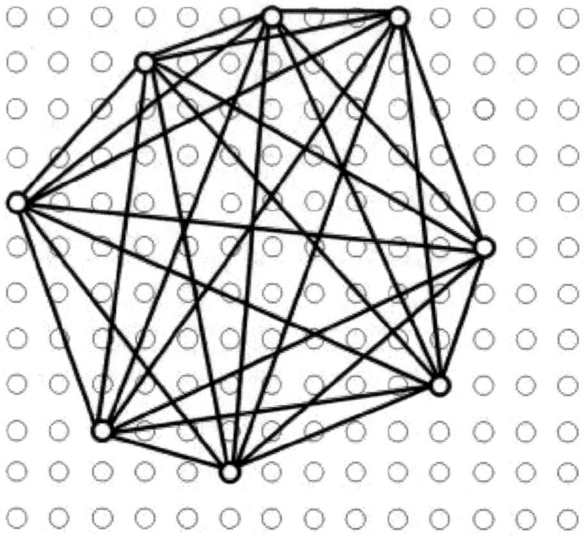

Figure 1. Willshaw network. A message composed of 8 nodes is displayed, the inter-connections being the means of its storage in the network.

The storage of this information element corresponds to the creation of connections with unitary weights between every two neurons in this message. The graphical pattern thus formed is termed "clique". The storing process of n binary vectors x^μ of length N, is equivalent to the modification of elements of the network's connection matrix W, according to the formula:

$$w_{ij} = \max_{\mu} x_i^\mu x_j^\mu \tag{3}$$

Note that here, the $\max$ operator is performed coefficient-wise. Equivalently, the connection weight between nodes i and j is equal to 1 if, and only if, those two nodes are both part of one of the n stored messages.

The network's density d is defined as the expected ratio of the number of ones in the matrix W to the number of ones it would contain if every possible message was stored. For cliques of order c, the number of connections they contain is $\binom{c}{2}$. Despite the correlation between these edges, the probability of a given connection to be picked when forming a message can be estimated as $\frac{\binom{c}{2}}{\binom{N}{2}}$. Provided that the n stored messages are uniformly distributed and independent, the density of the network equates to the probability for any given connection to belong to at least one of these messages. This leads to the formula:

$$d = 1 - \left(1 - \frac{\binom{c}{2}}{\binom{N}{2}}\right)^n. \tag{4}$$

The efficiency of a connectionist associative memory is defined as the ratio of the maximal amount of information carried by the messages it is capable of storing then retrieving with high probability, over the total information quantity represented by its set of connection weights. For a Willshaw network with N nodes, the number of potential connections, or binary resource, is

$$Q = \frac{N(N-1)}{2}[bits]. \quad (5)$$

After M messages have been stored in the network, the amount of information it contains is

$$B = M\left(log_2\left(\binom{N}{c}\right)\right)[bits]. \quad (6)$$

Hence the efficiency of a Willshaw network is

$$\eta = \frac{2M\left(log_2\left(\binom{N}{c}\right)\right)}{N(N-1)}. \quad (7)$$

The maximal attainable efficiency is $\ln(2)$ [22].

The stimulation of a Willshaw network with an input request can be performed as the product of the sparse input vector by the network's connection matrix. The resulting vector then contains the output scores of the network's neurons. The score of a neuron is thus the sum of unitary stimulations it receives from the request elements it is connected to. Neurons must then be selected based on their score. Algorithm 1 defines a procedure that can be used for the recovery of a complete message from a subpart of its content. The Global Winner-Takes-All step consists in discarding all active neurons with a score below the maximum.

Algorithm 1: Message retrieval procedure in a classic Willshaw network.

```
Data: Subpart x of a stored message
Result: Set of nodes z active after treatment
z = x
Repeat
    y = Wz
    z = GlobalWinnerTakesAll(y)
while (convergence not reached
    and max. nb. of iterations not reached)
Return z
```

The probability of error in the retrieval of a message from the Willshaw network can be calculated when the process uses a single iteration. If only one vertex of the clique to complete has been erased, the probability of completing the message accurately after one iteration is the probability that no unit is connected to all elements of the query, other than the missing one:

$$P_{retrieve} = (1 - d^{c-1})^{N-c}. \quad (8)$$

The probability of error is then:

$$P_e = 1 - P_{retrieve} = 1 - (1 - d^{c-1})^{N-c}. \quad (9)$$

Knowing (4), this gives:

$$P_e = 1 - \left(1 - \left[1 - \left(1 - \frac{\binom{c}{2}}{\binom{N}{2}}\right)^{n}\right]^{c-1}\right)^{N-c}. \quad (10)$$

This holds for numbers of erasures c_e superior to 1, bringing:

$$P_e = 1 - \left(1 - \left[1 - \left(1 - \frac{\binom{c}{2}}{\binom{N}{2}}\right)^{n}\right]^{c-c_e}\right)^{N-c}. \quad (11)$$

We aim to modify classic Willshaw networks in a way that is relevant in regard to biological observations. Emphasis is put on lateral inhibition, a phenomenon that has been observed in several areas of the brain. It is notably present in sensory channels. For vision, it operates at the level of retinal cells and allows an increase in contrast and sharpness of signals relayed to the upper parts of the visual cortex [7] [23]. In the primary somatosensory area of the parietal cortex, neurons receive influx coming from overlapping receptive fields. The Winner-Takes-All operation resulting from the action of inhibitory lateral connections allows localizing precisely tactile stimuli, despite the redundancy present in the received information [24]. The same scheme of redundancy among sensory channels, and filtering via lateral inhibition, is present in the auditory system [6]. WTA is observed in the inferior colliculus and in upper levels of the auditory processing channel.

IV. Preventing Connections Between Neighbor Neurons

Classic Willshaw networks have no topology. Their material is constituted with a list of neurons each having an index as sole referent. There is neither a notion of spatial position in these networks, nor, a fortiori, of spatial distance. We get closer here to a real neural network, by arranging them on a two-dimensional map. In the model we propose, the respective positions of two neurons impact the possibility for them to get connected together. The considered network is composed of a number N of nodes evenly distributed along a square grid, of side $S = \sqrt{N}$. Stored messages are of constant order, meaning they are all constituted of the same number of neurons. We forbid connections between nearby neurons. To this end, we apply a threshold σ on the spatial length of a connection. Stored messages must necessarily be conform to this constraint. Each message is formed in a random manner, units are chosen iteratively. Each new element of the message is picked from the positions left available after the removal of the neighbors of the formerly selected nodes, as indicated in Figure 2. One can consider the introduced constraint as applied on the network's material, as the weights of a predetermined set of short-range connections will be enforced to stay null all along the network's life. During the formation of a message, it is practical to pick neurons to satisfy this constraint in a sequential manner, with a local inhibition applied on a neuron's neighborhood from the moment it is selected until the message generation is complete.

A link can be drawn between this approach and Kohonen Self-Organizing Maps, where close-by neurons encode more similar information [25]. Therefore, long-range distance separates information elements that are different in nature, whereas

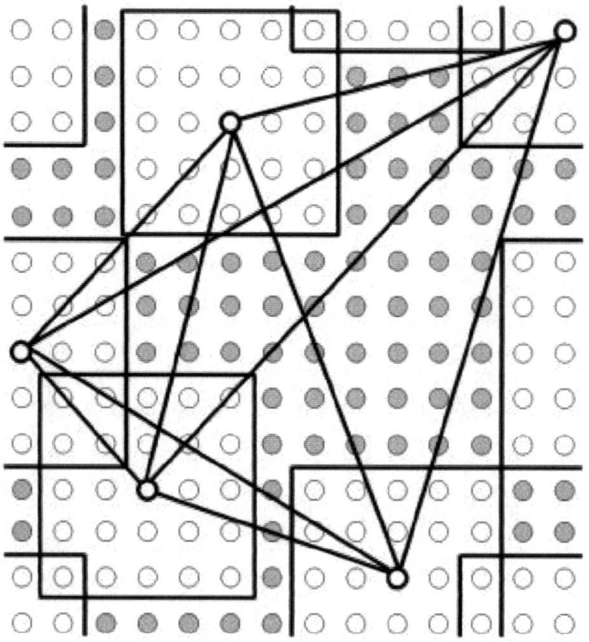

Figure 2. Cyclic Willshaw network with a constraint on local connections.

shorter-range distance depicts a difference in degree. Local competition is particularly relevant in this scheme.

During retrieval, the network is stimulated iteratively with a request that will most often change from one iteration to the next. Each node of the request will first stimulate every other element it is connected to. Scores are initialized with zero at the start of every iteration, and each stimulation is a unitary increment to the score of the receiver unit. For the first iteration, after the stimulation we apply a global Winner-Takes-All rule, which consists in excluding from the research scope all units that do not achieve the maximal score observed in the network. We know indeed that the neurons from the searched message will all have the maximum possible score, equal to the number of elements in the request. Once non-maximum elements are put to zero, we only pay interest in the remaining neurons during the rest of the retrieval process. Moreover, for every iteration after the first one, neurons in the new request are the only ones that can receive stimulation as the algorithm proceeds to only discard neurons from then on.

Thereafter, we can keep using the global WTA principle iteratively, but other algorithms such as Global Winners-Take-All (GWsTA) or Global Losers-Kicked-Out (GLsKO) [26] are more efficient in discriminating the right nodes from the spurious ones that can appear during retrieval.

GWsTA relies on the calculation of a threshold score to select winner neurons. This threshold is chosen such that neurons with an activity above it are in number at least as large as the order of stored messages.

GLsKO consists in putting off, at each iteration, all the units that do not have the highest score, or a subgroup sampled randomly in this ensemble.

These two algorithmic techniques allow to getting rid of an important proportion of false-positives. In the clique-based CCN, clusters play a similar role.

The iterative nature of the process means that a message retrieved as output from the network is typically reinjected in it until input and output no longer differ. A limited number of iterations is applied in the case where the network would not converge to a stable solution, an observable case in which it can oscillate between two states.

In addition to these two stopping criteria that are the maximum number of iterations and convergence, comes a third one which is the identification of a clique. Indeed, if we observe that the units still active after an iteration are in number equal to the order of stored messages, and that they all have the same score, this means it is a stored message. This ensemble is then retained as the response given by the network for the current request.

Algorithm 2 shows the message retrieval procedure used in the results we present. Phase II uses GLsKO.

Algorithm 2: Message retrieval procedure in the modified Willshaw network with spacing constraint.

Data: Subpart x of a stored message
Result: Set of nodes z active after treatment
Phase I
  $y = Wx$
  z = GlobalWinnerTakesAll(y)
Phase II
  Repeat
    $y = Wz$
    a = active nodes in y
    m = nodes in a with minimal score
    $z = a - m$
  while (convergence not reached
    and max. nb. of iterations not reached)
Return z

We experiment the storage of messages of order c in the connection matrix of the network. Messages are formed with the constraint of a minimal space between connected nodes. Two units in a message must be spaced apart at a distance superior to a minimum σ. In order to ease computations and avoid edge effects, we choose to use the L_1 distance, even though we believe this method should work using any distance. This way, when picking a node x for a message, all nodes located in a square grid centered on x, of side $2\sigma+1$, are excluded from the possible choices for the elements of the message remaining to be filled. Moreover, this distance is applied in a cyclic way, meaning a node located on the right edge of the grid will be considered a direct neighbor of the element located at the crossing between the same line and the left edge of the grid. All four corners of the grid will also be neighbors to one another. We call the network so described a torus.

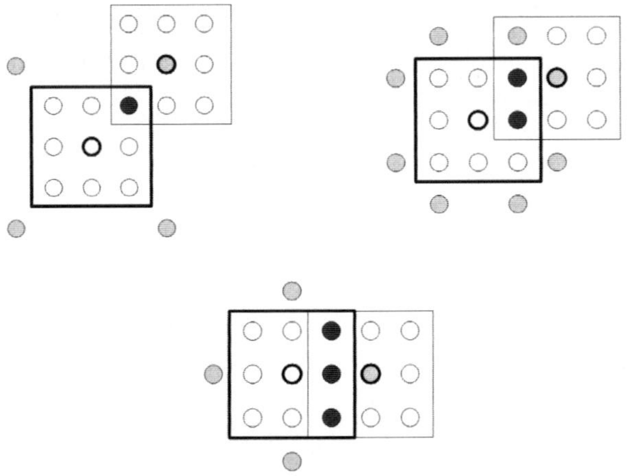

Figure 3. Different overlapping configurations between the inhibition areas around two neurons. Here with $\sigma = 1$, the union of the two overlapping areas can contain 17, 16, or 15 neurons.

V. EFFICIENT USE OF THE MEMORY RESOURCE

When applying a constraint σ on the minimal spacing between connected neurons, the number of potential connections in the modified Willshaw network becomes

$$Q = \frac{N(N-(2\sigma+1)^2)}{2}[bits]. \tag{12}$$

Let the total number of messages one can form in it under the spatial constraint, be denoted $\overline{M}$. The entropy per message b is given by

$$b = \left(log_2\left(\overline{M}\right)\right)[bits]. \tag{13}$$

The amount of information contained in the network after the storage of M messages is then

$$B = bM = M\left(log_2\left(\overline{M}\right)\right)[bits]. \tag{14}$$

Hence the efficiency of the network with lateral inhibition is

$$\eta = \frac{2M\left(log_2\left(\overline{M}\right)\right)}{N(N-(2\sigma+1)^2)}. \tag{15}$$

Predicting the diversity of the spatially constrained Willshaw network given N, c and σ, is not trivial for most values of these parameters. Indeed, once one neuron has been picked among N, the remaining choice for the second element of a message is naturally $N-(2\sigma+1)^2$. However, there are then several possibilities for the number of remaining allowed components, as the inhibition areas around the first two elements can overlap. Figure 3 shows this behavior with $\sigma = 1$. In this case, two overlapping inhibition areas can either share one, two or three neurons. Given a fixed position for the first neuron, there are four positions for the second one that will give an intersection of one, eight that will give an intersection of two, and four positions will give intersections of three neurons.

For $N-(4\sigma+1)^2$ neurons, the two inhibition areas do not overlap, and there are $N-2\left(2\sigma+1\right)^2$ remaining possible choices for the third node.

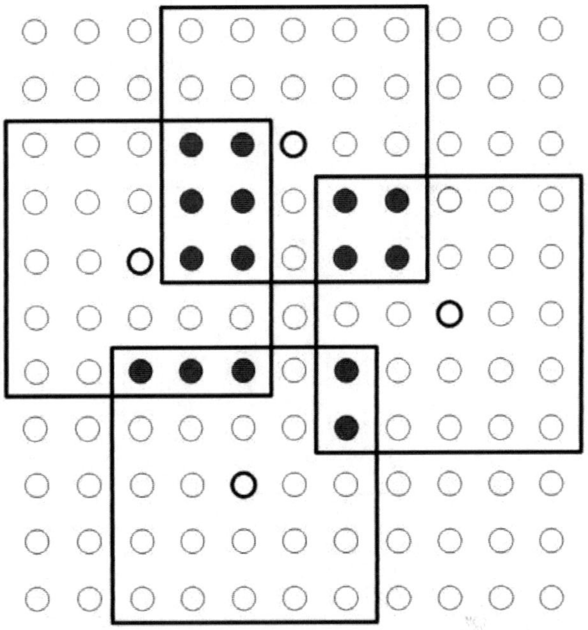

Figure 4. Overlapping inhibition areas for a combination of 4 selected neurons satisfying the spatial constraint $\sigma = 2$. Intersections are made of 2, 3, 4 and 6 neurons.

The possible number of ordered arrangements of three nodes respecting the spacing constraint $\sigma = 1$ is thus

$$\begin{aligned}|\mathcal{A}_{\sigma=1}(N,3)| = N((N-25)(N-18)+4(N-17)\\ +8(N-16)+4(N-15)).\end{aligned} \tag{16}$$

Finally, as the considered messages are unordered, we divide this formula by $c!$, here six, to get the number of allowed messages:

$$\overline{M} = \frac{N^3 - 27N^2 + 194N}{6}. \tag{17}$$

Similarly, for other sets of parameters, the diversity of the spatially constrained network will be obtained by sum-product compositions [27] of polynomials with integer roots. As c grows, the number of overlapping configurations between inhibition areas for neurons in a message is multiplied, and finding a formula to predict the diversity becomes increasingly complex.

Figure 4 shows a possible configuration for the intersections of the inhibition areas of four neurons that can be part of the same message, when $\sigma = 2$. This illustrates the variety of intersection configurations that arise as c and σ grow larger.

For some values of σ and c, the total number of legitimate messages can be predicted by polynomial formulas that can be easy to find experimentally. Table I shows a list of such formulas, for the diversity of the constrained network as well as its number of prohibited messages, for different values of c and σ. $\sigma = 0$ corresponds to the unconstrained Willshaw

Table I:
POLYNOMIAL FORMULAS FOR THE NUMBERS OF ALLOWED AND FORBIDDEN MESSAGES UNDER THE SPATIALITY CONSTRAINT ON CONNECTIONS, AS A FUNCTION OF THE NUMBER N OF NEURONS, FOR DIFFERENT CLIQUE ORDERS AND INHIBITION RADII. CASES WITH $\sigma = 0$ ARE EQUIVALENT TO WILLSHAW NETWORKS. THE VALIDITY OF THESE FORMULAS HOLDS FOR NETWORK SIZES SUFFICIENTLY LARGE BEFORE σ.

c	σ	Allowed messages	Forbidden messages
2	0	$\frac{N^2-N}{2}$	0
	1	$\frac{N^2-9N}{2}$	$4N$
	2	$\frac{N^2-25N}{2}$	$12N$
	3	$\frac{N^2-49N}{2}$	$24N$
3	0	$\frac{N^3-3N^2+2N}{6}$	0
	1	$\frac{N^3-27N^2+194N}{6}$	$4N^2-32N$
	2	$\frac{N^3-75N^2+1514N}{6}$	$12N^2-252N$
	3	$\frac{N^3-147N^2+5834N}{6}$	$24N^2-972N$
4	0	$\frac{N^4-6N^3+11N^2-6N}{24}$	0
	1	$\frac{N^4-54N^3+1019N^2-6798N}{24}$	$2N^3-42N^2+283N$
	2	$\frac{N^4-150N^3+7931N^2-149550N}{24}$	$6N^3-330N^2+6231N$
	3	$\frac{N^4-294N^3+30539N^2-1133958N}{24}$	$12N^3-1272N^2+47248N$

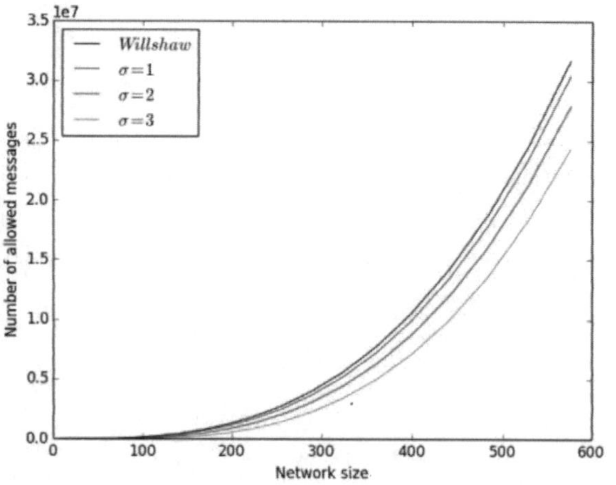

Figure 5. Number of allowed messages as a function of the total number of neurons, under constraints $\sigma = 1, 2$, and 3 compared with a Willshaw network, for $c = 3$.

network. Figure 5 shows the evolution of the number of legitimate messages for different constraints on connection length, for $c = 3$.

With $c = 3$ and $\sigma = 1$, the network's efficiency is given by

$$\eta = \frac{2M\left(log_2\left(\frac{N^3-27N^2+194N}{6}\right)\right)}{N(N-9)}, \qquad (18)$$

while the efficiency of the corresponding Willshaw network is

$$\eta = \frac{2M\left(log_2\left(\frac{N^3-3N^2+2N}{6}\right)\right)}{N(N-1)}. \qquad (19)$$

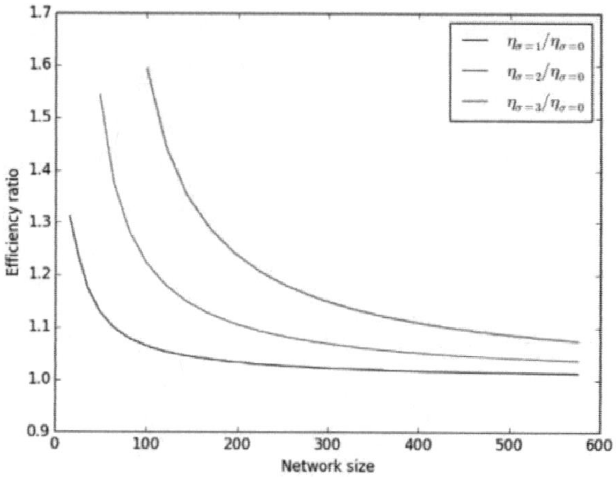

Figure 6. Evolution of the ratios of the efficiency of the networks with spatial constraints $\sigma = 1, 2$, and 3 over the efficiency of a Willshaw network, with increasing network size, for $c = 3$.

The ratio of these two efficiencies is thus given by

$$\frac{\eta_{\sigma=1}}{\eta_{\sigma=0}} = \frac{(N-1)\left(log_2\left(\frac{N^3-27N^2+194N}{6}\right)\right)}{(N-9)\left(log_2\left(\frac{N^3-3N^2+2N}{6}\right)\right)}, \qquad (20)$$

and is independent of the number M of stored messages. Figure 6 displays the evolution of the efficiency ratios between spatially constrained networks with $\sigma = 1, 2, 3$, and the unconstrained Willshaw network, as a function of the number of neurons in the network, for messages made of three neurons. It shows that, although the spacing constraint can not be applied on too small networks, for sufficiently large networks, prohibiting shorter connections comes with an increase in efficiency, as the ratio is superior to one. Larger values of σ are associated with larger gain in efficiency, and for a given σ the best improvement over a Willshaw network comes for the smaller networks where this constraint applies, i.e., where it does not block the vast majority of Willshaw messages. The decay of the ratio when the network size increases is due to the fact that the constraint then prohibits a smaller proportion of connections, making the difference with a Willshaw network less noticeable. Figure 7 shows the evolution of the same efficiency ratios for cliques of order 4. The improvement is then slightly lower for constant σ and network size, in comparison to the case $c = 3$. The ratio remains superior to one however. This tends to show that the spacing constraint can be more beneficial, in terms of gained efficiency, for shorter messages.

The efficiency is a measure of the amount of information one can store for a given amount of available memory. The improvement in efficiency brought by the spacing constraint on connections means that the reduction of used material is more significant than that of the quantity of information carried by messages.

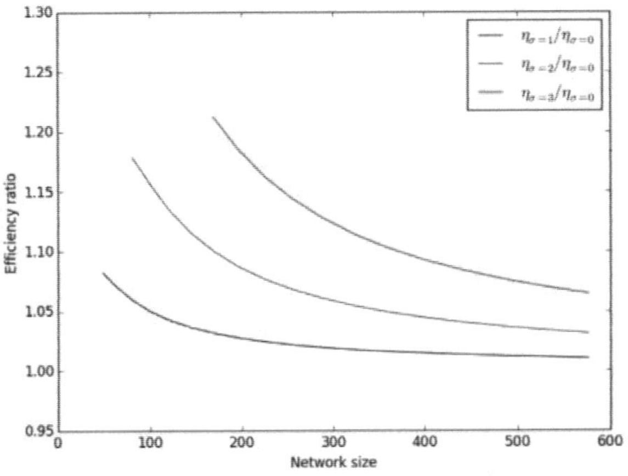

Figure 7. Evolution of the efficiency ratios for $\sigma = 1, 2$, and 3 over the Willshaw efficiency as a function of the number of neurons in the network, for $c = 4$.

VI. PATTERN COMPLETION

We now pay interest specifically in the pattern retrieval ability of the modified Willshaw network, as compared to the classical model. During retrieval, only a sample from the nodes of the complete message are stimulated, the inputs are subparts of stored messages. Units that are close to elements of an input will not reach the maximum score in the network, and will therefore be ruled out after the first WTA operation. During the second phase of the algorithm, nodes in the vicinity of input neurons will also be more likely to reach a low score if they are activated, and to be discarded. Hence, the local inhibition used initially during the creation of messages impacts the retrieval process as well.

We pay interest in the network's ability to return the exact memory associated with a request. Hence every difference, even marked by a single unit, between the expected pattern and the network's output is counted as an error.

We measure the performance of the network as the ratio of the number of successfully retrieved messages over the total number of requests.

Various parameters can impact this performance, albeit to different degrees:
- the length S of the grid's side
- the number M of stored messages
- the minimal space σ between two elements of a message
- the order c of stored messages
- the number of erasures c_e applied on stored messages to obtain the corresponding request messages

The behavior of this network is interesting in relation to Willshaw networks and clustered cliques networks, as it is close to a classic Willshaw network and displays the added feature of prohibited connections as observed in CCNs. This modification can be viewed as a form of sliding-window clustering.

Figures 8 and 9 represent the matrices of allowed and forbidden connections in a clustered clique network with

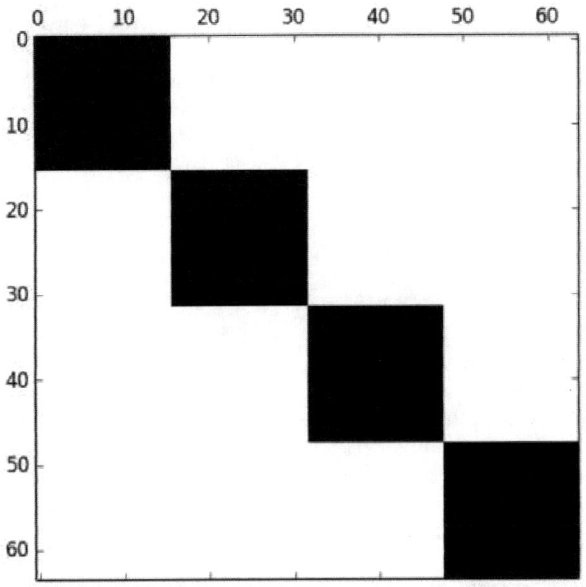

Figure 8. Matrix of the potential and forbidden connections in a clustered clique network with 4 clusters of 16 fanals each. The element at the crossing of a line i and a column j represents the connection between neurons n_i and n_j. White cells represent allowed connections, black cells correspond to forbidden ones.

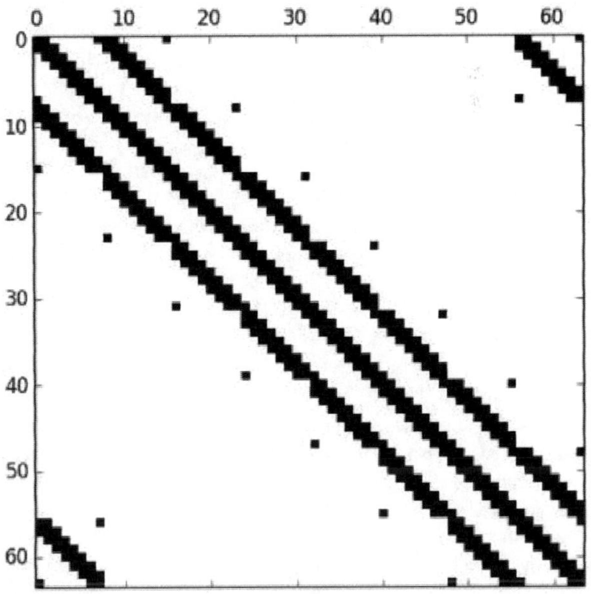

Figure 9. Connection matrix for a modified Willshaw network with side length $S = 8$ and $\sigma = 1$.

$\chi = 4$ clusters comprising $l = 16$ neurons each, and in a modified Willshaw network of side length $S = 8$ under constraint $\sigma = 1$, respectively. The two networks have the same number of neurons, and comparable numbers of allowed and forbidden connections. Indeed, the number of potential non-oriented connections in the clustered clique network is given by

$$Q = \frac{\chi(\chi-1)l^2}{2}[bits], \tag{21}$$

which here gives 1,536 allowed connections, against 480 ones forbidden due to the clustering constraint. On the other hand, from (12) we have 1,760 allowed connections in the spatially constrained Willshaw networks, and 256 forbidden ones. The major difference is the potential overlapping between the inhibition areas around different neurons in the modified Willshaw network. It follows that in this network, two neurons n_i and n_j can be both prohibited from connecting to a third one n_k, and yet be allowed to connect together. To the contrary, if in a clustered clique network, connection weights w_{ik} and w_{jk} are forced to remain at 0, then necessarily w_{ij} will be as well.

In a first series of experiments, we focus on the ability of the network to first store independent, identically distributed messages, and then complete them properly when probed with partial cues. For every configuration of the network, messages and requests we test, we store a set of thousands of messages in the network. These messages are generated randomly following the local inhibition pattern described in section IV. We then request it with the full set of queries associated with stored messages.

For each network size, we observe that there is an optimal value of the minimal distance σ, that lowers the most significantly the error rate, as compared to the corresponding Willshaw network without constraint on local connections. For a given minimal distance, the reduction in error rate depends on the number of stored messages, with an optimal number of messages which is a function of the network size. For cliques of order four and with two erasures, the maximal reachable improvement is close to 15%, and seems to be the same for all network sizes. In this configuration, the minimal distance bringing the best performance is approximately the third of the network side.

The evolution of the retrieval error rate as a function of the number of stored messages is slower with the appropriate constraint on connections than for a classic Willshaw network, as can be seen in Figure 10.

Figure 11 shows a similar comparison, this time between the modified Willshaw network with constraint $\sigma = 5$ made of 400 neurons, and an unconstrained Willshaw network with 335 neurons. Because of the reduction in the number of connections when $\sigma = 5$, the two networks have almost the same binary resource. Indeed, the Willshaw network has 55,945 connections while the constrained one has 55,800 possible connections, despite having more neurons. Even though the modified network has a slightly lower footprint, the improvement is even more noticeable than for the comparison with equal size of the neurons sets. In fact for $1,500$ stored messages, the modified network gains around 50% in error rate over the Willshaw network. Like the CCNs approach, this shows the interest of spreading the binary resource across a larger set of units with constrained connectivity, as opposed to allowing any two neurons to link.

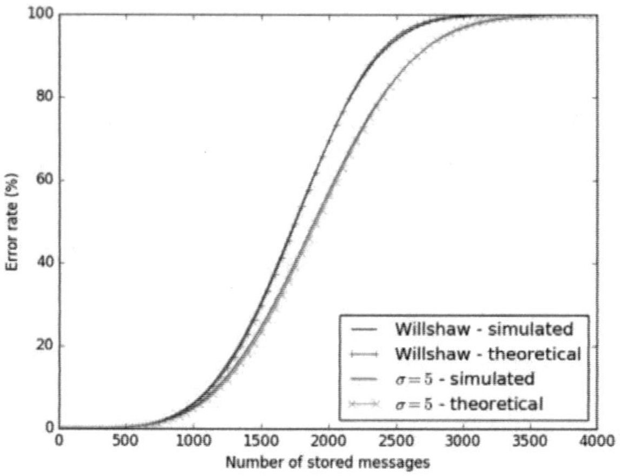

Figure 10. Evolution of the retrieval error rate with and without constraint $\sigma = 5$ in a network of side length 20 with 400 neurons, stored messages of order 6 and 1 erasure applied to form corresponding requests, with 1 iteration.

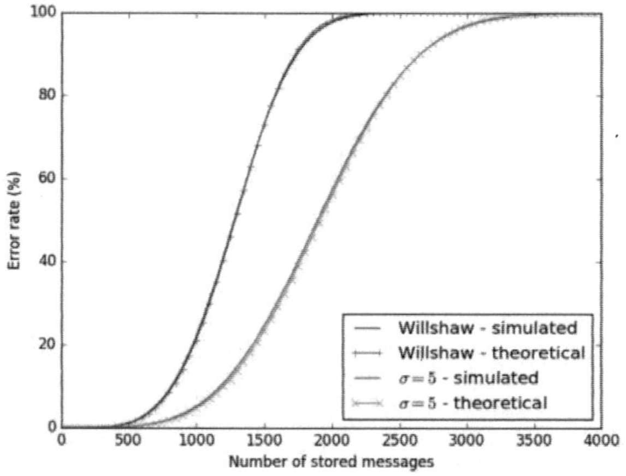

Figure 11. Evolution of the retrieval error rate after one iteration as a function of the number of stored messages, in a classic Willshaw network with 335 neurons, and in a modified Willshaw network of side length 20 with 400 neurons, with constraint $\sigma = 5$. Stored messages have order $c = 6$ and associated queries are obtained by erasing one vertex. The two networks have close numbers of possible connections.

For a constant number of stored messages, the graph of the error rate as a function of σ is characterized by a progressive decay down to a minimum, followed by a rapid growth for upper values of σ, as shown in Figure 12.

This can be explained by two phenomena. On the one hand, the prohibition of a growing part of the possible connections gradually decreases the probability of a "false message", characterized by the intrusion of a spurious node in the output. The existence of a node that is connected to all elements in

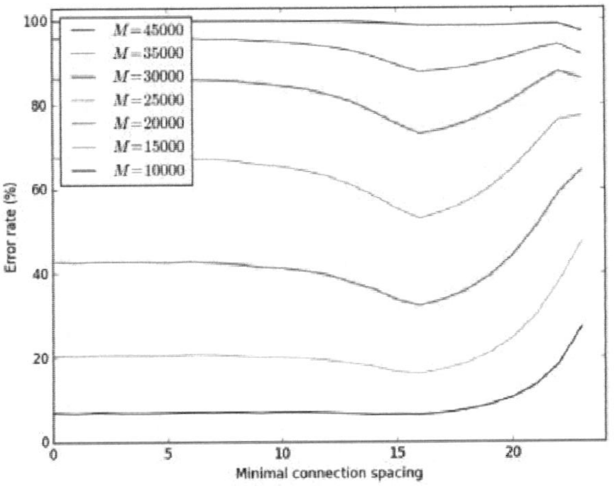

Figure 12. Minimal connection distance effect on performance in a modified Willshaw network with local inhibition, made of 2500 neurons. Stored messages are of order $c = 4$, and $c_e = 2$ erasures are applied to form corresponding requests. Seven different numbers of stored messages are tested. The case where minimal spacing $\sigma = 0$ corresponds to a classic Willshaw network.

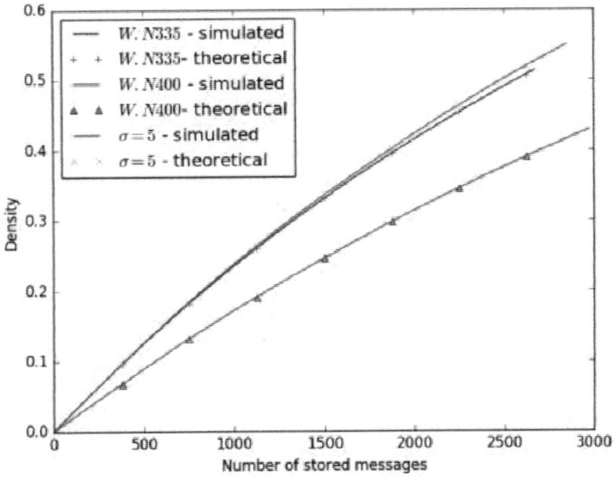

Figure 13. Evolution of the density for ordinary Willshaw networks with 335 and 400 neurons, and for a modified Willshaw network of side length 20 with 400 neurons and $\sigma = 5$, with stored messages of order 6.

a request, yet is not part of the corresponding message, will potentially cause an error. In fact, forbidding some connections has the effect of reducing the number of concurrent nodes susceptible to cause errors. We can estimate the mean number of concurrent nodes remaining after the choice of k neurons of a message:

$$N_c(\sigma, k) = N\left(1 - \left(\frac{(2\sigma+1)^2}{N}\right)\right)^k. \tag{22}$$

The corresponding number of nodes blocked by the constraint on connections is, on average:

$$N_b(\sigma, k) = N\left(1 - \left(1 - \left(\frac{(2\sigma+1)^2}{N}\right)\right)^k\right). \tag{23}$$

This explains the decay phase in error rate observed for the first values of σ. Let us note that it comes with a decrease in the diversity of messages, namely the total number of different messages that can be stored in the network. Following this decay, the decrease in the number of concurrent nodes has another effect: the reuse of connections by different messages becomes more frequent as the choice for possible connections gets reduced. This comes to counteract the former phenomenon and raises the error rate.

The density of the modified network after the storage of n messages can be calculated by

$$d = 1 - \left(1 - \frac{\binom{c}{2}}{Q}\right)^n, \tag{24}$$

that is

$$d = 1 - \left(1 - \frac{c(c-1)}{N\left(N - (2\sigma+1)^2\right)}\right)^n. \tag{25}$$

As for the Willshaw network, we can calculate the probability of error of the modified network after one iteration of decoding. Given queries where c_e nodes have been removed, the probability of retrieval after one iteration can be estimated by:

$$P_{retrieve} = (1 - d^{c-c_e})^{N_c(\sigma, c-c_e)-c_e}. \tag{26}$$

One can then deduce the probability of error:

$$P_e = 1 - (1 - d^{c-c_e})^{N_c(\sigma, c-c_e)-ce}. \tag{27}$$

Now referring to (24), this leads to:

$$P_e = 1 - \left(1 - \left[1 - \left(1 - \frac{\binom{c}{2}}{Q}\right)^n\right]^{c-c_e}\right)^{N_c(\sigma, c-c_e)-c_e}, \tag{28}$$

which holds for lower values of σ. For higher values, however, the global density is no longer a proper estimator of the probability of spurious connections. Given $c - c_e$ message elements, the local connection density between these nodes and the restricted ensemble of allowed neighbors they can all be potentially connected to, is then higher than the average density over the whole network.

Figure 13 shows how the network density grows faster as messages are stored in the network, than for a classic Willshaw network with equal number of neurons. This is because of the decrease in number of possible connections due to the spacing constraint. When the number of connections of the compared classic and modified networks is close, the two densities evolve at a similar rate however.

Besides, we observe that the maximal improvement in performance, for given values of c and c_e, does not considerably vary as a function of the network size. This can be explained by the fact that the minimal distance giving the

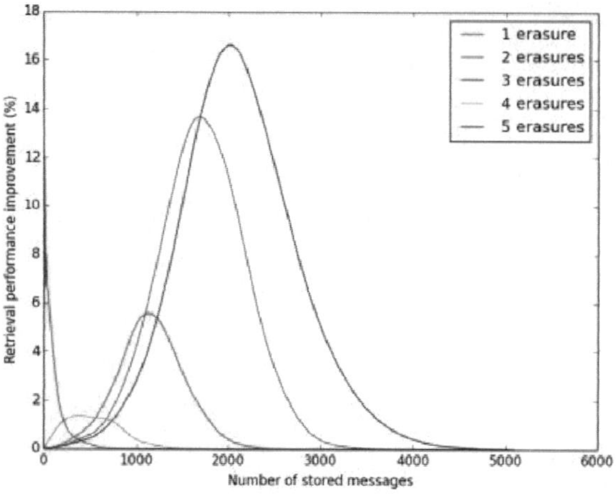

Figure 14. Maximal improvement obtained over a classic Willshaw network made of 400 units with messages of order 6, using a single iteration of decoding.

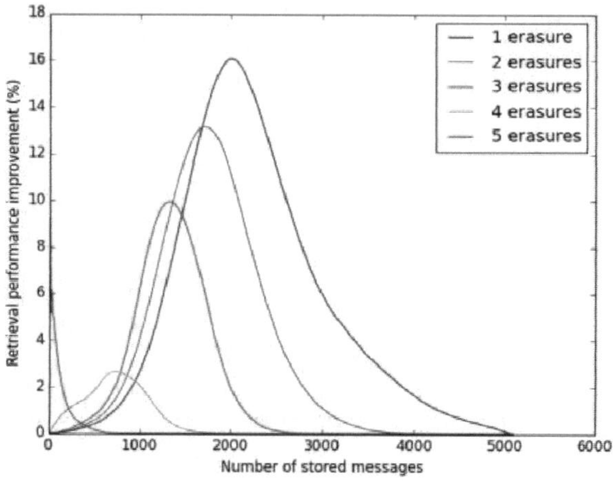

Figure 15. Maximal improvement obtained over a classic Willshaw network of 400 neurons with messages of order 6, using a maximum of 3 iterations.

best performance is approximately proportional to the side of the network. Consequently, the proportion of neurons in the network that cannot be connected to the $c - c_e$ neurons in the request remains more or less the same for different network sizes, with the optimal minimal distance.

The benefits brought by the constraint on connections tends to be stronger for smaller numbers of erasures. For erasures of about half the units of the messages, the maximum gain will be lower, yet for a high amount of erasures the performance may be more noticeably enhanced by the added constraint. The performance improvement over a classic Willshaw network also depends on the number of messages stored in the network. It reaches a peak for a certain number of stored messages, and then decays when additional messages get stored. The maximal improvement tends to be reached earlier during storage for higher numbers of erasures, as illustrated by Figures 14 and 15 for a network with 400 neurons, messages of order $c = 6$, and numbers of erasures c_e ranging from one to five. Figure 16 shows that for a larger network with 900 neurons, this arrangement is respected for the most part, with the exception of the case where $c_e = 1$, for which the peak in performance improvement occurs for a lower number of stored messages than for $c_e = 2$ or $c_e = 3$. Figures 14 and 15 also show that the maximum number of iterations applied during retrieval has a varying effect on performance improvement, depending on the number of erasures applied to form requests. With 400 neurons, increasing the number of iterations has a clear effect on performance for $c_e = 3$ and $c_e = 4$, more so than for $c_e = 1$ and $c_e = 2$.

When comparing networks with the same number of neurons, the greatest performance improvement is most often observed over a classic Willshaw network and a number of stored messages originally giving an error rate ranging from about 40%, up to 70%. The performance gain is then often close to 15%.

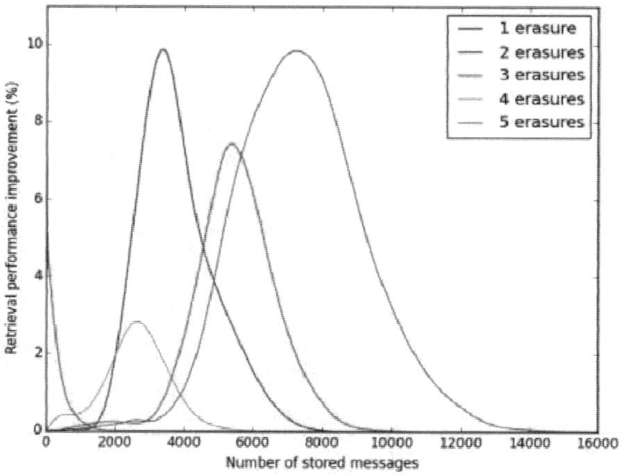

Figure 16. Maximal improvement obtained over a classic Willshaw network of 900 neurons with messages of order 6, using a maximum of 3 iterations.

VII. Conclusion and Perspectives

We introduced a modified version of Willshaw neural networks that has interesting properties regarding storage capacity and retrieval performance. By prohibiting certain types of connections in the network, we observe that the retrieval ability can be enhanced, and that the value of the threshold on inter-neuron connection spacing has a direct impact on performance. As a result, this constraint can be used to enhance pattern completion performance without modifying the size of the pool of neurons. In addition, the applied constraint comes with an improved efficiency, as the quantity of information carried by each single connection is made higher. Also, given a fixed amount of allotted memory to store connection weights, a constrained network with more neurons but a comparable number of connections can be created, which will display even better performances.

This is relevant to observations on clustered cliques neural networks, in that it shows constraining connections in a Willshaw network modifies its capacity and efficiency in a way that depends on the nature of the applied constraint. It is a step forward in understanding why the use of clusters in CCNs brings significantly higher capacity as compared to Willshaw networks. To some extent, it also emulates biological observations of lateral inhibition in the brain and sensory channels, as we prevent neighbor neurons from connecting and therefore let them compete for activity. This makes sense with a framework in which close-by neurons encode patterns that differ only in degree and where only one unit that resonates most with input stimuli must activate.

Future work may involve experimenting with other constraints on connections based on the relative locations of neurons, such as variable values of the spacing constraint applied in different subregions of the graph, or preventing random subsets of connections. It may also focus on theoretical advances in the understanding of the relationship between the nature of a connectivity constraint and its influence on efficiency and performance.

Acknowledgements

This work was supported by the European Research Council under Grant ERC-AdG2011 290901 NEUCOD.

The authors would like to thank NVIDIA for providing us with a free graphics card allowing to speed up computations for the experiments performed during this work.

References

[1] P. Tigreat, V. Gripon, and P. H. Horrein, "Improved Willshaw networks with local inhibition," Proceedings of the Eighth International Conference on Advanced Cognitive Technologies and Applications, 2016, pp. 96–101.

[2] H. Jarollahi *et al.*, "Algorithm and architecture for a multiple-field context-driven search engine using fully-parallel clustered associative memories," In IEEE Workshop on Signal Processing Systems (SiPS), 2014, pp. 1–6.

[3] H. Gazit *et al.*, " Network device architecture to support algorithmic content addressable memory (CAM)," U.S. Patent No. 9,367,645. Washington, DC: U.S. Patent and Trademark Office, 2016.

[4] L. Y. Liu *et al.*, "CAM-based VLSI architectures for dynamic Huffman coding," IEEE Trans. Consum. Electron., vol. 40, no. 3, 1994, pp. 282–289.

[5] M. E. Valle, "A class of sparsely connected autoassociative morphological memories for large color images," IEEE Transactions on Neural Networks, vol. 20, no. 6, 2009, pp. 1045–1050.

[6] S. A. Shamma, "Speech processing in the auditory system II: Lateral inhibition and the central processing of speech evoked activity in the auditory nerve," The Journal of the Acoustical Society of America, vol. 78, no. 5, 1985, pp. 1622–1632.

[7] H. Ozeki *et al.*, "Inhibitory stabilization of the cortical network underlies visual surround suppression," Neuron, vol. 62, no. 4, 2009, pp. 578–592.

[8] D. J. Willshaw, O. P. Buneman, and H. C. Longuet-Higgins, "Non-holographic associative memory," Nature, 1969, pp. 960–962.

[9] J. J. Hopfield, "Neural networks and physical systems with emergent collective computational abilities," Proceedings of the national academy of sciences, vol. 79, no. 8, 1982, pp. 2554–2558.

[10] P. Chossat, and M. Krupa, "Consecutive and non-consecutive heteroclinic cycles in Hopfield networks," Dynamical Systems, 2016, pp. 1–15.

[11] R. J. McEliece *et al.*, "Neural networks and physical systems with emergent collective computational abilities," Proceedings of the national academy of sciences, vol. 33, no. 4, 1987, pp. 461–482.

[12] M. Laiho *et al.*, "High-dimensional computing with sparse vectors," Biomedical Circuits and Systems Conference (BioCAS), 2015, pp. 1–4.

[13] H. C. Longuet-Higgins, D. J. Willshaw, and O. P. Buneman, "Theories of associative recall," Quarterly reviews of biophysics, 3(02), 1970, pp. 223-244.

[14] J. D. Keeler, "Comparison between kanerva's sdm and hopfield-type neural networks," Cognitive Science, vol. 12, no. 3, 1988, pp. 299–329.

[15] D. Willshaw and P. Dayan, "Optimal plasticity from matrix memories: What goes up must come down," Neural Computation, vol. 2, no. 1, 1990, pp. 85–93.

[16] A. Knoblauch, G. Palm, and F. T. Sommer "Memory capacities for synaptic and structural plasticity," Neural Computation, vol. 22, no. 2, 2010, pp. 289–341.

[17] V. Gripon and C. Berrou, "Sparse neural networks with large learning diversity," Neural Networks, IEEE Transactions on, vol. 22, no. 7, 2011, pp. 1087–1096.

[18] A. A. Mofrad *et al.*, "Clique-based neural associative memories with local coding and precoding," Neural Computation, 2016

[19] ——, "A simple and efficient way to store many messages using neural cliques," Computational Intelligence, Cognitive Algorithms, Mind, and Brain (CCMB), 2011 IEEE Symposium on, 2011, pp. 1–5.

[20] ——, "Nearly-optimal associative memories based on distributed constant weight codes," Information Theory and Applications Workshop (ITA), 2012, pp. 269–273.

[21] B. K. Aliabadi, C. Berrou, and V. Gripon, "Storing sparse messages in networks of neural cliques," Proceedings of the national academy of sciences, vol. 25, no. 5, 2014, pp. 461–482.

[22] G. Palm, "Neural associative memories and sparse coding," Neural Networks, vol. 37, 1987, pp. 165–171.

[23] J. R. Cavanaugh, W. Bair, and J. A. Movshon, "Nature and interaction of signals from the receptive field center and surround in macaque v1 neurons," Journal of neurophysiology, vol. 88, no. 5, 2002, pp. 2530–2546.

[24] M. A. Heller and E. Gentaz, "Psychology of touch and blindness," Psychology Press, 2013.

[25] T. Kohonen, "The self-organizing map," Neurocomputing, vol. 21, no. 1, 1998, pp. 1–6.

[26] A. Aboudib, V. Gripon, and X. Jiang "A study of retrieval algorithms of sparse messages in networks of neural cliques," Proceedings of the Sixth International Conference on Advanced Cognitive Technologies and Applications, 2014, pp. 140–146.

[27] M. J. Wainwright, T. S. Jaakkola, and A. S. Willsky, "Tree-based reparameterization framework for analysis of sum-product and related algorithms," IEEE Transactions on information theory, vol. 49, no. 5, 2003, pp. 1120-1146.

Good telecare: on accessible mental health care

Annemarie van Hout, Ruud Janssen, Marike Hettinga
Research Group ICT-innovations in Health Care
Windesheim University of Applied Sciences
Zwolle, The Netherlands
Email: a.van.hout@windesheim.nl,
twjm.janssen@windesheim.nl, m.hettinga@windesheim.nl

Jeannette Pols, Dick Willems
Medical Ethics, department of General Practice
Academic Medical Centre
Amsterdam, The Netherlands
a.j.pols@amc.uva.nl, d.l.willems@amc.uva.nl

Abstract — **Mental health care is increasingly given at a distance, supported by technology. In this article, we focus on whether care, when technology comes in, still counts as good care. Therefore, we looked into a mental health nursing telecare practice for patients that live at home. The telecare team offers 24/7 unplanned webcam contact. We observed and interviewed nurses whilst they were having webcam contact. In our analysis we focused on frictions in care. We found different examples, that relate to an overall dilemma in mental health care: how does the policy of reinforcing self-reliant patients relate to 24/7 care? The dilemma is reinforced through the webcam, as it makes care much more accessible. We used theories on good care, which show how good care is situational and established when enacted. We think professionals should look for complex examples and confer on frictions in order to improve good care. Exchange and discussion between care professionals, derived from their understandings with patients, will lead to applied knowledge, or even better, artisanal knowledge of good care.**

Keywords - good care; e-mental health; nursing telecare; ethnography.

I. INTRODUCTION

Telecare provides an alternative form of contact between patients and nurses [1]. It is part of a growing movement of technology in mental health care, in which web-based systems like serious games, self-management instruments and online therapies have become regular interventions [2][3][4][5][6]. For mental health care institutions, telecare is mostly used to introduce new care interventions, enhance flexibility and reduce face-to-face contacts [7][8]. Flexibility is regarded as an important asset and the online interventions are often available for patients to use at their convenience. This fits the prevailing views on self-managing patients: request for support at the moment you need it or when it fits you. Not all 'e-mental health' technologies make use of a webcam, but they do share the capacity for care at a distance. The webcam has an extra feature: it makes the care professional accessible to patients in sight and sound at the push of a button.

The webcam is bringing obvious changes to care. How to know if this care through a webcam is good care? Let us start with what comes to mind first: care at a distance leads to absent care professionals (or if you wish: absent patients). We can safely state that care at a distance will not be the same as care in person. There is the absent body, which can be cared for at a distance using information on the body that the patient provides through the technology [9]. Mental health care has more focus on non-physical care, with nurses guiding and supporting patients. For mental health care organizations, telecare seems a good alternative or addition to regular care, based on the presupposition that bodily presence might not always be necessary in mental health care [8][10][11].

The webcam does not only make the bodies absent, the webcam itself comes into the caring relationship: it is a technological object that asks for operation [12][13][14][15]. There is a rich literature in Science and Technology Studies on dealing with technology. Buttons need to be pushed, results read out, numbers configured and settings tailored for individual patients [16][17]. The work that needs to be done before the technology works, becomes part of the caring relationship, as technology is worked together, but is also a form of relationship with the technology. People get attached to devices, as technology becomes part of their day-to-day life [18][19][20][21]. Technology takes hard work and subtle tinkering to make it fit in daily practices. In the process of tinkering, relations are established. The technology becomes part of the caring relation, bringing along changes.

And that is our main concern: if care changes with the technology, is that care still good care? Good care is a widespread, presumably understandable idiom. We all have ideas on what good care is. Some are easy to name, but many aspects of care are in our actions. Care is about patients, but often also about the patients' spouses and everything that surrounds them. Care is for houses, pets, administration, food, plants and trinkets [22]. That we care not 'just' for patients, but also for what surrounds them (and us), shows how infinite care is and also how difficult to define. Care is not only about who and what we care for, but also why. Care is also about the relation and responsibilities of caregivers and caretakers. In prevailing health care policy, self-management is seen as a form of good care. The ideal of self-management is about patients who are in the lead, controlling their own lives, instead of being led by health care organizations. In day-to-day lives of patients though, a lot of practical stuff needs care, in which patients are often very dependent, next to dependency on physical care [23][21]. Care is also about rituals, in which the good is established [24].What binds the various aspects of care we just named, in which we were far from complete, is the establishment of the good along the way. Nurses and patients, patients by themselves, spouses, pets and stuff: in shared practices, or in the practice they are in, they all define good care by doing it. They make good care along the way, so to speak. The outcomes of the 'good' in care are different of course: in different practices good care will be establish in different ways. As the 'good' is constituted when care is given and depends on its environment, the 'good' is extremely fluid [25][23][26].

There is no uniform description of the good in care, but there is some agreement: care should be experienced as good by all involved. Therefore, good becomes something that needs to be negotiated. But will that what is negotiated as good, be renegotiated for its goodness when parts of it change? Or will this lead to the idea that the care is still good, because its seems the same, when in fact its merits might have changed? Making good care therefore demands tuning between nurses and patients, to discuss and adjust care when and where necessary. However, to know if there is something to discuss, nurses have to watch out for where the tuning fails. Following Mol [26] this means searching for moments and situations, for circumstances or signals that show the 'bads'. Often patients will be very capable of saying so themselves, but many cases they are not. Nurses have to look for those bads, because part of making care good (again) is to be sensitive to friction and to be able to fix it. Nurses should therefore search for friction to maintain good care.

In this article, we want to add to knowledge on how good telecare can function in a changed practice. We studied a mental health nursing care practice that uses a webcam to give patients the opportunity to make video contact with a nurse whenever they feel it is necessary. We specifically wanted to learn about the limits and constraints of good care and posed the following research question: What dilemmas and frictions on good care arise in a mental health nursing telecare practice?

II. METHOD

This article is based on our data collected from a nursing telecare team. This team consists of case managers from 18 different FACT teams. FACT stands for Flexible Assertive Community Treatment. The members of these multidisciplinary teams are mainly case managers (almost all nurses) as well as psychiatrists, psychologists and sometimes social workers [27]. FACT teams supply care to patients with severe mental illnesses who live at home. Home care varies from daily visits to once every month. FACT teams aim to be able to scale up care very quickly when necessary, and likewise to scale care down as well, therefore adapting the care to the patient's situation. The 18 FACT teams are part of one organization for mental health care that covers one region. Each team makes one case manager available to take one shift a week (from 8 am till 8 pm) at a health care post that provides care at a distance with a webcam. The telecare team handles all unscheduled webcam contacts ('calls') from patients throughout the region. After hours, calls are routed to various clinics in the region, which also have a webcam. Patients have a dedicated computer with touch screen and webcam. The system's hardware and software is very user friendly: patients can talk to the case managers at a push of a button. The telecare team is available 24/7, but most calls take place during office hours.

We followed the telecare team for nine months, conducting ethnographic research [28][29][30] to open up the practice, see and recognize changes and discuss these with nurses. In the process, we sought to become part of the care practice, or at least to get as close to it as possible so that we could recognize and understand it better. We became acquainted with the telecare practice by talking with and observing case managers in various mental health care settings [31][32]. We read project documents on various telecare projects and participated in team meetings. Our field work then concentrated on the telecare team. We joined case managers on their shift. We took field notes while observing webcam contacts of case managers and patients. We conducted interviews after the webcam calls, asking case managers to reflect on the call, and these interviews were taped and transcribed verbatim. We also interviewed two patients, and observed webcam calls from their homes. Patients were informed of our presence in writing beforehand and the researcher left the room if they had not consented. The independent ethics committee judged this project to be exempt from review [33]. The quotes in the results section of this article were translated from Dutch. We joined the team 27 times, for two to four hours each. We observed and talked to 11 case managers, who were in touch with 30 patients, some of them multiple times. Twice we were asked by patients to leave the room, as they did not consent to our presence.

The observations and analyses were led by sensitizing concepts, which were shaped by the theoretical notions on good care we discussed in the introduction, guiding the notes and the coding process. The researchers articulated these notions during the analytical process, and applied them in the second round of observations and interviews with the case managers [34]. Data analysis focused on what work is done in the telecare practice and how the case managers talk about it. We used theories on good care to interpret the findings and reveal the areas of friction in good care in the practice of telecare.

III. RESULTS

We will discuss three practices in which webcam use has changed care. We show how this has led to frictions concerning good care.

A. *What care do case managers address?*

This section deals with how telecare leads to uncertainty about who needs to follow up on particular questions from patients.

When a patient's call on the telecare system is not answered, the system registers the call and gives an engaged signal. There are two kinds of unanswered calls: the ones that arrive when a case manager is busy talking to another patient and the ones that come in after hours. For the first kind the custom is to return the call as soon as possible. Patients know that when there is no answer during daytime, this probably means the line is engaged. After hours calls are diverted to various clinics, but sometimes the staff in the clinic are too busy to answer. At the start of a working day, especially on Monday morning, when the team has been away for the weekend, the system shows a list of missed calls:

Rien contacts one of the patients on the list of missed calls.
Anja answers: I know you, but your hair looks different. They joke about hair gel. Then Anja asks: What's up?
Rien: I saw a missed call and knew you'd tried to reach the clinic.
Anja: Yes, I felt depressed. I worry about my cat a lot and it takes up my mind. The cat needs meat twice a day and I can't handle that. I've got to take it out of the fridge and give it at room temperature. I am all taken up by that cat.
Rien: Well, all the best.
Anja: Thanks. What's your name again?
Rien spells his name and tries to end the conversation. Anja talks a bit about the weather and finally says: Thanks for your interest in me and for returning my call.

There is no protocol for missed calls after hours. Rien feels he should find out what the matter was. After all the patient tried to reach a care professional. Some colleagues disagree, like Taco:

The telecare team doesn't do scheduled care. If you promise one [caller], you have to promise them all. Then we have to call back a lot. You'll see.. when you're returning one of those calls, you're in the middle of that conversation and then a new call comes in, so you have to return that one, and so on. So you're actually creating calls.

There are two routines for missed calls. The first is about returning only daytime missed calls, the ones caused by an ongoing call with another patient. This routine is not in dispute. The other one, returning after hours missed calls, is carried out very differently. Some case managers feel that each missed call needs following up. The idea is that telecare will 'only work' when it does as promised: provide a way to contact a care professional 24 hours a day. For Rien, good care means acting on the list of missed calls; for Taco only when the missed call appears when he engaged with an unplanned call.

Telecare conversations tend to be diverse. Sometimes, as soon as the technology establishes the connection through image and sound, patients tell their stories:

Case manager Taco talks about patient Tobias, who told very dark stories after his last admittance. For example, Tobias claimed that one of the nurses at the clinic had instructed him to 'go grab that borderline bitch'. Taco tells how much such calls affect him and how difficult these conversations are on a webcam. We discuss this for a bit, but do not seem to get to the heart of the matter. Taco says such contact seems like a stopgap, like it is not part of the process. I ask Taco if it would have been different if he were Tobias' case manager. Taco ponders on this, on how telecare is a part of the care offer and how it is part of the treatment, but that does not seem right to him after all.... It seems clear though that calls like the one with Tobias have more effect on Taco because they are by webcam.

The immediate contact established by the webcam and the fact that he is not Tobias' case manager make this situation hard for Taco to deal with. In some situations his colleagues find other solutions:

Wende: They [patients] should arrange these things with their own therapist. [A call] can get very substantive on medication as well. Then I say: I can't answer these questions!

The case managers on the telecare team feel that they should not be replacing the patient's regular care team (in which case they might feel obliged to answer all questions), but that the purpose of the telecare unit is to provide a first contact in (unscheduled) times of stress or social need. In the event of a crisis, the telecare team alerts the patient's regular case manager. A non-crisis situation is just recorded on the electronic system and patients are referred to their action plan or their own case manager. This demonstrates that not all questions can be asked, or put better, will be answered.

Another example of the issue on who takes care of what kind of questions from patients is the case of Maartje, one of the patients we interviewed. Case manager Hella told Maartje to use the webcam whenever she feels it is necessary. Maartje has a recurring belief that a man enters her house and leaves blood everywhere. Whenever she is frightened, she calls the team on screen and discusses her feelings and behavior. This service is very important for her, even when she is not delusional, because she can talk about what is going on in her life. A few times Maartje has discussed her delusion with the person on duty in the health care post, who in turn has called or mailed Hella, her regular case manager. Maartje's story about blood and violence is upsetting to case managers who do not know her, and that makes them mobilize Hella.

Maartje: I'm not supposed to go into details with other case managers, you know? When I do, Hella complains because she gets all these messages from her colleagues at the station. They report my questions and remarks, and then Hella gets telecare questions later on. So then the world's upside down.

Besides receiving the regular appointments at home, which steadily deal with her important issues, Maartje can mobilize help via telecare whenever she feels she needs it. This has a down side, though, as the issues that bother Maartje lead the case managers to warn Hella. Apparently Hella knows that these issues, although very serious, make the case managers undertake understandable, but mostly unnecessary, follow up. Hella might be used to regulating Maartje in their weekly talks at home, but now she feels she has to coordinate her telecare calls too. Maartje has noticed that Hella intervenes, because Maartje's questions are not supposed to come through the back door, as it were, and she even thinks that the telecare option upsets the logical process of care for Hella: it is an 'upside down' world.

Hella structured the logistics of Maartje's webcam use without consulting others. There is no clear rule and no communication on the practice:

Maartje called while Taco was engaged with another a screen call. Afterwards, he returns Maartje's call. Maartje immediately states that she 'does not want to go into the deep end'. She just wants to discuss social relations. 'I met someone at the fellow sufferers group. And I ate with this person I met at church'. Maartje talks about her plans.

After the call ends, Taco says he needs to examine whether it was Maartje's initiative 'not to go into the deep end' and not talk about everything that is bothering her. He respects her choice, but states that she used to discuss everything. Maartje and Hella's understanding on the content of telecare conversations has not traveled far.

We saw how telecare contact (or lack of contact in the case of missed calls) raises questions on who needs to follow up on what. What questions need answers and which ones can be passed on? Are telecare case managers supposed to answer all questions and deal with all issues that stem from the technology (like missed calls) just because they offer availability? Underneath these practical questions lies the issue if it is good care to address all questions. Unanswered questions create friction in the telecare situation, as presumably the system was installed for patients to use whenever they feel the need to talk about any and every subject and it turns out they cannot..

B. *A familiar difference in a new case*

Care professionals, of course, differ in their approach, opinions and knowledge. This section deals with the differences in webcam practices, revealing a dilemma related to good care.

Many patients use the webcam regularly and are thus known to all the case managers. Bob is one such patient, a middle-aged man with an anxiety disorder. Sometimes, when he is having a tough day, Bob calls several times:

Interviewer: What do you think is the purpose of telecare for people like Bob?
Mary (case manager): Well, it's for when people get stuck, for example. People who can't start the day by themselves, they call their case manager every five minutes. With the screen, I feel they can learn to give themselves a signal, like: I'm stuck, I have to do five things and I don't know where to start. Structuring your day, that's a perfect way of using it.

For Mary, Bob exemplifies the benefits of telecare. Bob is very insecure about many things, including organizing his days. The screen gives him an easy way to get in touch with a care professional so that he can ask for support for whatever is bothering him at that moment. However, there are other sides to this story too:

Bob calls. He says: I want to talk a bit.
Daniel: Why do you want to talk?
Bob: I want to get rid of my tension.
Daniel: You always do, but you have to talk to your psychiatrist, I can't help you.
Bob: I want to know what I can do about it.
Daniel: What do you think?
Bob: I think I'll go for a ride on my bike.
Daniel: Good idea!
Bob, terminating the call: I'm hanging up now.

Daniel has a different way of handling Bob's recurring requests for support. Daniel's intervention is based on the aim of letting patients develop their own resources. Bob has written an action plan of steps he can take when he is not feeling well. In this example, Bob suggests a bike ride, just as his action plan might indicate.

For case manager Rudi even this might not be enough, as Bob seems unable to use the plan as intended. Rudi says that Bob's frequent calls show that he cannot rely on his own resources. In Rudi's opinion telecare is not supporting Bob, but therapy might:

Field note: Bob has already called in once this morning. He knows what to do, but needs confirmation. Rudi thinks that Bob should be taught how to handle his thoughts himself, without the continuing intervention of others, for example with the help of cognitive behavior therapy. He does not know if that would be an option for Bob or if anything like that has been tried yet.

We learn from the example of Bob, Mary, Daniel and Rudi how a familiar issue in mental health care is reinforced by the webcam. It questions whether Bob should be allowed to call in whenever he feels it is necessary or should he be encouraged to rely on other resources than nursing care? The issue is not new, but the webcam renews it, as it makes care more accessible. Bob can ask for help whenever he likes; he just has to press a button and someone is there to support him. With telecare the team is available for unplanned contacts. The webcam puts forward a normative question related to telecare: should care be accessible on demand?

C. *Good platforms?*

In the wake of the issue of care on demand is the question whether patients can vent about anything on their mind? Is it important to share everything? And is therefore each call equally important, including ones that do not seem to be about care? This section explores a specific aspect of telecare: social talk.

Because most calls are unplanned, often the patient and care professional are unfamiliar with each other. Most case managers of team E find this unfamiliarity an asset in their work. It is a change from daily routine, they meet new patients and being separate from regular care gives them the chance to participate in social talk. The case managers recognize that many patients are very lonely and understand that some need to chitchat regularly. Case manager Wende:

Well, you've got the time to listen, you see, as you're here anyway. And because you are not in a therapeutic relationship, you don't have to do so much with these patients... That's what they do with their therapist, with their own care professional.

Wende regards telecare as an extra option for social talk, as patients do not have to 'work' with the members of the telecare team. Her colleague Taco adds another level of meaning to this. He actually sees that patients actively bond:

Some callers appeal to us differently than to their own case managers. They do more ... or sometimes less... It's definitely different. You can tell that they know they can ask us questions. Some have great confidence in us, they know us, our faces. They discuss everything. But some don't, they are more reluctant.

Not all case managers find social talk not part of their job though, or as Rudi puts it:

And what I do here, with telecare, is just show my face, chat a bit and listen to what someone says or wants. So that's more like, well, it's like being very understanding and not giving any old advice.... I don't fully understand the situation nor do I know where the patient is heading, so it's more like being available for a talk... Methodically it's not much... Perhaps it'll go somewhere, but that would still take a lot of work.

The question this raises is whether social talk is a pastime or a therapeutic intervention? For Rudi a webcam conversation without a care context becomes chitchat, while Wende and Taco feel that providing social contact is important as patients can be very isolated.

The matter of social talk is further complicated by an issue on the platform telecare offers. Is it a form of good care when patients get the opportunity of recurring conduct? Case manager Taco talks about the issue of handling the reappearing chitchats of the same patients:

Yes of course, it gives patients an opportunity to complicate things. Give an extra option and people will take it. Maybe not to the extent we want, but you do give people... When you look at it from the recovery perspective telecare is actually quite nice. That people can decide for themselves whether they want to talk to someone or not.

Sometimes it is difficult for case managers to handle, repeatedly listen to, or even look at the recurring stories. Taco tells about patient Sonja:

Taco: Sonja calls whenever she hears voices. She uses telecare to tell us how she used her own interventions successfully. But it's always the same conversation. Also, she always sits in the same way at her table. Always the same notification. Sometimes I find it a bit silly.

On one of the shifts we attended, patient Titia called at least a dozen times. It turned out that she had had a lot to drink and Rudi, who was on duty, cut off every subsequent call until she stopped calling. During the conversations, Titia complained about mental health care. Rudi knows her concern is exacerbated by alcohol, something he sees more often:

Rudi: You generate dysphoria with the screen... It becomes online grumbling.
Interviewer: So the screen paves the way to that?
Rudi: Yes, I think so. Many frequent users have an endless need to externalize everything without having any awareness of the part they play themselves. So with this lady, you can wonder if telecare is useful... But you don't know.

When is a call useful? Some case managers say that every form of contact is useful, whenever patients find it necessary. Others, like Taco and Rudi, have their doubts. Patients with a borderline disorder form a special group under debate. These patients often demand a great deal of attention and should be able to cope by themselves. Therefore, some case managers wonder if such patients benefit from a seemingly endless offer of care. Case manager Ab stresses this point by reminding us that the general vision of the organization is that caregivers should stimulate clients to undertake more things by themselves:

And then what do we do? We give them 24/7 [tele]care.

We have seen different ideas on the function of social talk in telecare. Some case managers find telecare a very good instrument for engaging in the important asset of social talk to prevent loneliness. Others find that talking without a care context, such as an action plan, is not very relevant. With the social talk comes a deeper friction: should telecare be a space in which everything can be discussed freely whenever necessary? And does telecare give stage to endless unnecessary chats, leading to a main dilemma: does 24/7 telecare relate to the self-recovery perspective on care that drives the care organization? Some case managers define social talk as good care, but they have not discussed this with others. The same goes for the dilemma of round-the-clock care.

IV. CONCLUSION

In this article we looked for dilemmas and frictions on good care that arise in a mental health nursing telecare practice and we discussed three different practices. In this section we discuss our main findings and relate them to theories on good care. We end by discussing how care professionals can deal with frictions in good telecare.

Telecare leads to frictions on what questions can be asked at what moments. According to some nurses, just because telecare makes care available, that does not mean that all questions need (or should) be answered. In fact, it might even be necessary to prevent some questions, which takes place when missed calls are not returned or particular topics are prohibited in telecare conversations. The moral questions are on when you offer 24/7 care, should you regulate or even answer every call that comes in? And is it good care to follow up missed calls, as they might represent unanswered questions for help. Or in other words: if a patient calls, missed or answered, does it represent self-control or does the stream of calls need any regulation? Case managers act differently in this situation, which can lead to uncertainties for patients.

The webcam renews a familiar issue while it makes care more accessible: should patients rely on themselves or on care? For Bob, the question of what good telecare is, depends on whether it supports his need to be self-reliant. Self-reliance is an important theme in mental health care. Patients are encouraged to solve things for themselves as much as possible. Does telecare, through its constant availability, really help Bob to develop self-reliance, when he can call for help whenever he wants? Or will Bob lean on (tele)care more, because it is constantly available? Does telecare create a missed opportunity for Bob to find his own solution first (on his action plan or his own social network)?

Social talk, which has always been an aspect of care practice, becomes more extensive when telecare is used. Whether or not social talk is good care is not a new question. Care professionals and patients usually handle this in their daily practices. However, the webcam adds friction to the practice, as it takes social talk out of the context of prevailing care and turns it in a care practice on its own. For some case managers it is unclear if this is a good care practice. The availability of telecare plays quite a role here, leading to questions on the amount or frequency of contact. Is a daily webcam chat about the weather a form of support or should that not be part of care? And what about its ceaseless accessibility? That can lead to uninhibited expression of feelings and ideas, even when that is not good for a patient. The webcam might facilitate a free space for unbridled expression, which some case managers think is the actual benefit of social talk.

We have seen patients using telecare to ask diverse questions or discuss subjects they regard as essential, whenever they feel it is necessary. Along the way they encounter care professionals with differing views on whether or not their needs should be met. Here we see where good care comes into being. The different practices described here could count as good care, as we stated that good care is situational and fluid. So why are the differences that we encountered a problem? Why do we call them dilemmas and frictions? Let us look at the case of Bob again. Let us say that nurse A and Bob have determined that for Bob good care means that he may call whenever he likes. Everything is fine during nurse A's shift but a few hours later, nurse B is on duty. She thinks differently and treats Bob differently. So Bob will not get good care, or he has to renegotiate it with every new shift. And even when he has established this with all the team members, he now needs a system to remember what he has agreed with whom (whomever is on duty). Differences between care professionals are not unusual, but Bob now encounters them far more often than he ordinarily would, had the care been given only by his regular case manager and a sole replacement during holidays. If we take this a bit further: how would this practice look like if Bob decided that good care means he wants to call in every three minutes? Mostly likely, all nurses would agree that is not feasible, and they would not think it is good care. The example of Bob is exemplary for all frictions and dilemmas we have seen. They are not the same, but they share that good care could be established, but is not, as the changes in the practice through the technology, are not in favor of the patients. Moreover, for all of them counts that outcomes are not discussed.

In the introduction we discussed how good care is established as it is being carried out. Conditional for good care is the intention, as care professionals strive for good care, as we saw when observing the case managers. When the telecare case manager on duty redirects a patient with queries about medication to his regular case manager, it is not because the case manager does not want to help. It is because she thinks she might not be the best person to answer these questions. Instead, she offers him contact and the opportunity to talk about other things, to give something extra to his day. She strives for good care, but it becomes a friction when the patient needs something else. Or when the dynamics of the telecare team, with an occupancy of two different case managers a day, creates too many differences.

In our examples the frictions are not discussed between either case managers and patients, nor between case managers. It is difficult to discuss what the 'bad' of care is [26], but not discussing it at all, risks aggravation of frictions. Just because the 'good' in care is not uniform (as we not share all our ideas, convictions, passions, experiences or desires), it is important to discuss what patients need when striving for the good [23]. To reestablish good care when the circumstances change, patients, professionals (by themselves and together) and society at large have to think and talk about, strive for, and provide the good of care while trying to limit the bad.

The different dilemmas and frictions on good telecare we have seen, all relate to an overall dilemma in mental health care: how does the policy of reinforcing self-reliant patients relate to 24/7 care? The webcam reinforces the dilemma as it enhances the accessibility of care. With the webcam come all new forms of the same dilemma to the front, leading to frictions on good care. It is difficult to recognize changes in good care in the changed care practice of telecare. In what way does this new practice contributes to what patients need and want?

Good care is complex, as it consists of various goods [26] that rely on each other or are at least bound to each other. As good care is situational and established when enacted, it is also subject to some consensus, and so it might be best to be discussed often. Following the ideas on the importance of discussing ethical issues [35] and alongside the continuing development of empirical ethics [36], we would encourage professionals to deliberate on the care they give. Case managers should look for complex examples, confer on the frictions, dilemmas and issues. In discussion the subject could be that following up on missed calls might turn out to be one of the goods of care. Or not. If it turns out not to help make care good, then it might as well be skipped. Case managers can also discuss if any boundaries for social talk are necessary. And for whom they are important. Taking into account the workload of most care professionals, we want to stress that discussing good care can be done in workable solutions, fitting the case managers' daily routines (think of phone calls, team meetings, forums, coffee breaks, corridor chats and lunches) and deliberate on the care given in daily telecare practices. The aim is to strive to uncover the (potential) frictions, in order to give good care. Exchange and discussion between care professionals, derived from their understandings with patients, will lead to applied knowledge, or even better, artisanal knowledge of good care.

ACKNOWLEDGMENTS

We are most grateful for the opportunity to observe and become part of the telecare practices of the professionals and patients we met. We would like to thank the care organizations that generously offered the settings for this research: Dimence, GGZ Drenthe, Lentis and GGZ Noord-Holland-Noord The research described in this paper was made possible by a grant from Nationaal Regieorgaan Praktijkgericht Onderzoek SIA (grant number 2014-01-52P). Special thanks also to our colleagues with whom we worked closely on this project: Marjolein Verburg, Marijke Span and Aranka Dol.

REFERENCES

[1] A. van Hout, R. Janssen, and M. Hettinga "Performing Telecare: Recognizing New Nursing Care Practices," Proc. eTELEMED 2016, Feb. 2016, pp. 247-251.

[2] L. van der Krieke, L. Wunderink, A.C. Emerencia, P. de Jonge, and S. Sytema, "E-mental health self-management for psychotic disorders: state of the art and future perspectives," Psychiatric Services. 2014 1:33-49.

[3] S. Lal and C. Adair, "E-Mental Health: A Rapid Review of the Literature," Psychiatric Services. 2014 1: 24-32.

[4] D. Ekers, R. Murphy, J. Archer, C. Ebenezer, D. Kemp, and S. Gilbody, "Nurse-delivered collaborative care for depression and long-term physical conditions: A systematic review and meta-analysis," Journal of Affective Disorders. 2013 149: 14–22.

[5] G. Clarke and B. Yarborough, "Evaluating the promise of health IT to enhance/expand the reach of mental health services," General Hospital Psychiatry. 2013 35: 339–344.

[6] F. A. Moreno, J. Chong, J. Dumbauld, M. Humke, and S. Byreddy, "Use of StandardWebcam and Internet Equipment for Telepsychiatry Treatment of Depression Among Underserved Hispanics," Psychiatric Services. 2012 12: 1213-1217.

[7] GGZ Nederland (Dutch Association of Mental Health and Addiction Care), "Een analyse van de verwachte kosten en baten van eHealth. Blended behandelen en begeleiden (An analysis of expected costs and profits of eHealth. Blended therapy)," Amersfoort. 2013.

[8] F. Garcia-Lizana and I. Munoz-Mayorga, "What about telepsychiatry? A systematic review," Prim Care Companion J Clin Psychiatry 2010: DOI 10.4088/PCC.09m00831whi.

[9] B. Lindberg, C. Nilsso, D. Zotterman. S. Söderberg, and L. Skär, "Using information and communication technology in home care for communication between patients, family members and healthcare professionals: a systematic review," International Journal of Telemedicine and Applications. 2013: DOI 10.1155/2013/461829.

[10] L. Barnes and T. Rudge, "Virtual reality or real virtuality: the space of flows and nursing practice," Nursing Inquiry.2005 12: 306-315.

[11] K. England and I. Dyck, "Managing the body work of home care," Sociology of Health and Illness. 2011 33: 206-219.

[12] B. Latour, Science in Action. Cambridge (MS): Harvard University Press, 1987.

[13] J. Law, "Actor-network theory and material semiotics," In: The New Blackwell Companion to Social Theory. BS Turner Ed. Oxford 2009: Blackwell, 141-158.

[14] M. Akrich, " The De-Scription of technical objects," in Shaping technology/building society. Studies in sociotechnical change. W.E. Bijker and J. Law, Eds. Cambridge (MS), London (GB) 1992: The MIT Press, 225-258.

[15] B. Latour, "Where are the missing masses? The sociology of a few mundane artefacts," in Shaping technology / building society. Studies in sociotechnical change. W.E. Bijker and J. Law, Eds. Cambridge (MS), London (GB) 1992: The MIT Press, 225-258.

[16] M. Winance, "Trying out the wheelchair: The mutual shaping of people and devices through adjustment," Science, Technology, & Human Values. 2006 31: 52-72.

[17] J. Pols, Care at a Distance: On the Closeness of Technology. Amsterdam, The Netherlands: Amsterdam University Press, 2012.

[18] I. Kamphof, "A Modest Art: Securing Privacy in Technologically Mediated Homecare," Foundations of Science. 2015: DOI 10.1007/s10699-015-9448-5.

[19] J. Pols, "How to Make Your Relationship Work? Aesthetic Relations with Technology," Foundations of Science. 2015: DOI 10.1007/s10699-015-9449-4.

[20] J. Pols and I. Moser, "Cold technologies versus warm care? On affective and social relations with and through care technologies," ALTER, European Journal of Disability Research. 2009 3: 159-178.

[21] I. Moser and J. Law, "Good passages, bad passages," in Actor Network Theory and After. J. Law and J. Hassard Eds. Oxford. 1999: Blackwell Publishers, 196-219.

[22] A. van Hout, J. Pols, and D. Willems, "Shining trinkets and unkempt gardens: on the materiality of care," Sociology of Health & Illness. 2015: 1-12.

[23] A. Mol, The logic of care. London: Routledge, 2008.

[24] J. Pols, "Washing the citizen: washing, cleanliness and citizenship in mentalhealth care," Culture, Medicine & Psychiatry. 2006 1: 77-104.

[25] J. Pols, "Chapter 7: Conclusion: Enacting improvement," in Good care. Enacting a complex ideal in long-term psychiatry. Utrecht 2004: Trimbos-reeks, 149-55.

[26] A. Mol, "Care and its values. Good food in the nursing home," in Care in Practice. On tinkering in clinics, homes and farms. A. Mol, I. Moser and J. Pols Eds. Bielefeld. 2010: Transcript Verlag, 215-34.

[27] J.R. van Veldhuizen and M. Bähler, "Manual Flexible Assertive Community Treatment," Groningen, 2013.

[28] T. Scheffer, "Event and Process: An Exercise in Analytical Ethnography," Human Studies. 2007 30:167–197.

[29] J. Law, "Seeing like a Survey," Cultural Sociology. 2009 2: 239-256.

[30] S. Hirschauer, "Putting Things into Words. Ethnographic Description and the Silence of the Social," Human Studies. 2006 29-4:413-441.

[31] A. Mol, "Missing Links, Making Links. The Performance of some Atheroscleroses," in Differences in Medicine: Unraveling Practices, Techniques and Bodies. M. Berg and A. Mol Eds. Durham, London. 1998: Duke University Press, 144-65.

[32] J. de Lange, "Vergeten in het verpleeghuis (Forgotten in the Nursing Home)," Utrecht. 1990: Trimbos-instituut.

[33] Central Committee on Research Involving Human Subjects, "Your research: does it fall under the WMO?," Available from: http://www.ccmo.nl/en/your-research-does-it-fall-under-the-wmo.

[34] H. Boeije, Analysis in qualitative research. London: Sage, 2010.

[35] T. Abma, B. Molenwijk, and G. Widdershoven, "Good Care in Ongoing Dialogue. Improving the Quality of Care Through Moral Deliberation and Responsive Evaluation," Health Care Analysis. 2009 17: 217-235.

[36] D. Willems, "Varieties of goodness in high-tech home care," in Care in practice. On tinkering in clinics, homes and farms. A. Mol, I. Moser and J. Pols Eds. Bielefeld 2010: Transcript Verlag 257-277.

The Real World is a Messy Place:

The Challenges of Technology Use in Care Provision

Anthony P. Glascock
Department of Anthropology
Drexel University
Philadelphia, PA 19104, USA
email: glascock@drexel.edu

Rene A. Burke, Sherri T. Portnoy, Shaleea Shields
NHS Human Services
Lafayette Hill, PA 19444, USA
email: rburke1@nhsonline.org, SPortnoy@nhsonline.org, sshields@nhsonline.org

Abstract—**This paper reports on the ongoing demonstration project to assess the effectiveness of three distinct eHealth technologies to contain costs, make the provision of care more efficient, and to contribute to the wellbeing of individuals with intellectual and developmental disabilities and severe mental illness by increasing their independence. The challenges of providing care for these populations is discussed, as well as the demographic trends that will, over time, not only result in a threefold increase in the number of individuals in populations, but dramatically impact the cost of their care. Data collected from 14 different locations over the past three years will be discussed in order to determine if the selected technologies resulted in an improvement in living conditions and wellbeing, and to judge the acceptance of the technologies by both staff and customers. Findings indicate that the technologies have largely exceeded expectations resulting in plans to expand the project to other facilities, to build a new residence in which all the tested technologies will be installed and to begin to move beyond company owned facilities into residences within the community.**

Keywords—real world; intellectual and developmental disabilities; severe and persistent mental illness; eHealth; demonstration project; multiple technologies; financial impact; business decisions

I. INTRODUCTION

The impetus behind the inception of this Project was the recognition that the only way to provide cost effective care to individuals with disabilities and severe mental illness, is to incorporate appropriate technologies into existing care models which is the driving force for a multi-year, multi-site demonstration project being undertaken in the eastern United States. Findings from the first Stage of the Project [1] were so encouraging that a second Stage is underway and a third being planned.

The urgency to contain the cost of delivering care to a wide range of populations has seen the development of new technologies, as well as the innovative use of existing technologies, in an ever enlarging number of care models. Thus, it is not surprising that new and reconfigured eHealth technologies are increasingly being used to provide care and services to individuals with Intellectual and Developmental Disabilities (IDD) and those with Severe and Persistent Mental Illness (SMI) [2][3]. As in the use of new technologies, in each and every care model there are challenges to using innovative technologies in addressing the needs of these two populations, but the necessity to contain, and if possible, reduce the cost of providing care to these populations, makes the use of technology, in some form, inevitable. However, three different trends intersect in such a way that result in the urgency to develop more efficient care models for the IDD and SMI populations: the cost of care; the aging of the two populations, and a reduction in the number of qualified staff providing care to the IDD and SMI populations.

First, because of a wide-spread belief that very large state-run institutions resulted in dehumanizing individuals with disabilities, deinstitutionalization, in both the United States and Europe took place from the mid-1960's through the 1970's. This process of deinstitutionalization had a dramatic impact on the care models employed [4][5]. Between 1967 and 2008, the population of individuals with disabilities in large state-run institutions declined from a peak of 194,650 to 32,909 in 2009 [6]. The impact of this massive deinstitutionalization on the cost of care varied significantly, but the impact on the care model was significant as most individuals with IDD and SMI moved into the community, many living in some form of group home. The number of individuals living in any particular group home varies based upon the needs of the individuals. The most common number of residents is four with some group homes having as many as eight to ten residents. In the most severe cases, an individual lives alone in a residence with 24 hour supervision. Although costs of providing care to the residents in group homes vary based upon the needs and location, the average cost is between $40-50,000 per year per resident and if an individual needs to live alone, the cost can top $150,000 per year [5][7].

Second, the rapidly increasing number of such individuals brought about by the same demographic factors as for the general population is adding even greater cost to the care of IDD and SMI populations [8]. In particular, it is estimated that the number of adults with disabilities will almost double between 2000 and 2030 to over 1.2 million [9]. As individuals with IDD or SMI age, they are as susceptible to chronic illnesses as the general population, but the cost of caring for them is much greater. For example, care for an individual with IDD who has congestive heart failure costs approximately eight times more than for a person without IDD [5]. The ratios for other chronic diseases and individuals with IDD versus SMI vary somewhat, but the reasons are consistent. Many individuals with IDD and SMI make poor lifestyle decisions—use tobacco products and abuse alcohol and drugs. In addition, many are unable to self-manage disease, e.g., adhere to complicated medication regimes and follow complex health care instructions. Thus, greater cost of care for individuals with IDD and SMI when combined with the cost of residential care in general, results in a compelling reason for attempting to use technology to contain costs.

A third reason is that there is an increasing imbalance between the growing needs of the IDD and SMI populations and the number of qualified staff available to provide care. Projections from the federal government suggest that the need for trained staff will increase by over 30% in the next decade, while the supply of individuals who traditionally have filled these jobs is expected to increase only by 7% [10]. In addition, the high turnover rate for individuals caring for these populations adds another dimension to the staffing challenges. It is estimated that turnover for direct support professionals (DSPs) ranges between 50-70% depending on the specific jobs undertaken, e.g., residential care versus in-home care. This high turnover rate adds at least $2,500 in direct expenses plus a minimum of an additional $1,000 of indirect expenditures for an organization to replace a single DSP, thus adding to the ever increasing cost of providing care to these populations [11].

In the next section, a brief discussion of attempts at using technology to provide care for these populations is accompanied by a short description of the the overall goals and objectives to be achieved by the introduction of the new technologies. In Section III, the different Projects undertaken are described along with the care models employed and the technologies introduced. The next section discusses the methodologies employed in gathering data on the individuals with disabilities and mental illness and staff in order to assess the effectiveness of the technologies in care delivery, while Section V offers a discussion of findings for the first two Stages. What has been learned from the analysis is summarized in the Discussion Section, while plans for the roll-out of the technologies to other facilities and the financial model to pay for this expansion comprise the concluding section.

II. THE PROJECT

Even though NHS Human Services, through its subsidiaries, is one of the United States' leading non-profit providing education and human services to individuals with special needs., it was clear by 2012 that to be able to continue to provide cost-effective and high quality care to individuals with disabilities and severe mental illness, it was essential to introduce new and innovative technologies into existing care models. As a result, in the spring of that year, the organization created an Assistive Technology Executive Steering Committee to plan a demonstration project. One of the first tasks of this Committee was to assess the state of related work.

A. Related Work

It was quickly apparent that there had been few other attempts to utilize technologies in a similar manner. From its inception, the Project was viewed as a demonstration project to assess both the benefit that could be gained by incorporating specific technologies into existing care models, and the ability of the company to pay for the use of the technologies. As such, the Project was neither a pilot study of the use of a specific technology within a controlled environment [12], nor a short-term, funded research project [13], nor a study relying on a small number of volunteers to evaluate the effectiveness of a given approach to the use of a selected technology [14]. Thus, previous studies were of only limited help in the planning and implementation of the project.

Certainly, the fact that the selected technologies had all been tested and were all commercially available was a definite advantage for the Project and saved much time and money. However, the downside was that many of the technologies had been tested by a relatively small number of individuals, usually 20-50, and in controlled environments [15][12][13][14]. Given that the goal of the Project was for the selected technologies to be used in the provision of care for hundreds and, eventually, for thousands of individuals within the normal care model and not in a laboratory or controlled environment proved challenging. It is one thing for a technology to be effective in a controlled environment being used by carefully selected individuals, and another to be incorporating the technology into existing care models within the community.

Another limitation on the usefulness of related works was that the majority of studies of individuals with disabilities that exist focus on children within the context of school [16][17][18], rather than adults living in the community. Some studies of younger populations did prove to be of limited utility because of the convergence of similar technologies, i.e., iPad, but most were too focused on educational issues to be that helpful. A final limitation of the literature, even if the populations were similar, was that the rapid advances in technology made many of the studies, even if conducted less than a decade ago, out of date [19][20][21].

Nevertheless, the related work did confirm one of the basic premises that motivated the Project in the first place, that the utilization of technologies in the provision of care to individuals with disabilities and severe mental illness is limited [22]. Given the trends outlined in the Introduction, this underutilization of technology must be rectified. Consequently, the effort to use a series of technologies within the NHS care system should be viewed as more than just a localized demonstration project; it is, in reality, a test of the resolve to make these technologies widely available within the larger IDD and mental health communities [23].

B. The Strategic Plan

The planning process was inclusive and there was a recognition that in order to "do it right" it would take time to put everything in place. The first step was to encourage employees to propose sites at which new technologies could be used. During the remainder of 2012, proposals were received and evaluated based upon specific criteria: administrative and staff buy-in; existence of suitable technology; evidence that technology would enhance care provision; evidence that, if successful, the technology could be used at a large number of other care facilities within the organization; and a financial model showing that the technology was sustainable—the organization would be reimbursed for its use. Finally, there was an attempt to achieve a rough balance among the different care models employed throughout the organization. This process took over a year which afforded a thorough evaluation of the resources available at each of the selected sites. The final decision was confirmed at an all-day meeting of administrators and representatives from each of the chosen sites in the fall of 2013.

C. The Project Goals

As planning progressed, three main goals emerged: 1) to determine which, if any, of the technologies being tested can allow for an improvement in living conditions and the care being delivered in the selected facilities; 2) to judge the acceptance of the technologies by both the staff and individuals with disabilities and mental illness; and 3) to assess whether the technologies should be rolled out to other facilities with similar care models. In order for any of the Projects to be deemed successful, it was necessary to determine if the new technologies allowed for an improvement in living conditions and the care being delivered in that the individuals with disabilities and mental illness express that their lives are better after the introduction of the technologies than before. It was also necessary to determine if care delivered is more timely, efficient and cost effective than the care delivered without the technologies.

It was also essential to ascertain if staff could properly use the new technology, that they believed in its effectiveness, and accepted that the technology would require that they did their jobs differently. Likewise, it was necessary to determine if the individuals with disabilities and mental illness accepted the use of the new technologies in the care that they received, if they were intimidated or not by the technologies and if they would willingly comply with requirements for the use of the technologies. Finally, even if it was determined that the new technologies provided improved care, were accepted by staff and individuals with disabilities and mental illness, it was still vital to find out if the care provided with the new technology was reimbursable as a billable expense.

D. Project Timeline

Given the challenges of incorporating multiple technologies into existing care models, the Assistive Technology Executive Steering Committee decided that it was best to start slowly with a gradual roll-out, rather than trying to introduce multiple technologies simultaneously. Although this gradual approach meant that some components of the Project would begin before others, thus risking the loss of initial enthusiasm, the ability to focus on getting one technology up-and-running, instead of having to deal with potential problems with three technologies, appeared prudent.

In addition, a preliminary plan was developed for the roll-out of the technologies to other locations, if, of course, they proved successful in providing care and were cost effective. Once again, a measured approach was taken and a two-step roll-out envisioned. The first expansion would be to a small number of locations that mirrored the initial test sites in care model, size and staffing.

Rather than create a formal timeline for the Project, a three-stage timeline that relied on measured success, rather than arbitrary dates was formulated. This flexibility was possible because the Project was largely self-funded and because long-term success in incorporating the selected technologies into the organization's care models far outweighed any short-term success brought about by reaching arbitrary benchmarks.

III. THE TECHNOLOGIES AND LOCATIONS

After reviewing the submitted proposals, the Assistive Technology Executive Steering narrowed the proposals to three that were to be part of the first phase of the Project: 1) the Communication Technologies Project (CTP); 2) the Smarthome Project (SHP); and 3) the Biometrics Project (BMP). Based upon the desire to stagger the start dates of the three Projects, it was decided to begin the SHP early in the summer of 2014, followed by the CTP in late summer and the BMP later in the fall. Although the SHP began slightly before the CTP, the CTP has advanced at a faster rate than the other two Projects which were delayed by unanticipated problems in the renovation of facilities and the equipment to be used to monitor vital signs.

A. The Communication Technologies Project (CTP)

The CTP began in the late summer of 2014 with the selection of sites and upgrading of wireless routers. Work continued during that summer with the focus of training staff in the use of AbleLink software [24] that had been selected for use and during the fall of 2014, staff and individuals with disabilities were surveyed and the Glasgow Depression Scale (GDS) administered to all individuals with disabilities participating in the Project [25][26]. The main goals of the CTP was fourfold: to enable individuals with disabilities to stay in touch with family and friends; to allow a greater ability for them to communicate with members of the support services team; to encourage them to acquire basic computer skills; and to permit safe and secure access to the internet in order for them to pursue their particular interests. Seven sites were selected for inclusion in the first phase of the CTP. Five of the sites are group homes and two program centers, all in Western Pennsylvania. The group homes are single sex residences for between three and six individuals. In contrast, the two program centers serve between 90 and 130 individuals on any given day.

The hardware introduced into the five group homes were iPads and laptops while at the day programs, all-in-one desktop computers, laptops, tablets and iPads were made available. The hardware was customized to meet the needs of the IDD population, e.g., large keyboards, headphones. After much research, a software package designed specially for individuals with cognitive disabilities—AbleLink—was purchased and installed. AbleLink allowed individuals to experience a more self-determined and fulfilled life through an empowering technology characterized by a person-centered design philosophy. Several AbleLink applications were installed that allowed individuals to use email (voice activated), Skype and webcam broadcasts, along with providing prompts for tasks that increased independence.

In 2015, the CTP was expanded to include three additional locations in Pennsylvania and Virginia: a large congregate intermediate care facility (ICF) and two adult training facilities (ATF). The ICF has 103 residents, of whom 30 are included in the project while the Pennsylvania ATF has 29 individuals receiving services of whom 16 individuals are participating in the Project and the Virginia ATF has twenty-two individuals receiving services with 10 participants in the Project. The ATFs are non-residential centers which provide services in functional activities, assistance in meeting personal needs and assistance in performing basic daily activities to individuals who are 59 years of age or younger and who do not have a dementia-related disease as a primary diagnosis. Unlike the first Stage, rather than laptops, tablets and iPads being made available to the participants, AbleLink applications were enabled on all-in-one computers. This configuration was deemed to be the most useful equipment for their client base and size within these facilities.

Three additional sites have been selected for inclusion in the third Stage of the CTP: a large congregate intermediate care facility with an estimated 30 participants; a large adult training facility with an estimated 60 participants and a smaller adult training facility with an estimated 29 participants. Currently, funding is being secured to enhance the internet capabilities in the three facilities, and training of the identified staff has begun with the goal of all three facilities being up-and-running in the fall of 2016.

B. Smarthome Project (SHP)

The SHP required a remodeling of a residential unit which faced construction problems delaying the start of the Project several months. However, by the summer of 2014, the four residents were able to move into the remodeled facility and be administered surveys and the GDS. The main objectives of the SHP were to increase the independence of the four IDD residents and to conserve energy through the use of "green" appliances and more efficient heating and air conditioning systems (HVAC). To achieve the goal of increasing the independence of residents, a Smart TV was installed, iPads and remote controls for lighting and window blinds were made available to them. Additionally, motorized cabinetry and cook tops and sinks were installed in a lowered position to allow wheelchair access. Finally, to reduce the amount of energy consumed, remote control HVAC systems and smaller and more easily accessible dishwashers and refrigerators were installed.

NHS' second Smarthome Project is currently in the planning phase. Over $650,000 in funding has been secured to build a brand new facility, as opposed to retrofitting an existing residence. The building plans include a six bedroom residence to support the six individuals who will live in the wheelchair accessible home. The home will be designed to include: a fully accessible kitchen; home automation system; tintable window technology/moveable blinds; green components; more accessible bathroom components; and the CTP technology and biometrics system used in the other two projects.

Given the cost of building new Smarthome facilities, there are no concrete plans to expand the SHP beyond the currently planned location. However, the goal is to gradually include smart technology in renovations of existing homes and build new facilities incorporating all previously used smart technologies that have proven to be sustainable and useful in the everyday lives of the residents, as well as additional technologies that can increase independence.

C Biometric Project (BMP)

The BMP began in the summer of 2014 with the development of protocols, the installation of the technology, training of staff, retrospective data collection, the creation of event and error forms and administering surveys to residents and staff and the GDS to the residents. The main goal of the Project was to use technology to reduce the number of emergency room visits and hospitalizations and thus, by doing so, curtail costs by delivering care in a more timely manner and at a lower level of care [27]. Two group homes with four residents each were selected for inclusion along with a Long Term Structured Residence (LTSR), a locked facility that served eight male individuals with serious and persistent mental illness. The technology installed was a basic vital signs monitoring system including a digital scale, blood pressure cuff and pulse oximeters. The software included with the system allowed data to be sent to an external location and was configured to send alerts when the data collected went outside preconfigured parameters.

Planning for Stage 2 of the BMP began in 2015, but the project has yet to get underway because of the complexity of the facility selected. The location is a large adult behavioral health outpatient clinic serving over 400 individuals. The goal is to use a vital signs kiosk for the purpose of capturing basic biometric measures, initially blood pressure and weight, during the intake procedure. The plan is to electronically transfer the collected data to both the individual's psychiatrist and general practitioner in real time and be available in the individual electronic record. In addition, the goal is to be able to display a longitudinal record on the vital signs and have a warning system that indicates when any one of the vital signs is outside the normal range.

Preliminary plans for a Third Stage for the BMP are under way, in which a vital signs monitoring system is installed in the residences of individuals living in the community. The goal of the Stage is to have the data collected sent for review and evaluation by a clinical professional who could consult with the individual by phone or video for the purpose of providing reassurance and/or guidance for the next step in receiving additional care. It is anticipated that the initial systems will be installed by the first quarter of 2017.

IV. METHODS

There were several challenges to the selection of the methods to use to collect data on the three Projects. First, the Projects were not a test of the technologies themselves, as it was already known that they worked. Instead, the objective of the Projects was to determine how the selected technologies could be used to enhance the provision of care, while at the same time curtailing the cost of that care. Thus, the methods had to capture specific data on various components of care delivery. This entailed collecting data on the staff at each of the sites, both the way they used the technologies, and their level of acceptance and willingness to change how they did their jobs. Data also had to be collected on the level of acceptance of the technologies by individuals at each of the sites. If individuals were uncomfortable with the use of the new technologies it would not be possible to roll out the technologies to other facilities. Secondly, although ideally the same methods of data collection would be used at each of the sites, this proved impossible because of the differences in the nature of the sites and the care models employed.

A. Communication Technologies Project

The main challenge was to develop questions that could be answered by individuals with disabilities and would, at the same time, provide the data necessary on which to make future decisions [28][29]. Achieving these twin goals necessitated the development of a Project-specific questionnaire for the CTP, which included simple straightforward questions and took no more than 15 minutes to administer. The questions asked included:

- Which of the following electronic devices do you use to communicate with friends, family or other people?
- How much help do you need to use these devices?
- When you want to communicate with friends, family or other people, how often is the device available?
- What devices do you use to play games or watch movies?
- How much help do you need to use these devices to play games or watch movies?
- When you want to play games or watch movies, how often is the device available?

The questions were to determine the amount of change that took place in both device use and amount of help needed by individuals with disabilities during the length of the Project. Staff were trained to administer the questionnaire to individuals with disabilities at each of the sites with the goal being that the same staff member at each of the sites would administer the questionnaire at initiation of the Project and at three, six and 12 month intervals. However, this proved to be difficult because of the high rate of staff turnover. Thus, in order to minimize the impact of staff turnover for the CTP and the SHP, specific staff at the program centers and the individuals' case workers were designated to conduct the surveys.

The staff questionnaire was self-administered and, similar to the questionnaire for individuals with disabilities, was repeated at three, six and 12 month intervals. The GDS was administered by staff members at the inception of the Project and six and 12 month intervals.

For reasons that are discussed below in Section E, the GDS was not administered to individuals in the second Stage of the CTP. All other instruments were used to collect data in the three facilities in Stage 2 and it is anticipated that they will be used in the locations in Stage 3.

B. Smarthome Project

The methods used for the SHP were closely matched to those used for the CTP: Project-specific questionnaires were given to the residents at the initiation of the Project and three, six and 12 month intervals; likewise the GDS was administered at the initiation of the Project and six and 12 month intervals. Questions focused on the ability of the residents to undertake basic tasks within the home, e.g., meal preparation, putting away groceries, controlling the lighting and blinds in their rooms, using computers and other electronic devices, using email to communicate with family and friends. Thus, it was possible to determine changes in both the residents' ability to use the new technologies and the impact on the technology of residents' well-being.

C. Biometric Project

The methods employed for the BMP, to a large extent, mirrored those for the other two Projects with a couple of exceptions. Staff were surveyed at the beginning of the Project and after six and 12 months. Questions for the staff focused on:

- The comfort level of staff members in the use of the biometric devices;
- The reliability of the devices;
- The acceptance of the devices by individuals with disabilities; and
- The perceived change in the quality of care with the use of the biometric devices.

Similarly residents at the three facilities were administered questionnaires and the GDS at the inception of the Project and at six and 12 month intervals. In addition to these instruments, event and error forms were developed for use. The event forms were used to record each event triggered by a biomedical alert, the actions taken by staff in response to the event and the outcome, e.g., a visit by a nurse, emergency room visit or hospitalization. The error forms were used to record problems with the various devices comprising the vital signs array, steps taken to correct the problem, the potential risk to the health/safety of the residents and how the problem was resolved.

D. Limitations of the Methods

There were several factors which limited the effectiveness of the data collection and the quality of the data. First, the fact that individuals at all the sites had either developmental and intellectual disabilities or were diagnosed with severe mental illness limited the type of questions that could be asked and often required prompting by the staff member administering the instrument. Secondly, although not optimal from a research perspective, given the scope of the three Projects, it was necessary for staff to administer the questionnaires and GDS. These staff members were para-professionals whose main responsibility was not research, but instead, the delivery of care. Staff turnover also impacted the ability of the questionnaires and scales being administered at the designated intervals. Finally, and perhaps most importantly, the Projects were not research per se, but a real world evaluation of the effectiveness of technology within challenging care models. In other words, the information collected was that which could help NHS determine whether the technology installed in the sites should be rolled out to other facilities, rather than what would necessarily be collected in a controlled research project.

E. Modifications in Methodology

The limitations outlined above have resulted in modifications in the methods used during the second and third Stages of the Project. The most significant was the discontinuing of the use of the Glasgow Depression Scale for three reasons: 1) no significant differences were obtained when the results from the pre-Project and the six and the 12 month intervals were compared, primarily, it is believed, because of the small sample size; 2) in the vast majority of cases, the GDS was not filled out by individuals receiving services, but instead by a caseworker, which violated the GDS protocol; and 3) the close contact between individuals and case workers provided a better guide to the psychological state of participants than the use of the GDS.

The inclusion of the outpatient clinic in Stage 2 of the BMP will also require significant changes in the methods employed. Brief questionnaires will be used to ascertain both the acceptance of the kiosk by individuals receiving services and the acceptance by staff. An event form, similar to that used during Stage 1 of the Project, will be used to track the care delivered when one or more of the vital signs for an individual are outside the normal range. The goal is through tracking body mass index and blood pressure to assist in the identification of two comorbid conditions most associated with mental illness—obesity and hypertension—so that appropriate care can be provided. Finally, the kiosk software should be able to create omission reports for missing data and an error form, modeled after the one used in Stage 1, will be used in order to track any problems that may arise with the recording and transfer of the vital signs data.

V. FINDINGS

In an ongoing, multi-year, multi-stage, multi-site, multi-technology Project, in some ways, all findings are preliminary, since data are continuously being generated and will continue to do so for the foreseeable future. However, there are sufficient findings available from the first Stage of each of the three Projects to draw conclusions on the success of the use of the selected technologies. There is also adequate data from Stage 2 of the CTP to present some preliminary findings, but since the second Stages of the SHP and the BMP are still in the planning phase, no data are available for analysis. However, with the completion of the second Smarthome facility in 2017, and the anticipated roll-out of the outpatient clinic in the fourth quarter of 2016, and the home vital signs systems later that year, data will be available for on-going analysis.

A. Community Technology Project

The findings from a comparison of the data collected from the four sets of questionnaires are, from an organization perspective, very encouraging. Questions were asked about the use of electronic devices, both the number of devices used and the purpose for the use of the device. Answers to these questions showed a distinct pattern of the increase in both device use and the number and type of applications used. Sixteen of the 35 individuals (44%) for which data on all four sets of questionnaires are available were using more devices after 12 months than at the initiation of the Project, while fifteen (43%) were using more applications than in the prior 12 months. For the majority of individuals, the added device was a laptop that was made available in their residences. The pattern that emerged was quite clear. Individuals learned to use new applications on the desk-top computers at the day programs and then used the applications on the laptop when they returned to their residences.

The findings for the amount of help that individuals required to use the new devices and applications are a bit more complicated to interpret. The raw findings are: 2 (6%) of the individuals did not need help throughout the twelve months; 11 (31%) of them had no change in the level of help needed to access the devices and applications; 12 (34%) increased the level of help needed to access the devices and applications; and 10 (29%) decreased the level of help needed to access the devices and applications. These data are confusing enough, but in addition, there is no distinct relationship between the individuals who increased their use of devices and applications and the need for help. The amount of staff time required to train staff in the use of the technologies and to help individuals with new devices and applications is a key factor in the decision to expand this Project to other facilities and therefore, having more usable findings is extremely important.

Similarly, the findings from the GDS are ambivalent. Although there is a slight overall decrease in the number of answers that reflect a depressive state for over one-third of the individuals with disabilities, there is no apparent relationship between an increase in the use of devices and applications and a decrease in a depressive state.

Although the main conclusion that can be drawn from the staff surveys is that the staff believes strongly that the introduced technologies have been greatly beneficial, the findings did expose some problems. A full quarter of staff believed that the technology was not useful for all individuals with disabilities. In particular, those individuals who had problems with reading grew frustrated when attempting to use the various applications. Secondly, almost half of staff reported that there were problems with the applications periodically crashing and/or having difficulty in getting the applications to work properly. However, the data did indicate that over time, the technological problems decreased significantly. Finally, the data confirmed the high rate of staff turnover, as only 12 of the 50 staff who completed at least one survey completed all three. In fact, an equal number—12—of staff completed only the last survey as those who had completed all three.

Findings from the three locations in Stage 2 of the CTP confirm the findings from the analysis of data from Stage 1. Since the findings from administering the questionnaire during Stage 1 at three and six month intervals contributed little if anything to the analysis, it was decided to only survey individuals receiving services and staff at the initiation of the Project and after 12 months. Questions were, once again, asked about the use of electronic devices, both the number of devices used and the purpose for the use of the device and, once again, the answers followed the same pattern: thirty-one of the 38 individuals (82%) reported that their use of electronic devices had increased over the twelve months and likewise, the same number reported that their use of applications had increased. There were two differences between the findings for Stages 1 and 2. The first was that fewer than 25% of the individuals included in the surveys reported that they used no electronic devices at the beginning of the Project, whereas approximately 50% of individuals in Stage 1 reported that they used no devices at the beginning of the project. The second difference between the findings for the two Stages is that a full 23% of individuals in Stage 2 were using no electronic devices at the end of the 12 months. In other words, their behavior had not changed, even though a direct effort had been made to encourage these individuals to engage in the project and utilize the readily available devices.

Another major difference in the findings from the two Stages is the amount of help from staff needed by individuals using the devices. Unlike the findings in Stage 1, the findings from Stage 2 indicate that the individuals who had increased the electronic device and application use needed "a lot of help"; from 29% to almost 90%. This dramatic increase in the help needed appears to be the result of the fact that such a small number of individuals in Stage 2 indicated that they used no devices at the beginning of the Project. Thus, the vast majority of individuals in Stage 2 were starting from ground zero resulting in the help of staff members to get them "up to speed". It will be interesting to see if the amount of help needed declines during the second year of the Project.

As was the case in the findings from Stage 1, the vast majority, 18 of 20 (90%) of staff at all three locations in Stage 2 viewed the Project as successful and they believed that the individuals who used the electronic devices benefited. Also, similarly to the results of the staff surveys in Stage 1, some staff in Stage 2 expressed a level of frustration with the amount and type of training they received in the use of the technology and various applications. However, the number of staff who expressed this frustration declined from almost 50% in Stage 1 to only 20% in Stage 2, thus indicating that the change in the type of training had been, to a certain degree, successful.

B. Smarthome Project

There were no problems with data collection for the SHP. All four individuals with disabilities completed the three questionnaires and GDS administered upon initiation and three, six and 12 months into the Project. Nevertheless, the simple fact that there were only four residents in the study does limit the ability to generalize and reach firm conclusions about the wisdom of expanding the Project to other facilities.

The findings are largely positive, as three out of the four residents expressed that over the twelve months of the Project, their level of independence had increased: three out of the four residents expressed an increase in the ability to operate blinds and lights without help; two out of the four residents expressed an increase in the ability to undertake chores in the kitchen without help; and one out of the four residents recorded a greater ability to communicate with family. Answers on the GDS indicated that two of the four residents experienced a slight decrease in their level of depression. Staff also filled out the GDS for the residents and once again, it appeared that the same two residents experienced a decline in their level of depression.

In addition to the quantitative data collected, more informal interviews with both residents and staff revealed a very high level of satisfaction with the modifications made in the residence and the addition of the Smart TV, iPads and remote controls for blinds and lights. In particular, staff indicated that the mood of the residents had become more positive and that residents are much more active in the kitchen and taking pride in their increased independence.

C. Biometric Project

The findings for the BMP are the most problematic of the three Projects, primarily because of equipment issues that delayed its start and continued during the entire data gathering period. Nevertheless, there are sufficient data to draw some conclusions that can be used as the Project is expanded to other facilities. For this analysis the two residential facilities serving IDD residents will be lumped together, while the findings for the LTSR are presented separately.

Staff surveys at the two IDD facilities showed that at the beginning of the Project over one-half of the staff did not know how to use at least one of the devices that was being installed. However, by the 12 month mark, all but one staff member, not only could use all of the devices, but were comfortable using them. The 12 month survey also indicated that, overall, staff were very positive about the use of the biometric equipment: a clear majority believed that care had improved with the use of the equipment; and all staff believed that residents had accepted the use of the equipment and were comfortable with its use. Over the 12 months of the Project, there were 10 instances when one or more vital sign reading was beyond the safe range. In seven cases, a physician was contacted and in three cases, a nurse was contacted. Although in none of these cases was hospitalization necessary, four residents were put on outpatient observation in order to more carefully track their vital signs.

The only negative finding was the number of problems with the equipment recorded in 32 error logs. Just over 50% of the errors were a failure of the data to upload from the device to the iPad, which was used to record and forward the data to the nursing staff. In one-third of the error logs, the problem was that the devices were not actually recording any data, e.g., the blood pressure cuff not indicating a reading. These findings have led to a reevaluation of the vital signs system being used.

All staff at the LTSR, when surveyed, expressed a high level of familiarity with all equipment used in the Project, both at the inception and twelve months later. Eight of the nine staff reported that the residents were comfortable with the use of the vital signs array, but one-third reported that the equipment was not as reliable as they would have liked. This unreliability was reflected in the nine error reports that indicated both problems with uploading data, and the blood pressure and oximeter cuffs not generating a reading. Finally, there were eight events when one or more vital sign reading was beyond the safe range. In four of the cases, the nurse was contacted and the resident more closely monitored for the next 24 hours.

VI. Discussion

As stated in the previous section, not only have the three Projects in Stage 1 been on different timelines, but Stage 2 of the CTP has produced findings from three additional sites with the result being that the discussion of the findings from the 14 different sites is complicated by this variation in timeline and amount of data available for analysis. However, there are sufficient findings for the steering Committee to make crucial decisions as to the success of the technologies in providing cost effective care to the different locations. In particular, it is possible to determine what has exceeded expectations, what has worked as hoped and what has not worked as well as hoped. Additionally, valuable lessons have been learned that can be used to move the SHP and BMP to Stage 2 and the CTP to Stage 3.

A. Successful Implementation

Findings from both Stages clearly indicate that the CTP has been a tremendous success and has far exceeded expectations. The vast majority of individuals in all ten locations embraced the new technology and the applications made available through the Project. These individuals were able, in a relatively short period of time, to use the technology to communicate with family and friends, safely surf the internet in order to pursue their individual interests and to play games and watch movies—none of which they could do on their own before the Project. Although not all individuals at the ten facilities were able to utilize the technology, the majority could and they were able, over time, to do so with less staff help. Staff were equally pleased with the introduction of the technology and consistently reported that individuals were happy with their increased independence.

Perhaps the most positive finding is that the technology and applications employed in the CTP proved equally effective in quite different types of facilities with dissimilar care models. Five of the locations were small single sex group homes, with between three and six individuals; two were Adult Training Facilities serving between 80 and 130 individuals who varied in level of ability and had been brought to the facility from their group homes; one was a large congregate ICF in which over 100 people lived of whom 30 participated in the CTP; while the final two locations were small adult training facilities to which individuals with disabilities were transported for skills training. The success of the CTP in these varied locations has allowed the Steering Committee to, not only plan for Stage 3 of the Project, but to begin planning for a much larger roll-out in the near future. This is exactly the success that was hoped for, in that a single technology and application (AbleLink) can be used throughout the organization without having to find, test and use a variety of different applications in facilities with varied care models.

The SHP also exceeded expectations. Even though the number of residents impacted by the introduction was small, four, the findings clearly show that they benefited from having the new technologies. Their independence increased over the duration of the Project because they were able to undertake tasks that they could not accomplish prior to the introduction of the new technology. The simple ability to control the lights and blinds in their own rooms, not only increased their level of independence, but staff reported that the residents' mood became increasingly positive over time.

This does not mean that everything went smoothly, even though the assistive technologies installed in the Smarthome were all well tested and appeared appropriate for the residents. One example of a technology not working as intended was an adjustable sink that could be lowered to a height that made it usable for an individual in a wheelchair. The original bowl of the sink was much shallower than normal kitchen sinks to allow for the space underneath for wheelchair accessibility. Unfortunately, this design made it much more likely that water was splashed during usage resulting in the individual using the sink frequently being soaked by the splash of water from the faucet. This led, not only to the necessity for the DSP to change the resident's clothes, but often the resident becoming agitated, thus disrupting the home's routine. Also, several of the individuals living in the residence have motor skill and spasticity issues which made it even more difficult for them to operate the water controls placed at the rear of the sink. The solution was to reconfigure the sink with more sensitive controls placed at the front of the sink, rather than at the back. This reconfigured sink will be installed in future Smarthomes.

B. Meeting Implementation Objectives

Although the BMP did not exceed expectations, it certainly succeeded in meeting the objectives set out at the beginning of the Project. The vital signs system is able to record, upload and send data to an external location as was hoped. In addition, the system was able to determine when readings are outside established norms and this information was used at all locations to take action, e.g., notify nurses, inform physicians. It is too early to determine if the use of the technology has reduced emergency room visits and hospitalizations, but staff believe that the system is able to allow more timely care and, as a result, the well-being of individuals has increased. However, even this degree of success was sufficient to begin planning for the use of a vital signs monitoring system in an outpatient clinic, as well as an even more ambitious plans to use such a system in the residence of selected individuals.

C. Hardware and Software Problems

Even though the CTP far exceeded expectations, there have been some issues surrounding the reliability of both the hardware and applications used and staff concerns. Most of the technology problems were resolved during Stage 1 of the Project, but the need for better and more continuous staff training lingered as Stage 2 commenced. Many staff expressed the need for better hands-on training at the beginning of the Project and for on-going training, once individuals receiving services began to explore additional capabilities of the AbleLink applications. The high rate of staff turnover indicated that a more thorough on-going training program than was anticipated at the inception of the Project was necessary. In response to these findings, members of the Steering Committee periodically conducted Skype sessions during the year with all staff responsible for Stage 1 and Stage 2. These sessions reinforced the goals of the Project, allowed the sharing of resources/ideas and encouraged on-going peer led discussions at each of the Project locations. In addition, newly identified staff for Stage 3 have been included in these Skype sessions during the spring and summer of 2016.

The biggest issue with the technology occurred in the BMP. The number of error reports filed at the three facilities, confirms the overall impression that a different vital signs system needed to be used as the Project moved forward. This has resulted in several vital signs monitoring systems being evaluated and a much more robust and flexible system being selected for use at the outpatient clinic. The encouraging conclusion is that, even with the problems with the technology, the results were sufficiently encouraging to plan for the Project's expansion.

D. What Has Been Learned

In the last three and one-half years, many lessons have been learned about the process of incorporating new technologies into NHS's various care models Among the most important are: 1) detailed planning is indispensable; 2) there must be buy-in at all levels—board, C.E.O, upper administration, management and line personnel; 3) one technology must be working before the next one is introduced; and 4) everything takes longer than originally thought.

In the early stages of the Project, many individuals at NHS believed that things were moving too slowly; they were anxious to "get-on-with-it". This urge to move quickly is natural, especially from individuals who have been recruited because they are enthusiastic about the introduction of new technologies. Nevertheless, taking the time to plan every step of the Project was vital to success. Even with careful planning, mistakes were made and problems encountered. Likewise, there must be buy-in at every level of the organization and this also takes time. Without buy-in and commitment, there is the tendency to "cut the losses" when problems arise. The buy-in of the NHS Board, C.E.O. and upper administrators was key to the continuation of the Project when things went wrong.

The incremental approach to the introduction of the technologies also proved to be a wise decision. Once again, there was a push to introduce "everything" at once, but the plan to make sure that one technology worked before installing a second, allowed staff and individuals with disabilities and severe mental illness to adjust to the first change before a second was introduced. Finally, although initially people involved with the Project were confident that the timeline for their slow and cautious approach was realistic, as the three Projects got under way there was a realization that the amount of time necessary to get 14 different sites up-and-running was going to take longer than anyone had anticipated. Fortunately, the fact that there was agreement among all individuals in the Project that the long-term goal of the implementation of the selected technologies throughout the organization, rather than success in one or two locations, allowed the Projects to progress at the slower pace required.

VII. CONCLUSION AND FUTURE WORK

When the Assistive Technology Project was being planned in 2012 there were doubts whether, because of its scale and complexity, it could achieve its objectives. Even as the technologies were being installed in the first Stage and problems emerged, there were concerns that trying to evaluate three distinct technologies in 11 locations was just too ambitious. However, the Project leadership persevered and with the commencing of the second Stage and planning for a third, it is impossible to conclude otherwise than that the Project has been a success.

A. Improvements for Next Stage

Based on an analysis of the results from the three Projects in Stage 1, several adjustments have been made to both the technologies employed and the implementation protocols for the next two Stages. First, problems with the biometric instruments used in the BMP have led to a switch to more robust products from companies that offer greater technical support. In addition, the needs inherent in collecting vital signs at an outpatient clinic have led to the decision to employ a kiosk that collects a series of vital signs. The advantage of the kiosk is threefold: first, it is a self-contained unit that can be easily placed in a small room at the clinic; second, it allows for as few as one vital sign to be collected and up to six, thus allowing the ability to start off slowly by collecting two vital signs—blood pressure and weight—and adding more as individuals receiving services and staff become comfortable; and third, it allows the information collected to be sent electronically to psychiatrists and physicians, prior to face-to-face meetings. The switch to a more robust system has also enabled concrete plans to collect vital signs in the residences of individuals receiving services from NHS and to send this data to care providers at a remote site in Stage 3 of the BMP.

Similarly, dissatisfaction on the part of staff and individuals with disabilities with several products in the SHP has resulted in the selection of more sophisticated equipment that incorporates more sensor technology. In addition, the positive results from the CTP and BMP will allow for these technologies to be incorporated into the Smarthome currently being constructed.

An unexpected finding from the CTP has led the Steering Committee to dramatically alter the selection of facilities for Stage 2. It was thought that expansion would be to small group homes, but the overwhelming success of the AbleLink technology in the larger day facilities has resulted in the technology being installed in a large congregate ICF and two adult training centers and plans for further installations in Stage 3 in similar facilities. Longer term plans include the installation of AbleLink in group homes, but the use of the technology in larger facilities has proven to be more cost effective because of the more efficient use of staff.

B. Expansion

The best measure of this success is that NHS has made the decision to extend all three technologies to additional locations. The CTP has already been expanded to three new locations bringing the total number of participants to approximately 200 and there are plans to include three additional facilities before the end of 2016. NHS is also working with AbleLink to develop new applications specifically targeting individuals with disabilities, as well as working to refine and enhance existing applications.

The Delaware County Adult Behavioral Health Outpatient Clinic will be added to the BMP by the end of 2016, and vital sign monitoring systems will be installed in the residences of selected individuals with disabilities during 2017. The complexity of incorporating vital signs monitoring into an outpatient clinic is, from NHS's perspective, outweighed by the opportunity to effectively offer psychiatric services, primary care physicians and pharmacy services in one location along with an integrated medical record. Likewise, the difficulty of recruiting participants, installing systems and establishing a remote monitoring location is balanced by the benefits of being able to track vital signs and respond to alerts.

Perhaps the most significant indication of the success of the overall Project is that there are plans to build, ground up, a new facility in Western Pennsylvania which will include technologies from all three Projects.

C. Financial Model

From the inception, one of the key components of the Project was to construct a financial model that would allow NHS to be reimbursed for the care delivered by the use of the new technologies. The Project itself, costing over $200,000 in real money and much more when the amount of staff time expended is included, has been financed by grants. Although grant funding is satisfactory for a project whose goal is to evaluate the appropriateness of new technologies in the delivery of care, it is not a satisfactory means for developing a sustainable financial model. A sustainable financial model can only exist if the care delivered with the use of the new technologies is reimbursable by Medicare (the health insurance program for people in the United States who are 65 or older. Medicare Part B covers certain doctors' services, outpatient care, medical supplies, and preventive services) and, especially, Medicaid (the U.S. government program, financed by federal, state, and local funds, of hospitalization and medical insurance for low income persons of all ages) as billable services that can be reimbursed to NHS. Without the ability to be reimbursed for the services provided through the use of the technologies, a hoped for roll-out to a large number of locations will be impossible.

The problem is that, currently, most of the care delivered in the three Projects is not billable and thus, not reimbursable, but this situation is changing. A number of states have granted Medicaid waivers that can now allow reimbursement for care delivered with some of the selected technologies, but not all of the services delivered through the use of the selected technologies are reimbursable. For example, one of the challenges faced by the Steering Committee in using a kiosk to collect vital signs at the out-patient clinic is how to pay the staff member who will help customers in using the system. It is anticipated that this staff member will need to be available to help with the kiosk all day, five days a week. If NHS is unable to bill Medicare/Medicaid for her time, the organization will be unable to continue providing this service and expanding to other facilities. If the three Projects have done nothing else, they have confirmed that the use of technology to aid in the delivery of care to individuals with disabilities and severe mental illness is inevitable. The increasing numbers, along with the aging of both populations, is increasing the cost of care exponentially, at the same time as the number of people available to deliver the care is stagnating. The only way to maintain, let alone enhance, the level of care to these populations is through the innovative use of technology and the only way to make this happen is to develop a means of reimbursing this care. This must and will occur; the only question remaining is when?

REFERENCES

[1] A. Glascock, R. Burke, S. Portnoy, and S. Shields, "Planning for sustainability: Multiple technologies for different care models," Proc. eTELEMED 2016, pp. 68-76, 2016.

[2] G. Demiris, B. Hensel, M. Skubic, and M. Rantz, "Senior residents' perceived need of and preferences 'smart home' sensor technologies," Inter J. Tech Ass in Health Care, vol. 24, no. 1, pp. 120-124, 2008.

[3] E. E. Stip and V. Rialle, "Environmental cognitive remediation in schizophrenia: Ethical implications of 'smart home' technology," Canadian J. Psychiatry, vol. 50, no. 5, pp. 281-291, 2005.

[4] E. Novella, "Mental health care and the politics of inclusion: a social systems account of psychiatric deinstitutionalization," Theor Med Bioeth, vol. 31, pp. 411-427, 2010.

[5] M. Knapp, J. Beecham, D. McDaid, T. Matosevic, and M. Smith, "The economic consequences of deinstitutionalization of mental health services: Lessons from a systematic review of European experience," Health and Social Care in the Community, vol. 19, no. 2, pp. 113-125, 2011.

[6] National Council on Disability. [Online]. Available from: http://www.ncd.gov/publications/2012/DIToolkit/Costs/inDetail/ retrieved August 2016.

[7] National Institute of Mental Health. [Online]. Available from: http://www.nimh.nih.gov/statistics/index.shtml, retrieved August 2016.

[8] Administration on Aging—Projected Future Growth of Older Population. [Online]. Available from: http://www.aoa.gov/ Aging_Statistics/future_growth/future_growth.aspx, retrieved August 2016.

[9] C. Tinglin, "Adults with intellectual and developmental disabilities: A unique population," Today's Geriatric Medicine, vol 6, no 3, pp. 2-26, 2013.

[10] U.S. Department of Health and Human Services Office of the Assistant Secretary for Planning and Evaluation. [Online]. Available from: aspe.hhs.gov/.../supply-direct-support-professionals-serving-individuals- intellectual-disabilities-and-other-developmental-disabilities-report-c, retrieved August 2016.

[11] The New Jersey Council on Developmental Disabilities Recommendations on the Direct Support Professional Workforce. [Online]. Available from: https://www.njcdd.org/ 2012-04-23.../direct-support-professionals, retrieved August 2016.

[12] M. Miza and J. Gammel, "Consumer directed goal planning in the delivery of assistive technology for people who are aging with intellectual disabilities," J. Applied Research in Intellectual Disabilities, vol. 22, pp. 445-457, 2009.

[13] S. Stock, D. Davies, M. Wehmeyer, and S. Palmer, "Evaluation of cognitively-accessible software to increase independent access to cell phone technology for people with ID," J. Intellectual Disability Research, vol. 52, pp. 1155-1164, 2008.

[14] D. Bryen, A. Carey, M. Friedman, and S. Taylor, "Cell phone use by adults with intellectual disabilities," Intellectual and Developmental Disabilities, vol. 45, pp. 1-2, 2007.

[15] C. Li-Tsang, S. Yeung, and C. Hui-Chan, "Factors affecting people with IDD in learning to use computer technology," Inter J. Rehabilitation Research, vol. 28, pp. 127-133, 2005.

[16] H. Parette, "The importance of technology in the education and training of persons with mental retardation," Education and Training in Mental Retardation, vol. 26, pp. 165-178, 1991.

[17] D. Hammond, D. Whatley, K. Ayres, and D. Gast, "Effectiveness of video modelling to teach iPad use to students with moderate intellectual disabilities," Education and Training in Autism and Intellectual Disabilities, vol. 45, no. 4, pp. 525-538, 2010.

[18] B. Bryant, R. Byant, M. Shih, and S. Seok, "The role of assistive technology in support of needs assessment for children with disabilities," Exceptionality, vol. 18, no. 4, pp. 203-213, 2010.

[19] A. Carey, M. Friedman, and D. Byrant, "Use of electronic devices by people with intellectual disabilities," Mental Retardation, vol. 43, no. 5, pp. 322-333, 2005.

[20] B. Weinberg, and A. Kjellberg, "Participation when using cognitive assistive devices from the perspective of people with disabilities," Occupational Therapy Instruction, vol. 17, no. 4, pp. 168-176, 2010.

[21] Braddock, M. Rizzolo, M. Thompson, and R. Bell, "Emerging technologies and cognitive disability," J. Spec Ed and Technol, vol. 19, no. 4, pp. 49-56, 2004.

[22] E. Tanis, S. Palmer, M. Wehmeyer, D. Davies, S. Stock, K. Lobb, and B. Bishop, "Self-Report computer-based survey of technology use by people with intellectual and developmental disabilities," Intellectual and Developmental Disabilities, vol. 50, no. 1, pp. 53-68, 2012.

[23] B. Bryant, S. Seok, M. Ok, and D. Bryant, "Individuals with intellectual and/or developmental disabilities use of assistive technology devices in support provision," J. Spec Educ Technol, vol. 27, no. 2, pp. 41-57, 2012.

[24] Ablelink Technologies Home Page. [Online]. Available at http://www. ablelinktech.com, retrieved August 2016.

[25] J. Mindham and C. A. Espie, "Glasgow Anxiety Scale for people with an Intellectual Disability (GAS-ID): Development and psychometric properties of a new measure for use with people with mild intellectual disability," J. Intellectual Disability Research, vol. 47, no. 1, pp. 22-30, 2003.

[26] H. M. Evenhuis, "Characteristics of instruments screening for depression in adults with intellectual disabilities: Systematic review," Research in Developmental Disabilities, vol. 31, no. 6, pp. 1109-1120, 2010.

[27] Centers for Disease Control, Adults with Disabilities and Vital Signs. [Online]. Available from: http://www.cdc.gov /vitalsigns/disabilities, retrieved August 2016.

[28] M. Tasse, R. Schalock, J. Thompson, and M. Wehmeyer, Guidelines for interviewing people with disabilities. American Association on Intellectual and Developmental Disabilities. Washington, DC, 2005.

[29] K. Caldwell, "Dyadic interviewing: A technique valuing interdependence in interviews with individuals with intellectual disabilities," Qualitative Research, vol. 14, no. 4, pp. 488-507, 2013.

Developing a Personalised Virtual Coach 'Denk je zèlf!' for Emotional Eaters through the Design of Emotion-Enriched Personas

Aranka Dol[1], Olga Kulyk[2], Hugo Velthuijsen[1], Lisette van Gemert-Pijnen[2], Tatjana van Strien[3]
[1]Institute for Communication, Media & IT, Hanzehogeschool UAS, Groningen
[2]Department of Psychology, Health and Technology, University of Twente, Enschede
[3]Behavioural Science Institute and Institute for Gender Studies, Radboud University, Nijmegen
[3]Department of Health Sciences and the EMGO Institute for Health and Care Research, VU University, Amsterdam
e-mail: [1]{a.dol, h.velthuijsen}@pl.hanze.nl, [2]{o.a.kulyk, j.vangemert-pijnen}@utwente.nl, [3]t.vanstrien@psych.ru.nl

Abstract — **Obesity is a fast-growing societal threat, causing chronic conditions, physical and psychological health problems, as well as sickness absence and heavy healthcare costs. Despite numerous attempts to promote physical activity and healthy diet, existing interventions do not focus on the common emotional causes of obesity. There is a need for self-management support of this vulnerable target group: emotional eaters. This paper presents the results of the design case study focusing on a holistic design and evaluation of a personalised virtual mHealth coach that provides self-management training *'Denk je zèlf!'* (Dutch for *'Develop a wise mind and counsel yourself'*). The target group are young adults with an emotional eating disorder and who are obese. The contextual inquiry study was conducted to gain insights into the needs and experiences of the target users, including interviews and questionnaires with emotional eaters, patients undergoing obesity treatment, and healthcare practitioners. Personas and the use-case scenario were derived from these results and translated into the new *'Denk je zèlf!'* virtual coach, based on Dialectical Behaviour Therapy and experience sampling measures to capture user experience and emotional state. The main contributions of this paper are: (a) combining holistic eHealth design, behavior chain analysis, and dialectic behaviour therapy in one personalised virtual mHealth coaching application for emotional eaters; (b) applying emotion-enriched Personas to guide the design; (c) the results of the initial user evaluation. Preliminary results suggest that the *'Denk je zèlf!'* virtual coach is useful for helping the target group. Future research will be aimed at further iterative (re)-design and evaluation, as well as development of the dialectical dialogues for the virtual coach and content for the education and instruction modules.**

Keywords-obesity; emotional eating; Dialectical Behaviour Therapy; Persona; personalised care; virtual coach; persuasive design; young adults.

I. INTRODUCTION

The fast growth of obesity is a major threat to society. Treatment of obesity and obesity-related conditions imposes a heavy societal burden due to high healthcare costs, reduction of life spans and increased risk of developing other chronic conditions such as diabetes, heart disease, osteoarthritis, and certain cancers [1]-[4]. Obese patients often suffer from psychological comorbidities, such as depression and low self-esteem [5]. *Obesity* is defined as an abnormal or excessive fat accumulation that may impair health and is classified as such by a Body Mass Index (BMI) of 30 kg/m2 or higher [6]. Nearly 50% of the Dutch population suffer from being overweight and 12% from obesity [5].

Obesity is associated with poor eating habits and lack of physical activity, unhealthy family lifestyle and low socio-economic status. Increasing physical activity and reducing food intake (dieting) are considered cornerstones in the prevention and treatment of obesity. However, though many of the existing interventions are successful and help patients lose weight in the short run, long-term randomised studies demonstrate that "diets are not the answer" [7]. Existing interventions and online weight loss programs, such as Weight Watchers [8], My Diet Coach [9] and Lose it! [10] focus primarily on the 'Big Two' aspects, namely: eat less, exercise more. They do not provide the necessary support in the long run, as people are unable to maintain their bodyweight over a longer period.

A. Emotional eaters and obesity

Recent studies have shown that a considerable group (40%) of the obese population overeat due to negative emotions [11]. Emotional eating is an atypical stress reaction. A normal reaction to stress and negative emotions would be loss a of the appetite. Emotional eaters show this atypical behaviour because they confuse negative emotions with hunger. They have a narrow view of what happens in their bodies (poor interoceptive awareness) and they are experiencing difficulties identifying and describing emotions and feelings (alexithymia). Emotional eaters are facing problems with emotion regulation - the ability to keep one's emotional system in a healthy condition [12]. Diets and behaviour therapies do not help people with high degrees of emotional eating as they do not tackle the underlying emotional regulation difficulties, that lead to emotional eating [13][14].

Most emotional eaters have a long history of dieting, followed by the inevitable overeating and starting dieting again. They gain weight because of poor emotion regulation, not just due to poor eating habits or an insufficient level of physical activity [15]. Many times, they have tried to lose weight and when the emotional eating behaviour kicked in again, they ende up being heavier than when they started their previous dieting episode. This is an example of the so-called *'yo-yo effect'* in health behaviour [16][17]. It is highly conceivable that this is the cause of an accumulation of disappointments and a growing lack of confidence that one will ever succeed.

B. Emotional eaters and eHealth

Evidence suggests that eHealth and Cognitive Behaviour Therapy can be just as effective as face-to-face treatments. Evidence-based therapeutic procedures can be delivered online [18]. In addition, emotional eaters need personalised anonymous support that is always available. Not only do emotional eaters need moral support, but a personalised self-management support could also clear away obstacles that keep emotional eaters away from face-to-face contact with a therapist.

Obese emotional eaters form a particularly vulnerable group of people. To avoid further setbacks, they need support they can rely on, that is both trustworthy and promising, but realistic, and that matches with their needs. Such support needs to be accessible and comfortable so that one feels safe and secure. There is a need for supportive training programs for this specific target group of emotional eaters. This paper presents a design case study aimed at the development of a personalised virtual mHealth coach application for self-management support of young adult emotional eaters who are obese.

The paper is structured as follows. In Section II, related work on online eHealth interventions for emotional eaters and obesity is discussed. Next, in Section III the approach and methods are presented for developing Personas and applying them to guide the design process. In Section IV, the results are presented including a use case scenario and the first prototype of the 'Denk je zèlf!' virtual coach. Finally, a discussion and conclusions are presented in Section V.

II. RELATED WORK

A. Dialectical Behaviour Therapy

Dialectical Behaviour Therapy (DBT) is a new way of treating emotional eating behaviour. DBT was originally designed to help people who are suffering from Borderline Personality Disorder [19]. The therapy focuses on the process of 'reduction of ineffective action tendencies linked with dysregulated emotions' [20]. Recent research into the deployment of Dialectical Behaviour Therapy shows positive results in weight loss management and weight maintenance in obese emotional eaters [13]. DBT might be successful in patients where insufficient progress was achieved using Cognitive Behavioural Therapy (CBT) [12][21][22]. Currently, CBT is considered the state-of-the-art in treating eating disorders, aimed at treating eating disorders such as Boulimia Nervosa (BN), Anorexia Nervosa (AN) and Eating Disorders Not Otherwise Specified (EDNOS). The core of these eating disorders is the patient's over-evaluation of his control of weight, shape and/or eating. The efficacy of DBT for the treatment of Binge Eating Disorder (BED) and emotional eating behaviour has been demonstrated in the results of various studies and trials [23][24][25][26].

B. The dialectical focus

One of the most powerful 'mechanisms of change' or mediators in DBT is its dialectical focus. Since an invalidating environment plays an important role in the lives of emotional eaters, it is important that they are treated with a well-balanced mix of being validated in their perception of negative emotions and being confronted with a practical focus on changing problem behaviour. "Based in the biosocial theory, DBT has a unique approach to targeting behavioural dysfunction that is not typically seen among other cognitive-behavioural treatments; one key difference is the emphasis placed on emotions and emotion dysregulation." [20].

C. DBT and eHealth

There is a broad variety of eHealth self-management treatments available but the majority focus on weight loss and behaviour change. The discussion on the effectiveness of such interventions is progressing only slowly [27]-[33]. Little knowledge in the field of eHealth treatment using Dialectical Behaviour Therapy or even emotion regulation has been acquired so far, let alone about emotion regulation focused on emotional eating behaviour.

The results of one quasi-experimental study on the effectiveness of the mobile "DBT Coach", that focused only on one particular skill in DBT (Opposite Action), showed that emotion intensity decreased within each coaching session in participants suffering from Borderline Personality Disorder [34][35]. The target group uses the DBT Coach when it is needed most for them – after engaging in dysfunctional behaviour. One paper discusses the lack of user-friendliness of a DBT self-management mHealth application [36].

A small number of DBT-based self-management mHealth apps can be found in the Google Play Store and in the Apple App Store. However, they typically lack scientific grounding, user involvement in the design process, psychological aspects, and personalization.

D. Virtual coach and behaviour change

Substantial research has been dedicated to the employment of virtual coaches [37][38]. A virtual coach, as an interactive and self-learning persuasive system, can assist in attitude and/or long term behaviour change by providing immediate and personalised support [37][39]. Various definitions of virtual coaching exist in research studies, focusing either on mediated communication via internet or phone, or on the telemonitoring of health but not on the personalised feedback side of coaching [37][40]. In this research, we adopt the definition of eCoaching by Lentferink et al. [38, p.16]: *"...eCoaching is defined as the remote and automatic provision of just-in-time tailored feedback for healthy lifestyle management, by enabling users to set personal goals and encouraging to track personal progress towards their goals, adapting the feedback to the usage patterns and context, and encouraging long-term use."* Current eCoaching developments and studies often lack user and stakeholder involvement and are not grounded by the behavior change theories, which leads to low usability and therefore low adherence [38].

III. APPROACH AND METHODS

The objective of this research is to develop a personalised self-management intervention based on Dialectical Behaviour Therapy for young adult emotional eaters who are obese. We regard young adults as people in the age range of 18-44 years. Midlife transition, starting at the age of 40 to 45, might bring along other circumstances. Development is guided by the CeHRes roadmap (Center for eHealth & Wellbeing Research, University Twente) – a holistic eHealth framework for developing eHealth interventions based on a participatory design process and persuasive design approach to maximize the impact of the behavior change support intervention [41]. It is essential to emphasise the effect of the 'look and feel' of the intervention interface design on the adherence of the user. The interface design should be attractive and engaging to the user, otherwise he or she is unlikely to use it. User- centered design and an inter-disciplinary approach are therefore incorporated in the CeHRes roadmap framework [42]. In addition, it is important to determine which design features are used in successful eHealth interventions. Other essential aspects are suitable input modalities (such as speech, text, gestures, mouse, touch, haptic, and tangible) [40] and interactivity, information architecture, the degree to which information is updated, aesthetics, usability, credibility, and entertainment factor [43][44].

A. *eHealth Intervention Architecture*

The intervention consists of a series of education and instruction modules on emotional eating behaviour and emotion regulation. A personalised virtual coach will guide the user through four modules. First an 'intake procedure' will take place: the user will be invited to make a commitment never to lose him or herself in emotional eating behaviour again, followed by educational modules on mindfulness, emotion regulation, and stress tolerance (Fig. 1).

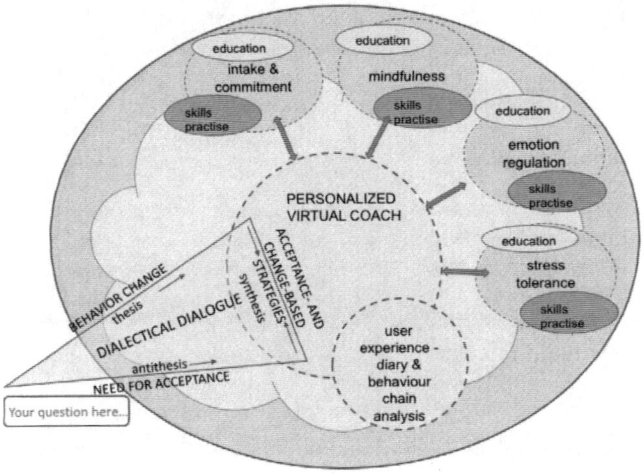

Figure 1. Schematic drawing of 'Denk je zèlf!' mHealth intervention

The training offers exercises based on practical experience in daily life. Modules are replaceable – they can be replaced by modules with content that might be focused on users with low socio-economic status. Users are invited to fill in their behaviour chain analysis (Fig. 2) and emotion diary on a daily basis. The behaviour chain analysis is to be performed at the moment a participant has given into cravings and poor eating behaviour or is just about to do so. The behavioural chain analysis is utilised to analyse problem behaviour and determine prompting events and vulnerability factors. People can also fill in new personal goals and consider ways to prevent prompting events and to think of solutions to reduce susceptibility in the future. Reminders to log in to the application and fill in the behaviour chain and the diary are sent out at the fixed times previously agreed with the user. Both components are considered indispensable in the face-to-face training being daily 'homework' for the participants.

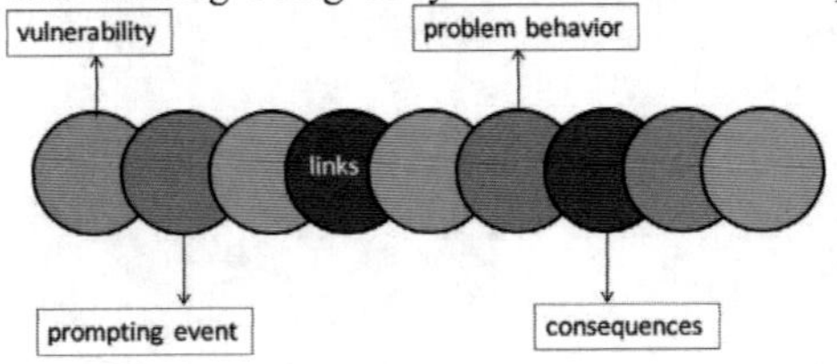

Figure 2. Behaviour chain analysis model by Linehan [19].

B. *Contextual inquiry*

This study focused on the contextual inquiry and early design phases. First, the contextual inquiry phase was carried out. To become familiar with what kind of support emotional eaters really need, it is important to understand the target group [45]. Even though interventions can be evaluated as positive in terms of effectiveness, if the target group is not captivated by its design and functionalities, they are not going to use it.

The use of user profiles and Personas as a tool to inform design is still rare in social sciences. We used the LeRouge classification model [45] and Van Velsen's additions to it [46] as a guideline to develop two Personas to guide the design of a virtual mHealth coaching intervention. Personas contain information on their *technology skills* and smartphone use, *demographic facts,* and *healthcare specifics* such as current practices in managing one's own healthcare, support network, and information seeking attitude. In addition to the standard Persona classification model [41], Persona Lisanne is enriched by emotions and feelings in second iteration of contextual inquiry.

To gather input for user profiles and Personas, questionnaires (N=321) were circulated via social media and the network of contacts. The target group was "young adults, 18-44 years of age" who are self-declared emotional eaters. Examples of questionnaire questions: "*For what purposes do you use your smartphone?* (social media, news gathering, mail, gaming)", "*At what specific moment in time would you like to/are you in need of contact with a help system?*", "*What kind of support do you expect from a smartphone application?*" We approached healthcare practitioners for expert interviews (N=13). In the next sections, we present the results of this design case study, including Personas, the use case scenario, and a description of the architecture and design of the *'Denk je zèlf!'* virtual coach.

The use-case scenario is developed based on Personas (Fig. 3) and the specific lifestyle characteristics of the emotional eaters from the literature and interviews.

Personas	
Lisanne, 25 years, high educated	Anita, 46 years, limited education

"I wish that I was more confident about myself…."	*"When I am stressed out I start snatching the cookie tin"*
Background	
Lives alone, single / High school graduate / Suffers from obesity / Co-morbidity unknown / l. 1.62 mtr., w. 91 kgs.	Lives with husband and two sons / low SES / Suffers from obesity / Co-morbidity unknown / l. 1.65 mtr., w. 85 kgs.
Attributes	
Feels insecure about her body / Is not able to sense the difference between hunger, appetite or emotion / Fears experiencing what she really feels.	Is worried about putting on more weight / Suffers from stress and finds comfort in food / Has a 'sweet tooth' / Hates to be patronized by authorities.
User needs	
Education on eating behaviour / Self-confidence / Help to set realistic goals / Support from peers.	Stop gaining more weight / Education about healthy food / Getting support from her family / Stop harassing thoughts.

Figure 3. Key Personas Lisanne and Anita

C. Design & prototyping

During the design phase, a first clickable prototype of the graphical user interface was designed, based on two Personas, namely Lisanne and Anita (Fig. 3), and the Gestalt design principles [47], Nielsen's usability heuristics [48], and Gomez's heuristics on mobile devices [49]. The wireframes were created in Adobe Illustrator and reshaped to screens in Adobe Photoshop. Interactivity was added using InVision. The prototype consists of loading screen (visible when the application is loading), a login screen and a link to Frequently Asked Questions. The user is shown an informative text about how to use this application. Navigation is divided into four buttons: 'coach', 'modules', 'diary' and 'profile'. The user can start a conversation with the coach through dialogues in a Whatsapp (messaging application for smartphone) layout style. The button 'modules' leads to subsets of learning modules on mindfulness, emotion regulation, and stress tolerance. In 'diary', the user can record feelings experienced. In 'profile', he or she can adjust settings and personal information such as uploading a personal picture. The user can also store encouraging items such as favorite pictures and quotes. Via the 'central' button the user can navigate back and forth to all the modules and functions available in this prototype.

To develop an architecture of the personalised virtual coach, including the two vital parts - the behaviour chain-analysis and the emotion diary, the Persona and use-case scenario were translated into user stories (example: *'As a user (Lisanne) I would like to get an overview of my diary so that I can see my history'*). Next, user stories were translated into a functional prototype of the architecture of the virtual coach.

D. Heuristic expert evaluation

Iterative user evaluation in the early design stage is essential to obtain early feedback from the potential users and so improve the protoype. An expert evaluation of the first version of the clickable prototype of the user interface was conducted by a usability expert (a lecturer in User Centered Design at Hanze University of Applied Sciences) This expert was asked to perform eight different tasks. The heuristics [48][49] were scored using a 5-point Likert-scale (1=does not fulfil requirements; 5=fulfils requirements). Examples of the applied heuristics include: *"Is there an option to navigate back (button)?"*; *"Is there a facility for horizontal scrolling?"*; "Do drop-down menus have a logical order?"; *"Is there sufficient contrast between text and background?"*

The expert was guided through a task scenario. Tasks were: *Log in to the app; Read the introduction about the app; Read the Frequently Asked Questions; Communicate with the Virtual Coach about your eating behaviour; Setup/configure your profile; Set the alarm to receive daily notifications;*

Figure 4. Screen capture from the usability test: eye tracking gaze plot video

Check on available modules and *Fill in your diary*. The results were used to modify and to adjust the prototype.

E. Usability evaluation with eye-gaze tracking

Next, a usability evaluation of the second improved version of the clickable prototype was conducted among the target group (N=10).

Participants, procedure & materials: Task scenarios were created. Tasks were presented as short stories to give more context. In total, ten participants [50] volunteered to perform a test with eight tasks. Among the ten participants, five were university students studying behavioural sciences, three were self-declared) emotional eaters and two were user interface designers. Participants were recruited via the Facebook social network used by the students. The usability test took around 30 minutes. Participants first received a pre-briefing on the goals of the usability test and the nature of volunteer participation. Next, users executed the task scenarios followed by a post-test interview.

During the usability tests (Fig. 4), participants were monitored by eye tracking, using a 'Tobii Mobile Device Stand' (MDS) [51], followed by an interview in accordance with the Retrospective Think Aloud Method [52]. A calibrated eye tracker (one that follows the eye pupils of a test user) registered what the user is looking at while performing each task using the second clickable prototype. The evaluator is able to see what parts of the interface the user is looking at, even if the test user scrolls, or zooms in while interacting with the user interface [51]. The results are plotted on so-called gaze plots and the evaluator is able to watch a slow-motion replay.

With participants' permission real-time gaze replay video is recorded of the eye-tracking sessions.

Measures: we measured the task performance by recording some quantitative measures, such as registering how much time is needed to complete each task, the number of times a particpant made 'wrong' choices/clicks, how many tasks were completed, and the number of taps or swipes that were needed to perform each task. The ways in which the designer performed the tasks was taken as a reference point. Tasks were measured on a scale 1-3 (1 = user failed to perform the task; 2=user performed the task but with errors; 3=user performed the task without any errors). Qualitative measures were extracted from the post-test interviews. The Retrospective Think Aloud (RTA) method [52] is used to collect additional information about the motivation of the choices users make while performing the tasks. After each user test session, post-interviews are conducted with each participant in order to give the user the opportunity to further explain why he or she made specific choices, as well as to obtain feedback on the user experience as additional input to improvement of the prototype. The RTA method - in combination with the eye-tracking allows a user to focus completely on the task. Research has shown that in retrospective think-aloud protocols, more problems were detected by means of verbalization, because the participants had more time to verbalize the problems [53]. Analysis of the gaze plots is not yet finalized and is work-in-progress.

IV. RESULTS

A. Questionnaires and interviews

In total, 321 responses were collected from the questionnaires and 13 interviews with healthcare practitioners (dieticians, physical therapists and psychologists) were conducted. The interviews were all transcribed and coded. We used a free coding style (no pre-set codes) so as not to lose the richness of the data. In addition, six obesity therapy patients were interviewed to obtain insight into the daily needs and experiences of the emotional eaters.

Data extraction from the questionnaires was processed for the purpose of creating user profiles and Personas according to the method proposed by LeRouge [45]: (a) personal and demographic information; (b) technical capabilities and limitations; and (c) needs and desires concerning support and care. Data derived from the interviews with experts and patients gave information about eating styles and the problems that emotional eaters encounter.

B. Personas and use-case scenarios

Two Personas were derived from the questionnaire data: Lisanne (25 years) and Anita (46 years). Figure 3 shows their personal profiles. Lisanne is a highly educated young woman. She is an obese emotional eater. Her eating behaviour is caused by a negative self-image. Eating gives her a feeling of comfort as long as the eating lasts. Afterwards, she feels guilty and depressed. Anita is 46 years old and a mother of two. She worries about the family's financial situation and overeats in stressful situations. Anita left school at an early age. She is from a low socio-economic background.

What we learned by creating the Personas is that we achieved a better understanding about the specific needs of the target group. We obtained a better comprehension of the moments and situations that cause emotional eating behaviour and the kind of intervention that might be helpful to them. During the design process, you can ask questions such as: "What would Lisanne think of this? Would she like it? Would she consider this as useful?"

To illustrate how Lisanne will benefit from the virtual coaching application, we created a use-case scenario (Fig. 5) in which she is about to give in to emotional eating craving. Use-case scenarios [54][55] are derived from Personas, interviews, and questionnaire results by describing the user goals, motivations, actions, and reactions while using the virtual mHealth coaching application.

In the "Day of Lisanne's life" scenario (Fig. 5) – the most common characteristics are mentioned, such as suffering from negative emotions and low self-esteem. Persona Lisanne is enriched with emotional characteristics and personal feelings. All her emotions are linked to realistic and possible causes, such as difficult situations at work or at home. Personas are distinguished from each other in terms of technical skills, demographic backgrounds and healthcare status specifics. As a result, they are realistic personifications of their very specific sub-target groups. The existing Personas Lisanne and Anita both are emotional eaters but the emotions and feelings that urge them to overeat, diverge widely.

Day of Lisanne's life – a Use Case Scenario

The alarmclock goes off; Lisanne presses the snooze button. She had a bad night. Probably because she spent a lot of time on her tablet and watching television last night, but she couldn't sleep anyway....
Lisanne gets out of bed eventually. She is standing in front of her drawer. What to wear? She slips into an oversized black sweater and a pair of jeans. She is way too fat to wear a fashionable dress.
"Oh bummer... no time for breakfast..." The bus is leaving within 5 minutes. The bus stops exactly in front of the building Lisanne is working. She meets colleague Esther at the coffee corner. Esther convinces Lisanne to come with her to the supermarket on the other side of the street. Esther didn't have any breakfast herself and she fancies a bite. Lisanne wants a freshly baked sandwich. That way she could make up for her missed breakfast. It is very busy in the store. She should focus on her presentation, scheduled for less then 10 minutes from now. Esther yells at her, waving, her hands full with donuts and soft drinks. Lisanne: "Oh well, I will get a healthy sandwich later in the day...."
Ok, one last sip of coke and off she goes. The meeting is about to start now. Lisanne is nervous – she hates presenting in front of a group. She thinks that everyone in the room is convinced that she is dumb and that she looks fat. Her clothes look ugly on her.
It is noisy in the hall. The door opens. A colleague is standing in the door opening. She is carrying a birthday cake and starts handing out cake to everyone in the room. Lisanne is upset. She hates to do presentations and now she is interrupted too.... The colleagues don't seem to care so Lisanne decided to keep her mouth shut.
Coffee time! Esther brings Lisanne a mug and nestles herself on Lisanne's desk. She fancies going out tonight to their favorite bar. Tonight is *Fry-night*: all-you-can-eat chips & snacks for a fixed price. Lisanne doesn't feel like going, she'd better have something more healthy to eat. But Esther persuades her, saying it will be fun. "Mm, I really should stop nagging.... why am I making all this fuss about nothing? Why care? Tomorrow is another day. Tomorrow we will start eating healthily!"
Esther turns around on her way out and throws a bag of chocolate peanuts on Lisanne's desk. Here, something to nibble while you are working. Now Lisanne is totally miserable... doesn't Esther realise what she is doing? The bag with chocolate peanuts is empty and the amount of work still high. She has got only two hours left and she starts panicking. She was not able to get a lot of work done. She is about to cry. Esther appears in the door opening. Time to go! Lisanne brings up some poor excuses - she really doesn't feel like going out. But Esther doesn't want to hear it and drags her along. The Fry-night has started of and both the girls are served a large plate with fries and chicken satay. Esther babbles along, as she usually does. In the meanwhile everybody in the bar is a bit tipsy. Lisanne feels lonely. Esther is standing at the counter, talking to some guys. She invites Lisanne to come over and join them. Lisanne grabs her handbag. She is going home. She waves and gestures to Esther that she feels nauseous and that she is leaving. On the one hand she feels relieved that she has left but on the other hand she is disappointed in herself. Why can't she just enjoy an evening out and have fun just like everybody else? Why is she so damn shy?

situation upcoming food craving:

Lisanne comes home and plumps down on to her couch, next to the pile of laundry that still needed to be done. She is hungry. She feels lonely. Nobody likes her...
Her stomach is rumbling and she takes a quick glance at the kitchen cupboard. Lisanne is about to giving in to her food craving. Butter biscuits and potato chips. She feels tempted to rip open the bag of potato chips and plunge into it, grabbing chips by the handful at the same time. Binge eating lies in ambush for attack and she senses a crying need for some kind of support, for someone who could provide her with advice to pull her through this situation. After finishing the bag of chips, she eats some leftover portions of ice cream she ordered at the take-away yesterday, and some mouldy cookies she found in the bread box.

situation upcoming food craving – consulting the 'Denk je zèlf!' virtual coach:

Lisanne comes home and plumps down on to her couch, next to the pile of laundry that still needed to be done. She is hungry. She feels lonely. Nobody likes her...
She reaches for her phone and activates the 'Denk je zèlf!' app. The virtual coach welcomes her with: "Hi Lisanne, how can I help you?" Lisanne starts typing:
-When I come home at night I start craving for snacks and chocolate... I just can't resist them...
-Hi Lisanne, I really feel sorry for you. Let me try to help you. The greater part of emotional eating occurs at night, due to feelings of loneliness or experiencing stress, but it can also happen because of irregular eating behaviour. Shall we give it a try to investigate this?
The virtual coach refers Lisanne to the *behaviour chain analysis*. Lisanne finds comfort in the reassuring words of the virtual coach and starts with the behaviour chain exercise.
After finishing the chain exercise Lisanne feels somewhat relieved. The coach urges her to reward herself. She treats herself with a hot bath. The perfumed bath water makes her feel calm and sleepy...

User Profile

Aspects of a sedentary lifestyle: computer, television; poor sleeping habits (sleep hygiene: going to bed too late); people with obesity often suffer from sleeping problems.

Obese people often suffer from low self-esteem and low body image.

Poor eating habits– no breakfast; no decent meals, just snacks, sugary soft drinks; little knowledge about what is healthy food.

What the hell effect: totally giving up on set goals (diet, exercise) after one single slip and feeling justified in giving in to poor habits (binging) because the intention is already 'ruined'.

Negative self image: people who are overweight suffer from a lack of self-confidence and have an inferiority complex.

(unrealistic) Assumptions about what others might think of you; not assertive, not able to stand up for oneself.

Mindless eating (mostly snacks and sweets); eating as a coping-strategy to cope with stress – emotion regulating eating behaviour.

Social introversion – quiet and reserved in large groups and around unfamiliar people.
Low self-esteem and a bad body image.

Having difficulties acknowledging and expressing feelings and emotions. Emotional eaters have trouble recognizing feelings of hunger and thirst, pleasure or discomfort. Most of the negative feelings are translated into appetite.

Figure 5. Persona Lisanne translated into use-case scenario aspects

Lisanne gives in to cravings, driven by disgust, while Anita binges when she is under a lot of stress.The next step in this research is to develop a use-case scenario for all the Personas next to Lisanne, each of them enriched with their own specific emotional characteristics.

C. Prototype of a "Denk je zèlf! personalised virtual Coach

The first prototype of the architecture of the virtual coach and a clickable prototype of the user interface were developed during the pre-design phase. The Personas Lisanne and Anita served as a starting point for a concept design (Fig. 5) of the user interface and content of *'Denk je zèlf!'* virtual coach modules: *intake*, *mindfulness*, *emotion regulation* and *stress tolerance*. The Java-based virtual assistant (developed on the Play framework) makes use of the *Alpino* open-source natural language parser. This Dutch linguistic language analyser [56] is self-learning and produces 'tree diagram' data in XML format. The output is stored in a graph database (NEO4J).

The virtual coach was developed to meet the needs of the user for immediate support. Every time a user is experiencing negative emotions, he or she can connect to the virtual coach and ask questions and start a dialogue. The virtual coach is the very heart of the e-DBT *'Denk je zèlf!'* training. It supplies users with so-called dialectical dialogues – providing answers to their need for change and to their need for acceptance. According to Lynch et al. [20], dialectical theory is defined as: *"The thesis (behaviour change) brought forth the antithesis (the need for acceptance), and both acceptance and change-based strategies were integrated into the treatment package (synthesis). Dialectical theory provides the theoretical undercurrent needed to balance and synthesize these strategies. Core acceptance-based strategies derive from client-centered approaches and Zen practice and these involve mindfulness skills, validation, and radical acceptance."* [20, pp. 463]. The goal of the training is to teach people how to develop their own *wise mind* and to learn to make decisions that have positive consequences for the quality of life. By providing dialectical dialogues, the virtual coach can help the user to practise this process of decision-making by identifying the possible consequences of making decisions. The output of the virtual coach is personalised by data derived from the behavioural chain analysis and the emotion diary. The virtual coach is a self-learning system. Ecological momentary assessment (EMA), often termed experience sampling measures (ESM), is applied within the virtual mHealth coaching application to assess behavioural aspects [57][58], for instance by assessing subjective momentary states several times a day via a user-experience diary integrated into a virtual mHealth coach application.

D. *Evaluation of the first prototype: 'Denk je zèlf!' virtual coach*

The expert user was asked to perform eight task scenarios based on 15 heuristics [44][45]. The overall impression, conveyed by the user expert was good, but he remarked that:

- the functionalities horizontal and vertical scrolling are not yet fully operational;
- no clarity about where to find oneself in the process;
- difficulties in locating the virtual coach;
- some modules do need additional explanation.

The feedback given was processed and the second clickable prototype was built. Ten test users performed the eye tracking test and a Retrospective Think Aloud session.

According to the results of a usability test (see Table 1), the virtual coach application scored highly on the usability, which in this case is defined as the accurate execution of a given task via a user interface. The accuracy of the tasks were assessed on a scale of 1-3. A task is graded 1 if the user fails to perform the task, grade 2 when the performance is done with (some) difficulties and a 3 for a problem-free performance. Two of the tasks scored as unsatisfactory: users found it difficult to navigate to and to find the biography, and to adjust the settings for a notification and to understand its purpose. Users tend to tap on the 'No logon code? Read this' link because they had no idea of how to log in to the application. They tried to utilize a non-active link in the text or to navigate back to the login page. Adjusting the personal profile was a difficult task because the corresponding button was labelled 'settings' and that made it hard to find. Some users had difficulties with setting up a notification. The routing via the start button and the 'hamburger menu' (icon with three parallel horizontal lines) was considered accurate but the alarm icon was not recognised. Test users delivered feedback such as: *'It is not easy to obtain a login code', "It is difficult to locate personal settings'* and *'I don't like the color orange.'* The designers used clickable prototypes. That means that the interface does not offer unlimited navigation and interactivity. This should have been communicated more accurately to the users.

TABLE I. TASKS AND THEIR SCORES

Task no.	Task description	No. of taps /swipes	T1	T2	T3	T4	T5	T6	T7	T8	T9	T10	researcher	Total grading
			grading scale 1-3											
1	Getting a login	1 swipe 1 tap	3	3	3	1[1] (4 taps)	1[1]	3	3	3	2[1]	3	3	28
2	Logging in	2 taps	3	3	3	3	3	3	3	3	3	3	3	33
3	read FAQ/ introduction	2 swipes 1 tap	3	3	3	3	3	3	3	3	3	3	3	33
4	send message to the virtual coach	3 taps	3	3	3	3	3	3	2[2]	3	3	3	3	32
5	adjust personal profile	2 taps	2[2]	3	2[2]	1[2]	2[2]	3	2[2]	2[2]	1[2]	1[2]	3	22
6	set up a notification	3 taps 1 swipe	2[3]	3	2[3]	2	3	2[3]	3	2[3]	1[3]	3	3	26
7	finish/close module	3 taps	3	3	3	3	3	2	1[4]	3	3	3	3	30
8	create item in diary	3 taps	3	3	3	3	3	3	1[4]	3	3	3	3	31
Total time needed (min.)			3:34	2:41	2:21	2:17	3:03	4:38	4:37	3:35	5:36	2:48	01:10	
T^x= test user			[1] didn't understand the concept 'swipe' [2] tapped on several buttons to navigate [3] couldn't find a way to confirm choice [4] didn't follow instructions					Grading scale 1-3: 1=user failed to perform the task, 2=user performed the task but with errors, 3 =user performed the task.						

Four users suggested that the layout color should be adjustable by the user. In the third clickable prototype the label of the button 'settings' is changed to 'profile'. The alarm functionality has a standard icon attached to it, and setting the alarm itself is made more explicit.

V. CONCLUSION AND DISCUSSION

The main contributions of this paper are: (a) emotion-enriched Personas Lisanne and Anita as a new approach to guide the empathic design, which contributes to a better understanding of the mental model of the target group; (b) use-case scenario describing the daily challenges of an emotional eater, namely Persona Lisanne, which can be used for designing other lifestyle support and eCoaching applications for this vulnerable target group; (c) a clickable prototype of the 'Denk je zèlf!' virtual eCoach user interface; (d) results of the first user evaluation with the usability expert via a heuristic evaluation. The usability of the second clickable visual prototype was tested with users via thinking aloud usability tests using eye tracking. First results demonstrated that users are positive towards the virtual coach prototype. However, the first results are rather limited due to a small number of participants and the incomplete analysis of the eye gaze tracking data.

As a next step, the user interface of the 'Denk je zèlf!' virtual coaching application will be modified, further developed and iteratively evaluated by a larger number of target users. Within this next design and evaluation phase, the preferred appearance of the virtual coach will be investigated, in order to ensure that the virtual character, whether a peer or a health professional, is visually appealing and motivating for the emotional eaters. In addition, the content of the dialectical dialogue feedback messages of a virtual coach will be derived and iteratively validated with users, based on handbooks for therapists and online user forums. The users will be given small assignments such as navigating to the virtual coach and starting a conversation. Participants of the next user evaluation session will be asked to judge the responses of a virtual coach in terms of their persuasiveness, faithfulness and truthfulness. The output will not only benefit the quality of the user experience and the virtual coach interface, but will also inform future design and summative evaluation.

Figure 6. clickable visual prototype of the user interface with screens virtual coach, module mindfulness and profile

ACKNOWLEDGMENT

Lisanne and Anita are imaginary persons. Photos are retrieved from Shutterstock.
We thank all participants of the user studies, as well as students of the minors Healthy Ageing, Applied Psychology, and Web & Mobile Services at Hanze University of Applied Sciences for collecting data and supporting this study.

REFERENCES

[1] A. Dol, "Denk je zèlf! Developing a Personalised Virtual Coach for Emotional Eaters using Personas," The Eighth International Conference on eHealth, Telemedicine, and Social Medicine (eTELEMED 2016), ISSN: 2308-4359, ISBN: 978-1-61208-470-1, pp. 42-47.

[2] National Institute for Public Health and the Environment (RIVM). Available at: https://www.volksgezondheidenzorg.info/onderwerp/overgewicht/cijfers-context/gezondheidsgevolgen#node-maatschappelijke-gevolgen 05-01-2016 [retrieved: November 25, 2016]

[3] L. Kristén, A. Ivarsson, J. Parker, and K. Ziegert, "Future challenges for intervention research in health and lifestyle research - A systematic meta-literature review. International," Journal of Qualitative Studies on Health and Well-Being, v. 10, 2015.

[4] T.I. Van den Berg, S.M. Alavinia, F.J. Bredt, D. Lindeboom, L.A. Elders, and A. Burdorf, "The influence of psychosocial factors at work and life style on health and work ability among professional workers," Int Arch Occup Environ Health, 81(8), 2008, pp. 1029-1036.

[5] World HO. WHO | Obesity and Overweight. January 2015; Available at: http://www.who.int/mediacentre/factsheets/fs311/en/ [retrieved: November 25, 2016]

[6] Statistics Netherlands (Centraal Bureau voor de Statistiek). Webmagazine, Available at: http://www.cbs.nl/nl-NL/menu/themas/gezondheid-welzijn/publicaties/artikelen/archief/2014/2014-3939-wm.htm?RefererType=RSSItem [retrieved: November 25, 2016]

[7] T. Mann, A.J. Tomiyama, E. Westling, A. Lew, B. Samuels, and J. Chatman, "Medicare's search for effective obesity treatments: Diets are not the answer," Am Psychol, 04;62(3), 2007, pp. 220-233.

[8] Weight Watchers. Available at https://www.weightwatchers.com/us/ [retrieved November 25, 2016].

[9] My Diet Coach. Available at http://www.mydietcoachapp.com [retrieved November 25, 2016].

[10] Lose It! Available at https://www.loseit.com [retrieved November 25, 2016].

[11] T. Van Strien, A. Cebolla, E. Etchemendy, J. Gutiérrez-Maldonado, M. Ferrer-García, C. Botella, and R. Baños, "Emotional eating and food intake after sadness and joy," Appetite, 7/1;66, 2013, pp. 20-25.

[12] T. Van Strien, C.P. Herman, and M.W. Verheijden, "Eating style, overeating and weight gain. A prospective 2-year follow-up study in a representative Dutch sample," Appetite, 12;59(3), 2012, pp. 782-789.

[13] M.A. Roosen, D. Safer, S. Adler, A. Cebolla, and T. van Strien, "Group dialectical behavior therapy adapted for obese emotional eaters; a pilot study," Nutr Hosp, 07/20;27(4), 2012, pp. 1141-1147.

[14] Z. Cooper et al. "Testing a new cognitive behavioural treatment for obesity: A randomized controlled trial with three-year follow-up," Behav Res Ther, 8;48(8), 2010, pp. 706-713.

[15] P.G. Koenders and T. van Strien, "Emotional eating, rather than lifestyle behavior, drives weight gain in a prospective study in 1562 employees," Journal of Occupational and Environmental Medicine, 11;53(11), 2011, pp. 1287-1293.

[16] A.I. Concepción, "Effects of diets and their role in weight control," Psychol Health Med, 05;12(3), 2007, 321-327.

[17] A.J. Blair, V.J. Lewis, and D.A. Booth, "Does emotional eating interfere with success in attempts at weight control?" Appetite, 10;15(2), 1990, pp. 151-157.

[18] J. Ruwaard, "The efficacy and effectiveness of online CBT," Department op Clinical Psychology, University of Amsterdam, 2012.

[19] M.M. Linehan, "Skills training manual for treating borderline personality disorder," New York, NY US: Guilford Press, 1993.

[20] T.R. Lynch, A.L. Chapman, M.Z. Rosenthal, J.R. Kuo, and M.M. Linehan, "Mechanisms of change in dialectical behavior therapy: Theoretical and empirical observations," J Clin Psychol, 04;62(4), 2006, pp. 459-480.

[21] A.H. Robinson and D.L. Safer, "Moderators of dialectical behavior therapy for binge eating disorder: Results from a randomized controlled trial," Int J Eat Disord, 05;45(4), 2012, pp. 597-602.

[22] P. Daansen, "Eetbuien en affectregulatie: het gebruik van dialectisch-gedragstherapeutische principes bij de behandeling van eetstoornissen," Directieve Therapie, 26, 2006, pp. 79-89.

[23] D.L. Safer, C.F. Telch, and E.Y. Chen, "Dialectical Behaviour Therapy for Binge Eating and Bulimia," New York: The Guilford Press, 2009.

[24] D.L. Safer, A.H. Robinson, and B. Jo, "Outcome from a randomized controlled trial of group therapy for binge eating disorder: Comparing dialectical behavior therapy adapted for binge eating to an active comparison group therapy," Behavior Therapy, 03;41(1), 2010, pp. 106-120.

[25] C.F. Telch, W.S. Agras, and M.M. Linehan, "Dialectical behavior therapy for binge eating disorder," J Consult Clin Psychol, 12;69(6), 2001, pp. 1061-1065.

[26] E.Y. Chen, L. Matthews, C. Allen, J.R. Kuo, and M.M. Linehan, "Dialectical behavior therapy for clients with binge-eating disorder or bulimia nervosa and borderline personality disorder," Int J Eat Disord, 41(6), 2008, pp. 505-512.

[27] M. Bardus, J. Smith, and C. Abraham, "Are eHealth interventions for obesity prevention effective? A systematic review of reviews," The European Journal of Public Health, v. 25 (suppl 3), 2015.

[28] B. Khokhar et al. "Effectiveness of mobile electronic devices in weight loss among overweight and obese populations: a systematic review and meta-analysis," BMC Obesity, 2014, 1:22.

[29] M. De Zwaan et al. "INTERBED: internet-based guided self-help for overweight and obese patients with full or subsyndromal binge eating disorder. A multicenter randomized controlled trial," Trials, 11/21;13, 2012, pp. 220-220.

[30] J. Harvey-Berino et al. "Internet delivered behavioral obesity treatment," Prev Med, 08;51(2), 2010, pp. 123-128.

[31] M.J. Hutchesson et al. "eHealth interventions for the prevention and treatment of overweight and obesity in adults: a systematic review with meta-analysis," Obesity Reviews, 16, 5, 2015, pp. 376-392.

[32] G.M. Manzoni, F. Pagnini, S. Corti, E. Molinari, and G. Castelnuovo, "Internet-based behavioral interventions for obesity: An updated systematic review," Clinical Practice and Epidemiology in Mental Health, 2011, 7, pp. 19-28.

[33] M. Svensson et al. "The Change in Eating Behaviors in a Web-Based Weight Loss Program: A Longitudinal Analysis of Study Completers," Journal of Medical Internet Research, 11;16(11), 2014, 1-1.

[34] L.A. Dimeff, S.L. Rizvi, I.S. Contreras, J.M. Skutch, and D. Carroll, "The mobile revolution and the DBT Coach," the Behavior Therapist, 09;34(6), 2011, pp. 104-110.

[35] S.L. Rizvi, L.A. Dimeff, J. Skutch, D. Carroll, and M.M. Linehan, "A Pilot Study of the DBT Coach: An Interactive Mobile Phone Application for Individuals with Borderline Personality Disorder and Substance Use Disorder," Behavior Therapy, 12/01;42(4), 2011, pp. 589-600.

[36] M. Washburn and D.E. Parrish, "DBT Self-Help Application for Mobile Devices," Journal of Technology in Human Services; 31(2), 2013, pp. 175-183.

[37] O.A. Kulyk, H.J.A. op den Akker, R. Klaassen, and J.E.W.C. van Gemert-Pijnen, "Personalized Virtual Coaching for Lifestyle Support: Principles for Design and Evaluation," International Journal on Advances in Life Sciences, 6;3&4, 2014, pp. 300-309.

[38] A. Lentferink et al., "Self-tracking and Persuasive eCoaching in Healthy Lifestyle Interventions: Work-in-progress Scoping Review of Key Components," Proceedings of the Fourth Int. Workshop on Behavior Change Support Systems (BCSS16), co-located with the 11th Int. Conference PERSUASIVE 2016, Salzburg, Austria, April 5, pp. 15-35.

[39] E. Hudlicka, "Virtual training and coaching of health behavior: Example from mindfulness meditation training," Patient Education and Counseling, 92, (2013), pp. 160–166.

[40] H. Geissler, M. Hasenbein, S. Kanatouri, and R. Wegener, "E-Coaching: Conceptual and Empirical Findings of a Virtual Coaching Programme International," Journal of Evidence Based Coaching and Mentoring, Vol 12 (2), pp. 165 – 187.

[41] E.J. Van Gemert-Pijnen et al. "A holistic framework to improve the uptake and impact of ehealth technologies," J Med Internet Res, vol. 13, no. 4, 2011, p. e111.

[42] D. Morris, T.S. Saponas, and T. Desney, "Emerging Input Technologies for Always-Available Mobile Interaction," Found. Trends Hum.-Comput. Interact. 4, 4 (April 2011), pp. 245-316. DOI=http://dx.doi.org/10.1561/1100000023.

[43] O. Kulyk, R. Kosara, J. Urquiza and I. Wassink Human-Centered Aspects. pp. 13-76. In Human-Centered Visualization Environments, A. Kerren, A. Ebert and J.-J.C. Meyer (eds), Springer Verlag, Berlin, 2007.

[44] L. M. Ritterband, F.P. Thorndike, D.J. Cox, B.P. Kovatchev, and L.A. Gonder-Frederick, "A Behavior Change Model for Internet Interventions," Annals of Behavioral Medicine, 2009, 07;38(1), pp.18-27.

[45] C. LeRouge, J. Ma, S. Sneha, and K. Tolle, "User profiles and personas in the design and development of consumer health technologies," Int J Med Inf, 82(11), 2013, pp. 251-268.

[46] L. Van Velsen, L. Van Gemert-Pijnen, N. Nijland, D. Beaujean, and J. Van Steenbergen, "Personas: The linking pin in holistic design for eHealth," The Fourth International Conference on eHealth, Telemedicine, and Social Medicine (eTELEMED 2012), IARIA, ISSN: 2308-4359, ISBN: 978-1-61208-179-3, pp. 128 – 133.

[47] L. Graham, Gestalt Theory in Interactive Media Design (ISSN 193407227) 2008

[48] Nielsen, J., and Molich, R. (1990). Heuristic evaluation of user interfaces, Proc. ACM CHI'90 Conf. (Seattle, WA, 1-5 April), 249-256.

[49] R. Gomez, D.C. Caballero and J. Sevillano, "Heuristic Evaluation on Mobile Interfaces: A New Checklist," 2014.

[50] Nielsen, Jakob, and Landauer, Thomas K.: "A mathematical model of the finding of usability problems," *Proceedings of ACM INTERCHI'93 Conference* (Amsterdam, The Netherlands, 24-29 April 1993), pp. 206-213.

[51] A. Rösler, "Using the Tobii Mobile Device Stand in Usability Testinhg on Mobile Devices," Whitepaper. 2012. Available at http://www.tobiipro.com/siteassets/tobii-pro/learn-and-support/whitepapers/tobii_pro_whitepaper_mobiledevicetesting.pdf [retrieved: November 25, 2016].

[52] Z. Guan, S. Lee, E. Cuddihy, and J. Ramey, "The Validity of the Stimulated Retrospective Think-Aloud Method as Measured by Eye Tracking," CHI 2006 Proceedings • Usability Methods.

[53] M. J. van den Haak, Menno de Jong and Peter Jan Schellens, "Retrospective vs. concurrent think-aloud protocols: the usability of an online library catalogue," Behaviour & Information Tecnology, sept-okt. 2003, vol 22, p339-351

[54] J.M. Carroll, "Five reasons for scenario-based design," Interacting with Computers, 13(1), 2000, pp. 43-60.

[55] A. Pommeranz, W-P. Brinkman, P. Wiggers, J. Broekens, and C.M. Jonker, "Design Guidelines for Negotiation Support Systems: An Expert Perspective Using Scenarios," VTT symposium, 258, 2009, pp. 323-330.

[56] University of Groningen, "PIONIER Project," Available at http://www.let.rug.nl/vannoord/alp/Alpino/ [retrieved: November 25, 2016].

[57] S. Shiffman, A.A. Stone, and M.R. Hufford, "Ecological momentary assessment," Annu Rev Clin Psychol, 2008, 4:1–3.6.

[58] I. Myin-Germeys et al. "Experience sampling research in psychopathology: opening the black box of daily life," Psychol. Med. 39, 2009, pp. 1533–1547.

Structuring the EPRs; The National development of Archetypes for Core Functionallity

Gro-Hilde Ulriksen
Norwegian Center for E-health Research – University hospital North Norway
Telemedicine and eHealth Research Group, Faculty of Health Sciences, Arctic University of Norway, Tromsø
e-mail: Gro-Hilde.Ulriksen@ehealthresearch.no

Rune Pedersen
University Hospital of North Norway, Section for E-health administration
Norwegian Center for E-health Research – University hospital North Norway
Telemedicine and eHealth Research Group, Faculty of Health Sciences, Arctic University of Norway, Tromsø
e-mail: Rune.Pedersen@unn.no

***Abstract*—The aim of this paper is to highlight the importance of introducing a set of core archetypes to succeed with the national consensus work on archetypes in Norway. This is illustrated by emphasizing on four important challenges identified through the national work with archetypes in Norway from 2012-2016. First, the process of establishing a well-functioning national archetype organization, including a network of competent healthcare personnel to participate in the consensus process. This comprises how to increase the pace of developing these national archetype. Second, the interdependence between the archetypes and the new electronic patient record system (EPR), focusing on the need for archetypes to enable developing an EPR system, at the same time as an EPR system is required to develop and test out high quality archetypes. Third, the development of local archetypes at a specialized hospital clinic, including the first attempts of using archetypes for clinical practice, and useful lessons learned for the national archetype work. Fourth, the need to define the number of archetypes to plan the future archetype work, including estimating resources for the work, and the challenges of governing archetypes in relation to who is responsible for what aspects of the archetypes. Establishing a prototype of an EPR system based on generic core archetypes may increase clinicians understanding of structured EPR. Our research questions are therefore: What has been the challenges in the national archetype process so far, and how could these challenges be met through developing a set of core archetypes? Empirically the paper focuses on the national archetype work, and the regional archetype organization in the North Norwegian health authority. Our qualitative case also includes other archetype initiatives in Norway like for instance the use of archetypes for clinical practice at the small clinic in the Southern and Eastern Norway Regional Health Authority. Except from being a status report on national archetype development, this paper also contributes to a longitudinal interpretive study, related to the development of a large-scale EPR system by the North Norwegian Health Authority.**

Keywords-electronic patient records; archetypes; openEHR; semantic interoperability; core archetypes.

I. INTRODUCTION

There has been an increased focus in Norwegian healthcare to improve the role of the electronic patient record system (EPR) [1]. This has been an important part of national strategies and visions for healthcare since the 1990's [2][3], due to the augmented focus to enable sharing and integrating healthcare, as well as organizing information in a more structured manner [4]. The increased emphasis on cost savings, patient safety, and efficiency in healthcare practices, has raised the focus on seamless integration and standardization, both within as well as across professional, departmental, and institutional boundaries [2][3][5]. Considerable amounts of resources have been directed towards establishing fully integrated healthcare infrastructures, both from the government, and healthcare providers [6]. A central element in achieving such goals are interconnected and interoperable EPR systems [6]. Significant improvements of the EPR have been necessary for these systems to evolve from being tools for information storage, into structured work tools supporting patient pathways and decision support [7]. EPR systems have been required to be flexible enough for representing specific medical knowledge, at the same time as ensuring the need for interoperability with other systems [8]. The EPR has been essential for coordinating hospital work processes, and there has been an extensive need for standardizing the EPR content, to facilitate sharing and comparing health data within and across health care practices, to reach semantic interoperability. This has been a key requirement for improving EPR communication [9], to ensure that both senders and recipients understand information and standards the same way [10]. Semantic interoperability: requires that the information system understands both requests information and the information sources. Semantics are defined as the meanings of different terms and expressions [11]. Hence, semantic interoperability is the way information systems exchange information on the terms of shared, pre-established and negotiated meanings of language and expressions [11].

Standards in hospitals are useful both to specify workpractices, and to define how technologies interact [12]. However, standardizing clinical workpractices and routines have been difficult to accomplish [13][14][15]. This paper builds on two important standardization processes in Norwegian healthcare, one regional and one national. In 2012 the North Norwegian Health Authority established a large information and communication technology (ICT) project named standardization of the regional ICT portfolio (FIKS), to standardize the regional ICT portfolio. One of FIKS's most important roles was to collaborate closely with the largest EPR vendor in Norway, on developing a new open Integrated Care Electronic Health Record (EHR) based EPR system, using archetypes as core elements for standardizing the clinical content. Archetypes are structured data elements of clinical concepts, envisioned to ensure technology-independent interoperability, easy reuse of information and efficient decision support [16]. They are clinical information models used to standardize the clinical content of an EPR system. The archetypes contain a maximum dataset, including evidence about knowledge objects, and relevant attributes [8] They also need to include rules, measurement intervals, data types, presentation formats, data representation conditions (codes, terminologies), etc. [8]. It is possible to combine archetypes into templates to create documents, messages, specific forms and reports, including referrals, radiology reports and discharge forms. Templates are often locally made, based on three things: the requirements for the form, available archetypes, and the local use of terminologies.

The openEHR framework built on a two-level modelling approach separating the clinical and technical development of the EPR system. The intention of the first level, the technical reference model, was to increase semantic interoperability, and secure a reuse of data [11][17][18]. The reference model is generic enough to store any type of clinical information, and it is a stable object model to build software and data on. These are used to specify how to organize and group clinical information, capture contextual information, query and update the EPR and so on [19]. The second level, contained archetypes and templates, as standards for the clinical content. This makes it possible for clinicians to be in charge of designing and defining the archetype standards, hence the clinical content of the EPR system. There are some important issues related to allocating the right resources for the national archetype work, recruiting participants for this work and the role clinical competence play in the future archetype development, these are not however addressed in this paper. The two-level model enabled making changes only to the clinical content of the archetypes, without having to alter the underlying open EHR information model. The openEHR framework allowed for archetype design at different levels of healthcare organizations.

The other standardization process we have focused on was the national standardization of archetypes as the clinical content of the EPR system. In Norway the primarily work with archetypes was conducted at a national level, coordinated by NRUA (National Editorial group for Archetype development in Norway) established by National ICT in 2013, National ICT is responsible for coordinating ICT-related initiatives in the Norwegian specialized health care service [20]. To design optimal archetypes to standardize the clinical content of EPR systems, it was necessary for clinicians to have a key role in both developing and approving the national archetypes. Therefore, one of the most important tasks for NRUA was to recruit enough clinicians to participate in standardizing archetypes. The clinical content of an archetype based EPR system had to contain numerous archetypes, to encompass all clinical practice. There were 42 nationally approved archetypes by May 2016, and even if more than 100 were in the process, this was not nearly enough to comprehend the total clinical content of an EPR system. One important question to address was how many archetypes standardizing the clinical content of an EPR system required? The absence of necessary archetypes complicated and delayed the development of the new archetype-based EPR system. Hence, this raised a question of when to start using archetypes for clinical practice, and the consequences of using archetypes before reaching national consensus.

In addition, it was challenging for the users included in developing the new EPR, to grasp the potential of this system based on the close relation between the archetypes and the technology in this entirely new technological solution. These factors might contribute to explain why the development process took much longer than expected. The aim of this paper is to highlight the importance of introducing a set of core archetypes to succeed with the national consensus work on archetypes in Norway. Establishing a prototype of an EPR system, based on generic core archetypes, is a promising way to increase the pace of the national archetype work. In addition, such prototype might provide clinicians with a better understanding of this new way of developing and using the EPR, at the same time as it is possible to test the archetypes in a production like environment. One important dilemma to solve before establishing such prototype is however, how many archetypes would such solution require? Our focus is on how to use a set of core archetypes to keep up and escalation the pace of the national archetype work in Norway, and to evaluate the establishment of this core set of archetypes through interaction with projects essential to the ongoing process. Our research question is therefore: What has been the challenges in the national archetype process so far, and how could these challenges be met through developing a set of core archetypes?

We followed different aspects of the archetype work in Norway describing how the development have progressed over time. In the paper some important experiences with, testing out archetypes and implementing them to clinical practice are addressed, in addition to how this has contributing to speeding up the work with archetypes in Norway. Being able to estimate the number of core archetypes needed for establishing such prototype would ease planning the future work with archetypes in Norway related to how estimating the number of resources the work requires and predicting when to expect having a complete set of clinical standards for an EPR system. Defining a set of core archetypes would contribute to increased clinical understanding of the archetype concept and the potentials openEHR system have for improving the clinical practice.

The rest of the paper has the following structure: Section two, has a presentation of the method. In section three, there is a description of the four aspects of working with archetypes in Norway. The discussion in the fourth section focuses on when to start using archetypes, the use of core archetypes for a prototype of a basic EPR system and how to plan the future archetype work. The fifth section concludes the paper.

II. METHOD

The work with archetypes in Norway is mainly conducted on a national level, however, also regional archetype organizations has gained increasing foothold over the last two years. Our site of research is therefore both NRUA, and the regional archetype organization in the North Norwegian health authority. Our qualitative case also includes other archetype initiatives in Norway like for instance the use of archetypes for clinical practice at the small clinic in the Southern and Eastern Norway Regional Health Authority (described in section C). Except from being a status report on national archetype development, this paper also contributes to a longitudinal interpretive study, related to the development of a large-scale EPR system by the North Norwegian Health Authority. The methodologically positioning of the study is within a qualitative interpretive paradigm. The focus is on evolving and improving the understanding of a studied phenomenon, by looking at it from different viewpoints, within a context [21][22]. An advantage of using a qualitative interpretive approach is enabling complex textual descriptions of how people experience a particular matter, by providing information about the human side of a given process [22].

The fieldwork draws on the first author's role working in FIKS for two years and afterwards continuing to follow activities in the project, by participating in workshops and meetings connected to the development of the new EPR systems. This author have also been an observer and participant in both the regional and national work with archetypes in Norway, participating in meetings, workshops and discussions. The second author has contributed in the regional and national work with the new EPR and archetypes for the last seven years, participating in meetings, discussions and observations. He has represented the North Norwegian health region in NRUA, and been the leader of the regional archetype organization. He has also recently become the leader of NRUA. The personal information protection commissionaire for research in the health region, and the Norwegian social science data service (NSD), approved the data collection for this study. All informants provided written consents for the interviews by e-mail.

The data analysis was accomplished in several stages with data and citations from numerous sources. A document study including different iterations and negotiations of several archetypes like problem/diagnosis, had a key role during the field work, and provided contextual and historical insight into the process of establishing archetypes in Norway. The Norwegian CKM repository consists of nationally approved archetypes including the documents from each review iteration, where clinicians discuss and approve the content of each part of the archetype in what they refer to as "consensus" processes. We have read all the logs several times, both separately and combined with the interview data as a whole, to enable extracting the most important topics. Interesting citations from the different participants, related to each of the topics, were translated into English. Second, we conducted participatory observations in NRUA meetings and workshops as well as vendor meetings and workshops over several years. We also observed a reviewer while he used the CKM to review an archetype, to understand better, how it was for clinicians to use this web-based tool. There was a highlighting of events and milestones from the observations, these became the starting point for questions in the interview guide. This part of the analysis was also an iterative process in, which analysed and transcribed data led to new questions in the next interview, and so on. Third, the authors conducted 30 open-ended interviews with participants in the consensus process, initially while guiding them to become users of the CKM, then after they had used it to work with archetypes for a while, related to the development of the new EPR, and the national work with archetypes. Some participants in the archetype work were interviewed more than once, based on their specific roles in terms of being identified as key informants. The purpose of using open-ended interviews is enabling informants to tell their story, without the author's pre-perceptions getting in the way. The interviewers still prepared some questions for the interviews, to make sure the interviews covered the topics they wanted to focus on. In addition, new interesting issues to include emerged in several interviews. The interviews were transcribed and analyzed both separately, and as a part of a whole [22]. Information infrastructure and infrastructuring theory were used to transcribe and analyzed the interviews. This was done to put it all together as a whole, to complement different perspectives of the situation.

In addition to interview data from key personnel of the processes and documents from the CKM, we used official reports from organizations such as the National ICT Health Trust, and other websites such as the official site of the openEHR organization. We have aimed at obtaining a historical and contextual understanding of the work involved in the development of archetypes. Table I describes the details of the data collection.

TABLE I. THE DATA COLLECTION

	Number of persons	Duration	Period
Interviews			
Archetype reviewers	17	30-90 min	2014-2016
NRUA members	5	60-90 min	2014-2015
Persons involved in the EPR development	8	60-120 min	2013-2015
Observations			
NRUA/regional resource group		200 hours	2014-2016
Development of EPR system		80 hours	2012-2016
Archetype review and CKM use		5 hours	2014-2015
Discussions		100 hours	2012-2016
Document studies			
		240 hours	2015-2016

III. DIFFERENT ASPECTS OF NORWEGIAN ARCHETYPE WORK

We have focused on four different aspects of working with national archetypes for Norwegian healthcare in the period 2012-2016. First, the process of establishing a well-functioning national archetype organization (NRUA), including a network of competent healthcare personnel to participate in the consensus process. Second, the relation between the archetypes and the new EPR system focusing on development of a new openEHR based EPR system using archetypes as standards for the clinical content. Third, the development of local archetypes at a specialized hospital clinic, including the first attempts of using archetypes for clinical practice and useful lessons learned for the national archetype work. Fourth the need to define the number of archetypes to plan the future archetype work, including estimating resources for the work, and the challenges of governing archetypes in relation to who is responsible for what aspects of the archetypes.

A. Organizing the National Archetype Work in Norway.

In Norway, one system vendor had gained more than 70% of the EPR marked [23]. This vendor was developing a new openEHR based EPR system, which three of the four health regions had committed themselves to use. Consequentially it made sense to organize the construction of the archetype standards for the new EPR system at a national level. Hence, National ICT instigated a national organization to work with archetypes in Norway. The established NRUA in 2013, to form a national archetype repository – a clinical knowledge manager (CKM). The overall goal of NRUA was to coordinate development and use of archetypes on a national level, both handling the national consensus process of reviewing and approving archetypes, as well as supporting local initiatives for archetype design and usage in Norway. NRUA had five part-time engaged employees, working with governing and modeling archetypes. In addition, 2-3 representatives from each of the four Regional Health Authorities in Norway, and representatives from the Norwegian directorate of health were members of NRUA.

NRUA established an editorial group to initiate archetype reviews, these reviews were highly depended on clinicians participating in standardizing the clinical content of healthcare practice, and organizing these standards as archetypes. Therefore, recruiting clinicians and activating them as archetype reviewers for the national consensus process, was an essential part of NRUA's work. In the consensus process clinicians used the web based CKM to review and approve archetypes, enabling flexible asynchronous communication between the different contributors. The archetype reviewers only communicated through the online CKM without participating in any face-to-face meetings. The first year NRUA focused primarily on establishing a well-functioning organization prepared to handle coordinating hundreds of participants from different professions, including a large network of clinicians, working with archetypes. They offered training and support for new CKM users, and established relations with the international CKM run by the openEHR foundation. In addition, NRUA imported existing archetypes from the international CKM, and translated them into Norwegian, for clinicians to review. One of the archetype reviewers stated: "*NRUA has members with a genuine interest in archetypes, and they have worked very hard to get this organization up and running.*" The organizational work also included defining the steps of the consensus process, and forming a priority scheme for how to organize working with the national archetypes. Prioritizing launching a well-functioning organization, the actual consensus work moved slowly the first year.

However, they still managed to approve the first national archetype as early as in June 2014, only six months after NRUA was established. Investing a considerable amount of time on the organizational concerns initially, enabled NRUA to increase the pace of the archetype development the next two years. In May 2016, there were 42 nationally approved archetypes in Norway. The population of archetypes approved was considered complex and essential for EPR functionality. In addition, more than 100 had started the consensus process. The goal was to have 200 archetypes approved by the end of 2016. In addition, NRUA gained valuable knowledge along the way, both on how to structure archetypes, and on how to run a national organization. Hence, NRUA has gradually become an accomplished organization for supporting and supervising local and regional archetype projects. They have also become competent to arrange information and modelling courses to expand the competence on archetypes within Norwegian healthcare.

It was yet challenging to recruit reviewers for the national consensus process, especially clinicians. This was a rather extensive process to participate in, and the reviewers were not in any way compensated for contributing to the national archetype work. When they first started as reviewers, it was also time consuming to learn how to use the CKM work tool. Several reviewers said that it took quite some time to understand the complex relation between the clinical and technical components of the archetypes, and to get used to the concept of archetypes as clinical standards. In addition, many archetypes went through more than one review iteration before being nationally approved, the average number was 2 iterations, however, one of the archetypes had as many as seven without reaching national consensus. Each review iteration took a considerable amount of time for the clinicians to finish and some reviewers had more than one archetype to review simultaneously. One clinician stated: *"A review iteration can take between 15min and 1,5 hour or even longer to finish, depending on the complexity of the archetype and whether I need to look things up, or consult with colleagues."* The time-consuming review process done at the clinician's free time, and the loosely committed online work process, both led to several dropouts from the archetype work, especially among clinicians. Since National ICT had recommended archetypes as standards for the clinical content of the EPR, it was important for clinicians to have an essential role in defining and designing these standards since they were the main users of the standards in the EPR system. Also the two-level model of openEHR related on domain experts (clinicians) as the main developers of the clinical content of the archetypes. One clinician said: *"It is crucial to include clinicians in this work; they have the clinical knowledge and know what is important to focus on, for the archetypes to be useful standards for clinical practice."* He also commented, *"If the archetypes are not designed by clinicians it will be very difficult to get clinicians to accept and use them."* Though it was difficult for NRUA to plan the work with archetypes ahead, included estimating the number of clinicians needed for the national work, what archetypes to prioritize working with, and how many archetypes the Norwegian CKM needed. It was also problematic to ensure that the archetypes fit the clinical requirements of Norwegian healthcare, since they did not have any way of testing then out in clinical practice.

B. Relations between Archetype Standards and the EPR System

National ICT gained an interest for archetypes as clinical standards several years, before the vendor started working with the openEHR based EPR system in 2011. In 2008, National ICT run an internal project translating variables for electronic medical charts into archetypes. However, deciding to purchase the new EPR system for most of Norwegian hospitals, the interest for archetypes expanded to several parts of Norwegian healthcare.

The vendor started developing the new EPR system in close collaboration with the North Norwegian Health Authority. After a bid for tender process in 2012, this health region decided to regionalize their new ICT portfolio. To complete this process, they established a regional project, FIKS, to run from 2012-2016. FIKS was one of the largest ICT investments in Norwegian healthcare, with a total cost expected to exceed €100 million [24]. The main goal of FIKS was to establish a regional ICT portfolio, as a foundation for regionally standardized patient pathways, decision support, and integrations between clinical ICT systems. A regionalization, including standardizing EPR work practice, was a necessary requirement to reach such goal, enabling the Health Authorities to better administrate and compare information from the hospitals in the region. The FIKS project run in close collaboration with system users from the hospitals and the EPR vendor. One of the most important goals of FIKS was to collaborate with the EPR vendor on developing the openEHR based EPR system for Norwegian healthcare. This new EPR was designed to improve the user's workdays, providing structured data including predefined content elements and schemes for documentation, enabling better overview and reuse of patient information. In addition, the possibilities to include patient pathways and increase the semantic interoperability were important to improving the EPR. This was enabled by using the international openEHR architecture, standardized by CEN/ISO [25]. The openEHR architecture built on standardized information models, open source components, and highly structured clinical content, with archetypes as core building blocks. Archetypes were structured data elements of clinical concepts, where observations, evaluations, instructions, and actions, formed the ongoing process of treatment and care [16]. Archetypes were used to define how clinical data was structured, seamlessly stored, and transferred between EPR systems [26]. The intention was for archetypes to contain a maximum dataset, including evidence about knowledge objects, and relevant attributes [8][27]. It was possible to design both widely reusable generic archetypes securing interoperability within and across healthcare institutions, as well as specialized ones intended for a distinct local setting [9][27][28][29][30]. In figure 1. the archetype problem/diagnosis is used as an example of an archetype

Figure 1. The archetype problem/diagnosis

The new EPR system required archetypes, as standards for structuring the clinical content it was necessary to have a common language to raise the semantic interoperability. According to the two-level model the openEHR architecture built on, it was possible to develop the archetypes standards and the reference model of the EPR system separately [16]. Still, splitting the development processes for the new EPR systems and the archetypes, turned out to be more complex and challenging than expected, due to close interrelations between these two processes. The vendor first became aware of this in 2012, when they started to develop functionality for the openEHR based EPR system. The development of the new EPR system included an extensive process of designing necessary functionality. They also decided to develop the new EPR module by module, to prevent a "big bang" implementation for the healthcare practice. This required including clinicians and other domain experts in the development process. More than 100 system users, from all the hospitals in the North Norwegian Health Authority, participated in an attempt of user centered system design, where an agile method - scrum was practiced. System users participated in workshops suggesting and prioritizing requirements for the new system. The vendor developed functionality based on the needs and requests identified by the end-users and domain experts. The users mainly based their requests on challenges, flaws, and limitations they had identified from using the existing EPR system. It was difficult to understand the potential and possibilities of the new system since it was based on a very different architecture from what they were used to. One clinician stated: *"When you ask clinicians, they will most likely point out the needs for changes based on their current workpractice. Their starting point is the more than 20 years old EPR system they use today. They are more likely to think small steps ahead to improve their current work, rather than focusing on large revolutionary changes, necessary for exploiting the potential of archetypes and openEHR."* It was very difficult for the system users to grasp the potential of the forthcoming EPR system, and the new type of clinical standards, as neither the archetypes nor the EPR system functionality was finished. Consequently, developing the EPR based on this approach was time consuming, and inefficient with a risk of ending up with a system unable to exploit the potential of openEHR based EPR systems. The original strategy was for the vendor to import a set of generic archetypes from the international CKM to have some basic ones to start working with. Then the system users themselves would continue developing archetypes necessary for their clinical practice, for example clinical observations such as blood pressure, body weight, clinical scorings, and schemes for procedures. One member of FIKS said: *"At first it seemed possible for clinicians to design archetype based schemes on the fly, I don't however think this will be the case."*

It seemed rather straightforward to import already existing archetypes, and adjust them to Norwegian conditions. According to the openEHR organization, the archetypes were system independent and could be downloaded and used freely for any purpose. However, after trying to implement archetypes for a year, without achieving the desired results, it became evident that this work was more complex than anticipated. Thus, developing and testing functionality for the EPR system became unmanageable without being able to include the necessary archetypes related to the modules of the EPR system developed.

Based on both the archetype initiative and the vendors experiences it was decided to establish one technological solutions for storing and reusing archetypes a Norwegian clinical knowledge manager (CKM) containing archetypes designed to fit Norwegian conditions, the CKM also including standardized methods for developing and maintaining national archetypes.

One example illustrating the relation between system development and the archetypes is the ongoing work with the pre-surgical planning module. FIKS started working with this module in 2012, yet, in May 2016, this module still was not implemented in clinical practice. One of the reasons was the lack of necessary archetypes to support the functionality. They started working with this module before the national archetype work had even started, and the definitions of the requirements for working with archetypes at a national level were not yet made. After including NRUA in the work with the necessary archetype requirements for this module it was possible to gradually increase the pace of the development process of the pre surgical planning module. The project defined together with NRUA the need for 18 national archetypes for this module. There is an ongoing process of finishing the necessary archetypes, six of them had reached national consensus in May 2016, and more of them had started the consensus process. The North Norwegian Health Authority had defined a principle of only using nationally approved archetypes. However, as time went by, this principle was severely challenges by the need to complete the pre surgical planning module. There was a dilemma whether to wait for the rest of the archetypes to reach national consensus, and further delay the implementation of the new EPR, or begin to use archetypes not yet nationally approved, enabling implementing the pre surgical planning module at the hospital, however risking the need for changing the archetypes later on.

In addition to defining when to start using the archetypes, another important question to address was if all archetypes had to be national. When they worked with the pre surgical planning module defining the necessary archetypes, they defined some national ones. Others were more system specific for this particular EPR system, and were possible for the vendors to develop themselves. It seemed unnecessary to approve archetypes only useful for this particular EPR system at a national level. Another issue might be how much of the EPR system to structure as archetypes. Was it only possible to use the archetypes to standardize the clinical content, or could they also be used for structuring other parts of the EPR system, like workflows and administrative routines. It was important to include NRUA in assessing archetype requests, due to the extensive competence on archetypes they had gained over the years. These questions might have been easier to answer if there were a way of testing out archetypes for clinical practice. Accordingly, some local projects have lately been trying out archetypes for real time use, for very small and restricted parts of clinical practice. The following section describes one such example.

C. *Using the Archetypes for Clinical Practice*

Based on the notion of archetypes being the currently used standards for communication within Norwegian healthcare, many projects eager to start using them. Since the national archetype design of archetypes used several years to gain foothold, some local initiatives began to develop and implement archetypes for clinical practice, before they had reached national consensus. Some even started developing local archetypes themselves. One member of NRUA stated: *"Systems that use archetypes today are not designed on nationally approved archetypes, or even international ones. They are mainly constructed by system users themselves."*

One example was a hospital clinic in the Southern and Eastern Norway Regional Health Authority were they developed a "self-check" registration form for patients to fill out when entering the clinic. This form built on the new openEHR based EPR system, and needed a number of highly specialized archetypes to meet the clinical requirements. The clinic worked with this registration form as a project between February and November 2015. When the clinic started working with structuring their EPR content, the Norwegian CKM only contained a few nationally approved archetypes (even if they had started the consensus process of several other archetypes as well), and NRUA was still a rather immature organization. Therefore, the project had to develop most of the necessary archetypes themselves, in close collaboration with the system vendor, and Ocean informatics (the international CKM community). They searched the international CKM and CKM's from other countries to find relevant archetypes for their practice, without any luck. They identified a need for 58 archetypes, in total, and they ended up with using six nationally approved archetypes in addition to developing 52 themselves. However, due to NRUA being a newly established organization they had not yet defined the final modelling patterns for national archetypes. The project had to start using an immature modelling pattern, which consequentially led to needs for structural changes to the archetypes after implemented them to the EPR system, generating extra work and interoperability problems for the project and the clinic using the archetypes.

This clinic had a clearly defined focus area, working within a very narrow clinical field, and thought they mainly needed specialized archetypes designed especially for their field of expertise. Still, while developing and implementing the archetypes, they became more and more aware that some of the archetypes they thought were suitable only for this particular practice, also were applicable for other specialties. Having to include the needs of other potential users made the development of the archetypes, more complex than first expected. After implementing the archetypes, they also identified a need for some of the locally developed archetypes to be included into already existing national archetypes instead of being stand-alone archetypes.

Developing archetypes locally became a time-consuming and challenging process for the project, in addition to the lack of modelling patterns, there were no national guidelines for developing national archetypes in Norway. It was very challenging for the system vendor to create high quality archetypes without having national procedures to follow, since the archetypes were complex data elements that could be potentially structure in numbers of different ways. One member from NRUA said: *"It is not difficult to create an archetype, though it is very demanding to construct high quality archetypes."* Comprehending the complexity of archetypes and the interrelation between different archetypes is a maturity process that takes time to grasp. Another challenge potentially leading to interoperability issues, were the problems related to versioning the archetypes. The local and the national archetypes were versioned the same way creating a risk of mixing up the different sorts of archetypes. IT was necessary to convert the locally designed archetypes to national ones at some point. By doing this there was a risk of changing the local archetypes so much that the two versions (local and national) were no longer compatible, hence there were a risk of losing historical data and having to spend lots of money to convert data to the national archetype.

Another challenge identified in this project related to the involved clinical resources. They had two clinicians working 20%, one nurse working 80% and one mercantile resource working 20 % in the project, this turned out to not be enough clinical resources for developing the 52 archetypes within the deadline of the project. This consequentially compromised the quality of the archetypes they developed, and representatives from the project underlined the importance of including enough clinicians to enable making them the main developers of archetype standards for the EPR system. In addition, the project experienced that creating local archetypes based on schemes from the old EPR were unpractical to use. There was a need for a new way of thinking to utilize the potential of the openEHR architecture. It took time to adjust the clinician's way of thinking to fit the openEHR and archetype based systems. The clinic still, six months after implementing the archetypes in the registration form, use a lot of time and resources on adjusting the archetypes and including them in the national consensus processes.

When this clinic started using the new EPR system including archetypes for clinical practice, this was one of the first attempts in Norway to try out archetypes in a clinical setting, and one of the first times using the new EPR, including archetypes, for a clinical setting. This provided the clinic, NRUA, system users, and the vendor, with important insight on the usability of archetypes at different levels of the system, especially compositions, evaluations, and cluster archetypes. Testing archetypes in an actual clinical setting enabled identifying necessary requirements for improving, not only the local, but also the national archetypes, to make them useful for both small and large-scale clinical usage. Gaining this important experience further underlined the need for NRUA to assist in similar future projects better enabling the local archetypes to conform to the national ones, and prevent projects from ending up with the extensive challenges this clinic experienced related to both developing the archetypes using them and adjusting them to the national ones after they were implemented.

This clinic creating a "self-check" registration form ended up needing 58 archetypes, and even if some of them were included as parts of existing archetypes and others were expanded to cover other specialties, this was a quite high number. It became important to try estimating the total number of archetypes needed for an EPR system.

D. Coordination and Distribution of Consensus Based Archetypes

Establishing NRUA for coordinating the national work with archetypes in Norway was an important step towards archetypes being a successful way of standardizing the clinical content of an EPR system. Defining the interrelation between the archetypes and the openEHR based EPR system, and trying out archetypes for clinical practice, also contributed to moving this process forward. However, there were still some important question to address in order for this work to continue expanding. One of them was to try estimating how many archetypes this standardization process required. This directly related to defining the number of resources needed to standardize all necessary clinical content. Another issue was governing the archetypes in relation to distributing the responsibilities and different aspects of archetypes and the CKM, including defining the borders between NRUA – developing archetypes, and system vendors – using the archetypes for openEHR based EPR systems.

Defining how many archetypes to develop and estimating the need for resources, turned out to be difficult, since the archetype work in Norway were quite immature and still mainly conducted at a theoretical level. To estimate the forthcoming work, what archetypes to prioritize, and estimate the number and specialties of necessary reviewers, it was important to try to predict the total archetype number. NRUA partially addressed this by defined a synthesis of using core archetypes. This was an attempt of defining what archetypes to prioritize working with [28]: *"90% of the journal functions in the electronic patient record including non-specialized examinations and procedures can be represented by using 30 core archetypes* [28]*."* The synthesis underlined that the core archetypes did not include data from any clinical specialties, just the basic structured data of an EPR system. An even though one member of NRUA made some modifications to the synthesis, expanding it to include between 30-50 archetypes, due to extended experience with establishing and modelling archetypes, this number was quite low, and should be possible to accomplish if the national work was structured around these archetypes. Prioritizing archetypes for the national consensus process, to get an overview of the basic structure of an openEHR base EPR system, seemed like a clever strategy for the archetype development. If NRUA had prioritized finishing the national consensus work on the core archetypes, it would have been possible to design a prototype of an EPR system including the necessary archetypes. This prototype would have enabled testing the clinical usability of the archetypes, and the relation between the clinical archetype standards and the reference model of the EPR system. Consequentially such prototype would contribute to increase the pace also of the national consensus work. One clinician stated: *"Having a prototype would ensure that the archetypes cover the necessary clinical content for the modules of the new EPR."* Another reviewer said: *"It is difficult for clinicians to imagine the possibilities of new EPR and not base their requirements on today's needs."* Moving the development from a theoretical to a practical level were the clinicians were able to test out archetypes in relation with functionality, would potentially make it easier to detect the possibilities and advantages of using a structured archetype base EPR system. Thus, we had to question why they did not prioritize the core archetypes for the national consensus process, enabling designing such prototype. Until May 2016, 20 of the original 30 core archetypes were nationally approved, and 4 more were in the consensus process, the last 6 had not yet started the process. One member of NRUA said: *"We started out prioritizing the defined core archetypes. However, the work with core archetypes takes time since these are very generic and extensive concept archetypes."* NRUA also had a policy to start the consensus work on an archetype only on request from the healthcare organization. This was partly bases on the lack of available resources to work with archetypes. The requested archetypes are often specialized ones covering specific areas of the EPR, there are rarely requests for generic archetypes, covering basic elements of the EPR system. Another reason why they did not prioritize the core archetypes was probably the uncertainty about the validity of the core archetype synthesis, since NRUA had already started to modify it by including more archetypes than originally intended. The reason for this was the fact that archetype development was very complex process due to both the archetypes themselves being complicated standards demanding extensive technical and clinical knowledge. In addition, there was a closer interrelation between the archetypes and the reference model established than first expected. As a result, in a national forum in May 2016 the leader of NRUA stated that it would most likely be necessary to include about 200 archetypes to cover 80% of the clinical content of an EPR system. This was extensively more than the original core archetype synthesis from 2013. This indicated that a successive awareness and increased knowledge about archetype standards have required the hypothesis of core archetypes to change over time. Standardizing an EPR system by using archetypes was more complex than firs expected, by both NRUA and the system vendor. The further the work with archetypes in Norway evolved, and the more experienced NRUS got, the more archetypes seems necessary to include, even for covering the basics of an EPR system. Another statement underlining the assumption that increased knowledge raises the number of archetypes required for standardizing the clinical content, was made by one of the members of Ocean informatics, a partner in the international openEHR community working with archetypes since 1999 [31]. The international CKM is a repository including about 500 archetypes. She said that between 1000-2000 archetypes were necessary for covering all the content of the EPR system. These diverging numbers made estimating resources and a timeline for the future work with archetypes very demanding.

In addition, it was important to outline some borders between the archetypes and the systems they were included in. It was a fine balance since the system vendor was included in the consensus work, (which is open for all vendors to participate in). In addition, this new EPR system was the only one conforming to openEHR and archetype standards, meaning that if the standards developed were not useful for this system, it was no other large-scale system in Norway to use them in. One dilemma was therefore how to get system independent standards that still were useful for the only EPR system using archetypes. The Western Norway Regional Health Authority recently brought a request to NRUA for changing the nationally approved archetype Observation.nutritional_risk_screening. After implementing the archetype to the EPR system, the health region realized that the modelling pattern was not optimal for clinical use. The request included altering some of the variables in the archetypes. The 15 members of NRUA discussed this request by email and members from all health regions were involved in the discussion. One of NRUA's main concerns was whether this change request really related to the quality of the archetype itself, or to limitations in the EPR system were the archetype was implemented. In the following discussion, members of NRUA underlined the need to address the dilemma of altering archetypes to fit systems requirements. Making such adjustments to fit system requests increase the risk of archetypes becoming too system specific. Several NRUA members advised against this, since it was against the intention of creating flexible and system independent archetypes useful for different purposes. Members from three of the four health regions participated in this discussion, in addition to NRUA's editorial board, indicating that the archetype work was a national initiative were the health regions were included in all important decision-making. It was important to conduct the important dependency discussion concerning the national archetypes and that all health regions agreed before making such requested changes. Another important issue to address was how to define governance standards based on archetypes (such as schemes, scorings, clinical processes) and how to distribute these between system vendors using different clinical Information Models (CIM). NRUA and the regional archetype organizations contributes to this, but the government, the regional health authorities have to be active decision makers

IV. DISCUSSION

The four examples described in part III, all indicates a need for improving and speeding up the national archetype process in Norway. The main issues to discuss are 1) the question of consensus and/or the clinical value of archetypes: When to start using the archetypes? The balance between only using nationally approved archetypes and the need for speeding up the process by using unapproved archetypes to test the archetypes in a production environment (the EPR). 2) The importance of testing the archetypes for clinical practice.

A. The Temporal Evolvement of the Archetype Work

The archetype strategy of the North Norwegian Health Authority is to use only nationally consensus made archetypes. The overall goals are to secure high quality structured archetypes, in line with the national standard, confirming that the archetypes they include in the new EPR system are compatible with other archetypes in Norway. They adopted this strategy, due to the unknown consequences of using archetypes that has not reached national consensus. A project leader in FIKS stated: *"Some of the consequences we dread from using unapproved archetypes are the lack of interoperability, the need for converting data, loss of historical data, all leading to increasing cost."* However, they did not want to stop developing the new EPR system completely, due to the lack of archetypes. *"There is a risk that if we are too cautious with starting to use archetypes it will make our development set to provide excessive profit related to reuse of data and clinical parameters fall way behind (project leader FIKS)."*

One problem is that they started working with the EPR system functionality in 2012 and the national archetype work did not start until 2014. The reason for this was the notion the vendor had of importing a basic set of archetypes, and letting the system users develop the specific ones themselves. When the vendor defined a need for working nationally with the archetypes due to both their complexity and the extensive use of the same openEHR based EPR system in Norway the process of developing functionality had already reached a stage where they needed archetypes for the system. One example is the pre surgical planning module, the vendor started working on this in collaboration with system users in 2012, and still in May 2016, only six of the necessary archetypes were approved nationally. Thus, the strategy of waiting for nationally approved archetypes led to delays in the development process. At this point, the health region had to assess their strategy and consider using unapproved archetypes to prevent further delays. However, it is decided that all archetypes have to go through the national consensus process at some point, and before this has been complete, it is always a risk of having to change the archetype both technically and clinically after they are implemented, potentially costing the health region a lot of money.

In the Southern and Eastern Norway Regional Health Authority, they have nearly the opposite archetype strategy. Their criteria for using archetypes is that the archetype have started the consensus process. Thus, occasionally, (as in the small clinic described in section C)), they use archetypes not yet included in the consensus process, and even developed new ones themselves. Starting to use unapproved archetypes in a clinical production environment provides the vendor, the clinicians, and NRUA with important insights on the actual usability of the national archetypes, as well as for how to implement them in and use them for an EPR system. The development initiated in the local clinic was important for the maturity of NRUA as an organization, the archetype development, as well as for the EPR vendors. If they had waited for the regional EPR project to have all their archetypes approved, for example for the surgical pre-planning module, before testing out archetypes in clinical practice, they would still not have this valuable knowledge gained on how the archetypes actually work in a clinical setting. The experiences achieved from this project, made it conceivable to improve the structure and content of the national archetypes.

On the other hand, using archetypes for clinical practice before nationally approving them, leads to several challenges as described in section C) related to e.g., the structure of the archetypes and the interoperability of local and national archetypes. As an example, the local archetypes were versioned following the same standard NRUA used. This consequentially led to both local and a national archetype with the same version number. Firstly, if the possibility to create a new version disappears there is a risk of losing historical clinical data when converting to a consensus made archetype. Secondly, mixing up the local and the national archetypes might be a secondary problem, since the two definitely are comparable. Further, since the national archetypes are developed after the local ones are implemented, there is a risk of a dissimilar structure of the local and national archetypes modelling patterns. If the deviations are too extensive, the local and the national archetypes might not be compatible. One of the archetype reviewers with a technological background described this potential problem: *"The local system will continue to work on its own, but if the structure of the archetypes is changed extensively to enable national consensus, they will no longer be able to communicate with the old version of the archetypes. Consequentially a system based on the local archetypes, cannot communicate with systems using national archetypes."* Accordingly, this might lead to losing data, or having to spend an extensive amount of time and money on converting all existing data to the new national archetype format. This indicates a need for an improved overall structure for the national archetypes including a guiding manual for developing such, if there should be any local archetype development in the future.

When to start using the archetypes and the consequences of using them before the consensus process is complete are not questions that are only relevant for this health region. This question is important to address at a national level of Norwegian healthcare. If the archetypes needed for a project or a system are not available nationally, is it better to wait for the potentially time consuming consensus process or is it necessary to develop some archetypes locally e.g., in collaboration with the vendor of the EPR? In addition, if local archetypes are developed, how is it possible to ensure the quality of these compared to national ones? To solve these complex issues, instead of trying out the archetypes in local projects with potentially complicated and expensive consequences, another approach would be to create a test version, or a prototype of the new EPR system to test archetypes for clinical practice.

B. Future Development of the National Archetype Work

As the example in section D from the Western Norway Regional Health Authority describes, there is a need for emphasizing the interrelation between the archetypes and the systems using them. In addition, some of the lessons learned from the archetypes so far, and the need for defining the extent of the future work includes estimating how many archetypes to make, how much resources this work requires and how long establishing archetype standards for an EPR system will take.

First, the work with developing archetypes has over the last years enabled the system vendor and NRUA to approach each other, since they both have experienced that this are two interrelated processes that cannot fully be separated the way the two level model of openEHR describes. Testing out archetypes for clinical practice and working with developing the EPR system have enabled both organizations to gain evidence based knowledge and experience of this work. However, these two development processes have been unsynchronized from the start. Establishing a national archetype organization and the work with national archetypes started almost two years after the development of the new EPR system. The reason for this was mainly the original notion that archetypes was easy both to implement from other CKM's, and to develop by the system users themselves. After both National ICT and the vendor had worked with archetypes for a while, it became clear that establishing a Norwegian CKM was necessary to develop archetypes that fit the conditions of Norwegian healthcare. Since nobody had experience from working with archetype standards and openEHR based EPR systems in Norway, a maturity process where necessary for both organizations, before establishing an organized way of working with these issues. Working with archetypes and the system as separate processes provided a risk for challenges when trying to combine them. That brought on the question of how close the archetypes and the system needed to be. There was an example in Norway described in section D were the Western Norway Regional Health Authority requested a change in one archetype after trying to implement it to the EPR system. When discussing the need for changing the archetype the main question was how to relate to such changes, and if the archetypes should be adjusted to fit specific systems vendor's requirements, or if the vendors should rather adjust their systems to fit the archetype standards. OpenEHR archetypes are systems independent, which is one of the great advantages of using archetypes as standards for the clinical content of EPR systems. Therefore, it is not expedient to change the structure or content of the archetypes to fit one particular vendor's needs. Adjusting to the requirements of one vendor risk compromise the archetype and there is a possibility that they end up as system dependent standards that just fits with the needs and requirements of one particular vendor. The previous leader of NRUA stated *"National archetypes should not be modelled from limitations in the software of the vendor using them"* However, the archetypes in the Norwegian CKM are very new, and have mostly not been used for clinical settings. Implementing them to clinical environments might therefore require some changes due to the structure and content of the archetypes. In addition, if the Norwegian archetypes do not fit the requirements of the only large-scale EPR system conforming to openEHR architecture in Norway, who will then use them? In addition, in the national consensus work with archetypes, the EPR vendor is strongly involved in establishing the national standards influencing both their content and structure. It is however possible for all vendors to join the national archetype work and it might even be positive that the vendors participates of this process. This way they can follow and participate in the national discussions related to archetypes and contribute to detect challenges related to archetypes. They may also understand the concept and construction of archetypes better, and how to create their EPR system to fit the requirements of these standards.

Due to the archetypes and the EPR system being closely interrelated, there is a need for defining some borders between them and what part of archetype governance NRUA, and what part the vendor are responsible for related to establishing and changing archetypes. For example if one nationally approved archetype is changed after it has been implemented in the EPR system who is responsible for ensuring that all relevant data is interoperable, and all relevant systems, schemes and templates using this archetype gets updated? There are different degrees of changing archetypes. Minor problems related to for instance data storage only requires a new version of the archetype without compromising the interoperability between the new and the old version. A major change in the data structure will require a new archetype and this may not be compatible with the old one.

Another important question for planning the future work with archetypes is how to define the number of archetypes needed for the EPR system and how many would be necessary for creating a prototype of a basic EPR system? Are the 30-50 core archetypes defined by NRUA [28] a sufficient basis for covering 90% of the clinical content of an EPR system, and thereby create a general prototype of the system, or does this require a more extensive set of clinical standards? It seems to be diverging opinions related to this matter. This was illustrated on the national forum in May 2016 the leader of NRUA said that it is likely that 200 archetypes are enough to cover 80% of the EPR systems clinical content. This might be why NRUA has set a goal of nationally approving 200 archetypes by the end of 2016. The question is however if the archetypes approved by then are generic enough to actually cover 80% of the EPR content, or if the practice of working only with archetypes requested by the healthcare practice leads to ending up with working with less generic archetypes more specified for one particular use like the core journal or "the patient security program." It might be necessary for NRUA to adjust the current strategy, if the goal is to develop archetypes to cover as much of an EPR system as quickly as possible. At the same forum, a representative from Ocean informatics stated that between 1000-2000 archetypes are necessary to have a complete set of standards for an EPR system. Is there a need for 30-50 archetypes to cover 90% of the EPR, 200 to cover 80% of the EPR or 2000 to cover all of the EPR that is correct? There is a vast different between these estimates, and it seems rather uncertain at this moment what the total requirements of archetypes really are. The difference in the estimates makes planning the future work, and assessing the necessary resources for participating in the work very challenging. One of the reasons why this is difficult is probably the absence of archetypes used in clinical practice. According to this, establishing a prototype of a basic EPR system based on the core archetypes would be a good way of testing how much of the EPR content the core archetypes really cover. This way it may be easier to estimate the total number of archetypes required and plan the future work with archetypes in more detail due to the time and resources needed. The progress in the national archetype work have increased the competence of all the involved actors, this has been necessary for increasing the pace of this work. As the national archetype work progress, the hypothesis of core archetypes also evolves including more archetypes than first predicted covering less of the EPR system that expected. Realizing this is a result of the clinical and technical development processes approaching each other, NRUA being an important part of the vendor's development projects, and the vendor being a significant contributor to the archetype work and national consensus process. They have both gained experience and competence from trying out the archetypes for clinical practice. In the project in the Southern and Eastern Norway Regional Health Authority the vendor got to experience the challenges related to developing archetypes and they were able to test how well the system was equipped to implement archetypes. NRUA gained experience on how the archetypes fit clinical practice moving the development of archetypes from a theoretical to a practical level. The system users also experienced archetypes for real clinical use, helping them better understand the potential of archetypes and the new structured EPR system.

In addition to presenting how much of the clinical content the core archetypes cover, creating a prototype would be very useful for the clinicians, enabling them to see and try out archetypes in a practice like environment. This could enable them to better grasp the potential of the new EPR, and identify what the necessary requirements for continuing the development process are. Such prototype might also provide useful information for the vendor on how to include the archetypes technically in their new system, and gain knowledge that is missing today on how to create and import templates into the EPR system based on archetypes. However, this would require reaching national consensus on the defined core archetypes. The reason for not having prioritized the core archetypes for the consensus process were many: Establishing generic archetypes within an immature organization was time consuming, since there were no experienced archetype designers, neither clinicians nor technical personnel in Norway. Hence, trying and failing was part of the process. In addition, the requests for archetypes from different parts of the healthcare sector like the core journal project and the "patient security program" led to down prioritizing the more general core archetypes. NRUA was an organization with few persons, set to do most of this national archetype work and several issues like prioritizing their time, how many clinicians to include, how to structure the archetypes, whether to translate international archetypes, or establish new ones from scratch, what type of archetypes to prioritize, had to be considered. As a result of more and more focus on archetypes in the healthcare sector, the last two years of working with archetypes in Norway on different levels of healthcare, NRUA has gained the necessary level of competence to fulfill their role as an organization that coordinate the national work with archetypes, as well as supporting local initiatives. On the other hand, the lack of organizational and clinical resources in the governance organization is slowing down a positive development. Having such competent organization to coordinate the archetype work might necessitate reconsidering how to work with archetypes in Norway. In the example of implementing archetypes for clinical use described in part C, this project developed in collaboration with the system vendor 52 archetypes in 9 months. This illustrates that intensifying the archetype work by defining a set of archetypes requested for a smaller part of the health care service, and work systematically with these in a project is a way of speeding up the archetype development. Maybe a similar approach would have been beneficial for the pre surgical planning project, to complete the necessary archetypes for this module, or for completing the core archetypes, to enable developing a prototype of a basic EPR system. There is of course a difference in developing archetypes locally in collaboration with a system vendor and establishing nationally approved archetypes, however it do not necessarily need to be impossible to establish projects to work with national archetypes as well. This way there would be a set of clinicians and other necessary reviewers working with a particular set of archetypes as a part of their workday. This way reviewing archetypes could be included in the project preventing the reviewers from having to do this at their free time. Consequentially the group of clinicians involved would have been more dependable, however, including all of them in the project would either require a very large project or review processes with fewer participants with the risk of compromising the notion of a maximum dataset for the archetypes. NRUA is however currently working towards the goal of having 200 consensus made archetypes by the end of 2016. The question is however, if the archetypes they are working with are the core archetypes requested to fulfill the basics of an EPR system or if the development of archetypes so far includes too many specialized archetypes requested by part of the healthcare service.

V. Conclusion

Working with archetypes in Norway has been a gradually expanding process starting with a project in National ICT in 2008. The work gained national focus when the largest EPR vendor decided to develop an openEHR based EPR system based on archetypes as clinical standards. Since then, three of four health regions in Norway conformed to this system vendor, hence they decided to work with the OpenEHR standards to standardize clinical information at a national level. It has been challenging to increase the pace of the national archetype work in Norway due to several factors. It is however essential for the progress of the archetype work and the new EPR system that the development of archetypes continues to gain momentum. One important factor to accomplish this, is to develop a set of core archetypes to cover the basics of the clinical information in the EPR system. These archetypes are set to cover 80-90% of the EPR content leaving only the most specific ones out. To accomplish this there are several issues to address

First, how to organize the national archetype work in Norway, and the need for a solid organization to coordinate this work including hundreds of system users participating in the review process. Second, the relations between the archetype standards and the EPR system. There turned out to be a closer interconnection between the two levels of the openEHR model than first expected. It was difficult to develop the system before the archetypes were finished, as well as it was challenging to develop high standard archetypes without having a system to test them in, leaving the development at a theoretical level. Third, using the archetypes for clinical practice, and the balance between establishing national archetypes and developing local ones. To have some practical experience related to archetypes in clinical use is important for NRUA, the system vendor and for the system users to get an impression of how the new archetype based openEHR EPR system will be, and the potential of this new way of structuring clinical information. There also have to be a balance between only using nationally approved archetypes and the need for speeding up the process by using unapproved archetypes to test the archetypes in a production environment (use in the EPR). Fourth, how to coordinate and distribute the consensus based archetypes. This relates to versioning the archetypes, how to relate them to the existing national archetypes and the general complexity of developing archetypes.

A need for defining a set of core archetypes to develop a prototype of a basic EPR to solve some of these challenges have been addressed including the difficulties related to define the number of archetypes required for such prototype. Planning the future work with archetypes including defining how many archetypes an EPR system need for standardizing the clinical content, in relation to how many resources the work requires, and when it can be possible to have a complete set of archetypes, covering the necessary clinical content of the EPR system. Another important issue is how to distribute the governance of the archetypes between NRUA and system vendors using them in relation to establishing and changing archetypes to ensure that the archetypes keeps as high quality as possible.

Using the core archetype as a basis is a promising tool for future accomplishments in standardization, and speeding up the development of the new EPR. Still the process described, and the following maturation of the national environment has been a necessary process. Now, NRUA and the national consensus work have reached the required maturity level, to exploit the possibilities that constructing a prototype/model by using core archetypes can provide for the future work with archetypes, and the new EPR system in Norwegian healthcare. Establishing archetypes as standards for the clinical content of an EPR system is a way of establishing a national governance of clinical standards in Norway. It is important to share the responsibility for maintaining this between the 4 health authorities, since they all have to contribute to this work to reach the goal of archetypes being reusable standards enabling process and decision support as well as interoperability between the EPR systems and the rest of the ICT portfolio.

References

[1] G. H. Ulriksen and R. Pedersen, "Core Archetypes, The Means to Build Confidence Around the Power of Structured EPR Systems?" The Eighth International Conference on eHealth, Telemedicine, and Social Medicine (eTELEMED), April 2016, Venice.

[2] Norwegian Ministry of Health and Care Services: White Paper, Report 9, "One citizent one journal,"2012.

[3] National ICT: Action 41, Use of archetype methodology for the definition, availabiliy and use of clinical information models in health-based Information Systems, 2012.

[4] T. Meum, G. Ellingsen, E. Monteiro, G. Wangensteen, and H. Igesund, "The interplay between global standards and local practice in nursing," International journal of medical informatics vol. 82, no. 12, pp. 364-74, 2013.

[5] National ICT: Action 48, Clinical documentation for overview and learning, 2012.

[6] M. Aanestad and T. Jensen, "Building nation-wide information infrastructures in healthcare through modular implementation strategies," The Journal of Strategic Information Systems, vol. 20, no. 2, pp. 161-176, 2011.

[7] The office of the Auditor General: The office of the Auditor General's investigation of the ICT in hospitals and electronic interaction in health services, 2008.

[8] M.R. Santos, M.P. Bax, and D. Kalra, "Dealing with the archetypes development process for a regional EHR system," Appl Clin Inform, vol. 3, no. 3, pp. 258-275, 2012.

[9] C. Rinner et al., "Creating ISO/EN 13606 archetypes based on clinical information needs." Proceedings of EFMI Special Topic Conference" e-Health, Across Borders Without Boundaries; pp. 14-15, 2011.

[10] P. Gibbons et al., "Coming to terms: Scoping interoperability for health care," Health level seven, 2007.

[11] S. Garde, P. Knaup, E. Hovenga, and S. Heard, "Towards semantic interoperability for electronic health records," Methods Inf Med, vol. 46, no. 3, pp. 332-343, 2007.

[12] S. Timmermans and M. Berg, The Gold Standard: The challenge of evidence based medicine and standardization in health care. Philadelphia, Temple University Press, 2003.

[13] S. Timmermans and M. Berg, "Standardization in Action: Achieving Local Universality through Medical Protocols," Social Studies of Science, vol. 27, no. 2, pp. 273-305, 1997.

[14] R. Pedersen and G. Ellingsen, "Standardizing nursing work: works in practice but not in theory?" Studies in health technology and informatics, vol. 157, pp. 91-96, 2010.

[15] G. Ellingsen, E. Monteiro, and G. Munkvold, "Standardization of work: Co-constructive practice," The Information Society, vol. 23, no. 5, pp. 309–26, 2007.

[16] R. Chen, P. Georgii-Hemming, and H. Åhlfeldt, "Representing a Chemotherapy Guideline Using openEHR

and Rules," Studies in Health Technology and Informatics, vol. 150, pp. 653–657, 2009.

[17] R. Chen et al., "An archetype-based testing framework," Studies in health technology and informatics, vol. 136, pp. 401-406, 2008.

[18] T. Beale and S. Heard, "Open EHR Release 1.0.2," Architecture Overview. London, 2008.

[19] L. Bird, A. Goodchild, and Z. Tun, "Experiences with a Two-Level Modelling Approach to Electronic Health Records," Journal of Research and Practice in Information Technology, vol. 35, no. 2, pp. 121-138, 2003.

[20] National ICT: *Welcome to national ICT*. [Online] Available from:http://www.nasjonalikt.no/no/om_nasjonal_ikt/ 2011 accessed 2016.02.01.

[21] H. K. Klein and M. D. Myers, "A set of principles for conducting and evaluating interpretive field studies in information systems," MIS Quarterly, vol. 23, no. 1, pp. 67-94, 1999.

[22] G. Walsham, "Interpretiv case studies in IS research: nature and methode," European Journal of information systems, vol. 4, pp. 74-81, 1995.

[23] DIPS-ASA: *About us*, Available from: http://dips.com/eng/about-us.2015, accessed: 2015.10.16.

[24] B. Christensen and G. Ellingsen, editors, "Standardizing Clinical Pathways for Surgery Patients through ICT," European Workshop on Practical Aspects of Health Informatics (PAHI), Edinburgh, Scotland. http://ceur-ws.org http://ceur-ws.org. 2013.

[25] ISO: *Health informatics - Electronic health record communication, Part 1 Reference model CEN*. Available from: http://www.iso.org/iso/catalogue_detail.htm?csnumber=40784.European comitte for standardization, 2007, accessed: 2015.12.07.

[26] A. Moreno-Conde et al., "Clinical information modeling processes for semantic interoperability of electronic health records: systematic review and inductive analysis," Journal of the American Medical Informatics Association, pp. 925-934, 2015.

[27] M. Marcos, JA. Maldonado, B. Martínez-Salvador, D. Boscá, and M. Robles, "Interoperability of clinical decision-support systems and electronic health records using archetypes: a case study in clinical trial eligibility," Journal of biomedical informatics, vol. 46, no. 4, pp. 676-689, 2013.

[28] National editorial board for archetypes (NRUA): *Core archetypes*. [Online]. Available from: http://wiki.arketyper.no/wiki/Kjernearketyper, 2015, accessed: 2015.12.09.

[29] E. Hovenga, S. Garde, T. Carr, and C. Hullin, "Innovative approaches and processes for capturing expert aged care knowledge for multiple purposes," Electronic Journal of Health Informatics, vol. 2, no. 1, 2007.

[30] H. Leslie and S. Heard, Archetypes 101, Ocean Informatics 2006.

[31] D. Ingram, *Origins of openEHR*, [Online]. Available from: http://openehr.org/about/origins, 2002, accessed: 2016.09.05.

Student Views on Academic Reading and its Future in the Design and Engineering Disciplines

Kimberly Anne Sheen and Yan Luximon
School of Design
The Hong Kong Polytechnic University
Hong Kong, SAR
e-mails: Kimberly.Sheen@connect.polyu.hk
yan.luximon@polyu.edu.hk

Abstract - **Electronic textbooks are becoming a common educational tool, but there is little research on the student desires, which will affect the effectiveness of this tool. This paper aims to add to the current research by outlining students' reading habits in physical and electronic textbooks and identifying what students feel they need to study using future electronic textbooks. This paper describes a series of focus groups with a total of thirty design and engineering students. Findings illustrated the different ways in which these disciplines approach their academic readings and that future electronic textbooks require some discipline specific components. There were some similarities in views and ideas, such as being able to insert their own images into the textbooks and the desire for less text and more interactive components to facilitate their learning. Identifying design criteria based on discipline needs and including student input based on their task needs will assist in designing future electronic textbooks that will meet academic reading requirements.**

***Keywords** - focus group; electronic textbooks; academic reading; design education; engineering education.*

I. Introduction

Electronic textbooks are becoming more prevalent in higher education. Still, students are not as excited about this trend as many universities. While 60% of students reported using electronic textbooks during their academic studies, with half being required to by their instructors, student preference for physical textbooks has not waned [1,2]. In fact, many studies have shown that student preferences of some components, such as search functions and long blocks of text, negatively impacted student's opinions on electronic textbooks [1,3].

While electronic textbooks are starting to evolve past simple Portable Document Format (PDF) representations of the physical text, they are in their infancy. It has been individual schools creating their own interactive electronic textbooks, which are shifting away from the textbook metaphor [4] and creating this evolution. This shift from the textbook metaphor will allow for additional materials and components, which will enhance and assist with the reading task [5]. Yet, creating this type of electronic textbook for individual courses is time consuming and impractical on a larger scale due to the number of courses offered worldwide and the ever changing course material. On the other hand, electronic textbooks coming from major publishers do not tend to use diverse components that would be more suitable for the disciplines they serve and instead use components that would be appropriate for all areas of study. Although different disciplines are known to approach their education in different ways [6], it is still in broad practice to create this type of electronic textbook. Regardless of creating electronic textbooks specifically tailored for courses or broader textbooks, there is still the challenge of selecting and creating new supplemental materials and components for this new type of electronic textbook [7].

Not only would academics and publishers find creating new content difficult, advancing technology and the use of electronic textbooks may have altered the ways in which students use textbooks. Students can now easily read in cafes or while travelling [8], moving away from the desks and tables that used to confine students. Being able to study in more locations may seem positive, but without normal study aides such as highlighters and notebooks, students may find themselves slipping from the deep reading required during revision, which allows for in-depth comprehension and recall [9] to surface reading, which provides students with a more limited understanding of the materials [10]. While some components included in current electronic textbooks seem similar to the support activities students employ during reading, they are noticeably different. For example, many students take notes in the margins of their physical textbooks to support their studying. While electronic textbooks commonly offer notation software, notes are typically not displayed on the screen and require clicking on a small icon to later revisit (see Fig. 1). This could cause the students to miss their notes or interrupt their reading process leading them to become distracted. In fact, electronic annotation software is used less often than traditional note taking done with a

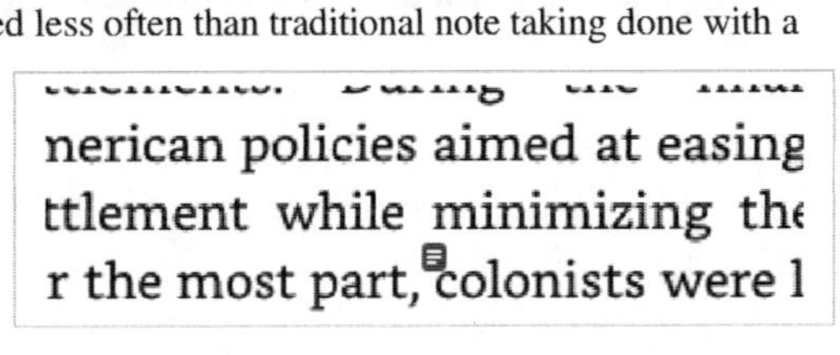
nerican policies aimed at easing
ttlement while minimizing the
r the most part, colonists were l

Figure 1. Example of note taken in a Kindle electronic textbook [15].

physical textbook [11]. The lack of tangibility associated with electronic textbooks also negatively affects the reading task [12]. Past research has stated that electronic textbooks should enhance current physical active reading activities while presenting an interface that is easy to use [13].

Currently, the components that are used to support academic reading are being designed for future electronic textbooks with limited understanding of the students and their habits [14]. While educators should be designing the content within electronic textbooks, students are the central user of electronic textbooks and know their own study behaviors and what additional tools they need to feel comfortable with the material. This could lead to new textbooks not being able to fully support students' study habits and not only failing the student but becoming something that is looked on with distain.

The purpose of the focus groups outlined in this paper was to identify components that are important to students during their studies from the task requirements reported by students, something that limited in the current discussion. Since each discipline has different approaches to studying and different needs, focus groups were separated based on the two disciplines studied: engineering and design. This allows for a better understanding of how these groups of students approach their studies. It also assists in identifying what type of supplemental content needs to be created and what tools need to be included to support academic reading in these different disciplines. This paper also aims to bring a deeper understanding to the data from an earlier survey released at The Hong Kong Polytechnic University [16]. It also provides insights into how students complete their academic readings in physical and electronic mediums and how they envision their future electronic textbooks based on their discipline specific needs. The rest of this paper is organized the following way. Section II describes the method employed in this paper. Section III presents the results of the focus groups. Section IV discusses the results within the literature and in a more general context. Section V presents the main conclusions and presents some future areas that should be explored.

II. METHOD

Focus group method was chosen to uncover current and future student needs and approaches related to academic reading, and was used identify task related design criteria for future electronic textbooks [7]. The focus group method allows for internal validity, a better understanding of the phenomenon that would not be possible through methods that use quantitative analysis, and assists in understanding truly complex issues [17], which are necessary in this type of research.

A. Participants

Students were recruited from The Hong Kong Polytechnic University. There were two requirements for participation. The first, students need to be enrolled in either an undergraduate level design or engineering program. The second, the students needed to have prior experience using electronic textbooks during their academic studies. Once students volunteered for participation, they were placed into three person focus groups made up of participants only from their discipline. While the disciplines of the students remained the same, the different programs within that discipline were mixed. For example, computer science, electronic engineering, and product engineering falls within the Engineering Discipline at the university, so all of those groups of students were included in engineering focus groups. Design students' programs also varied with students from programs such as product design, communication design, and multimedia design. Overall, five focus group sessions from the design programs and five from the engineering programs were conducted. Thus a total of thirty students participated in these focus groups. After three focus group sessions, homogeneity was reached [19, 20, 21] but sessions continued for two more focus groups per discipline so that findings would be more significant. Total student participants were 16 males and 14 females aged between 18 and 23. Engineering focus groups consisted of a total of 11 males and 4 females. While design focus groups consisted of 5 males and 10 females. The increased number of males in the engineering department and increased number of females in the design department reflect the distribution of genders in these faculties with design disciplines being more female heavy and engineering being more male heavy.

B. Session Design

Each focus group session was designed to last approximately one hour. The sessions were made up of sixteen semi-structured interview questions, which were followed up with unscripted questions related to the answers. Based on the similarities between the answers, many follow-up questions were the comparable. Students also participated in two activities during the session. The first activity asked them to express how they define current electronic textbooks. The second activity asked them to envision their future electronic textbooks, without considering the limitations of current technology. In this activity, students were asked to include components they wanted in their discipline specific electronic textbooks and then asked questions about how they would interact with these new textbooks. During both activities, students were given markers and paper and allowed to complete them with little oversight from the moderator. Each session was audio taped and later transcribed. The papers from the activities were kept for analysis and examples appear later in this article.

C. Data Analysis

Once each session was transcribed, the data was coded. The codes used in this research were grounded in the data [18] and used to organize the data into recurring topics and subtopics for easier analysis, description of the results and development of theory. Some of the codes, which emerged from the data are as follows: task requirements, technical requirements, preference, technical issues, ergonomics issues, and habits.

III. RESULTS

The semi-structured interview questions investigated the habits, task requirements, and preferences of students in regards to textbooks. The questions were broken up into three segments: one on physical textbooks, the next on electronic textbooks, and finally future electronic textbooks. The same questions were used for both design and engineering focus groups, although follow up questions differed slightly based on the responses given by students. During the future electronic textbook segment, students were also asked for feedback on ranking data gained from an earlier survey [16]. Two activities were also completed by students, one during the electronic textbook segment and one at the end of the future electronic textbook section, which wrapped up the focus group sessions.

A. *Physical Textbooks*

The questions regarding physical textbooks mostly dealt with student habits regarding physical textbook reading. Habits and preferences are diverse for many reasons; however, trends did emerge when analyzing the full transcriptions of the focus groups. When design students were asked about the frequency of their use of physical textbooks, answers ranged from 20% to 90% of their time reading. However, these initial responses are deceiving. Students later admitted during the sessions to underestimating their use of physical textbooks because they frequently printed out the electronic versions and initially included them in their estimates of electronic textbook usage. Overall, design students' usage of electronic textbooks was much closer to the higher percentage. Most often, these students reported to completing their academic readings in the physical form while at a desk at home in the morning before lectures or late night. Other locations design students reported competing their physical academic readings were home on the sofa, while traveling, and in the classroom. The majority of design students did not wish to read while traveling. When asked to expand on this, students reported the issue of dizziness as the main cause of their decision, although they also reported a dislike of carrying heavy books or a large amount of papers influencing the decision. They reported using the textbooks as the main source of learning concepts and reported looking past the required readings to find other resources regarding the concepts.

On average, engineering students reported using physical textbooks less than 50% of their time while doing academic readings, although two participants out of 15 claimed to use them almost 80% of the time and two state that they would go out of their way to use electronic textbooks as much as possible. How often engineering students did academic reading varied from only during exam times to one hour per day in the afternoons and evenings. This large discrepancy translated into their average time spent with a physical textbook. Those who reported only reading for revision would spend upwards of five hours reading per instance over the entire day. Most often, engineering students believed that reading should be done when it was required and not necessarily to look at concepts outside of what they are taught. Engineering students placed high emphasis on quiet when completing their studies. Those that had a quiet home environment reported to working at their desks at home, while the rest believed that the school library was the ideal quiet environment to keep them away from distractions. No engineering students reported using physical textbooks while travelling. All engineering students reported that the main purpose of their academic reading was to review what they had learned during the lecture and if required reading for their homework assignments.

Investigation into the task requirements of academic reading in a physical textbook was undertaken as a part of these focus groups. Students were questioned about what types of supporting activities they did during physical textbook reading to help them comprehend and engage with the material. Design students reported different supporting activities such as summarizing important points from the text into lists, highlighting, and searching for more information by keywords. These students make notes in the margins of the text, or if on a separate piece of paper, they attach it to the original text. They reported to using the margins of the text when the book was their own, whereas if it was a library book, they would use post-its or other paper. Most students reported that their notetaking was more visual in nature and included things like sketches and timelines. When searching for more information or other resources regarding the concepts, students reported using Google. Similarly, engineering students reported taking notes in the margins, underlining, highlighting, looking over drafts from class, and looking up definitions in the dictionary. Engineering students also reported doing practice exercises, something that based on the requirements of their discipline were novel.

Students also reported some ergonomics issues and other considerations when deciding to use physical textbooks. Both groups of students reported that physical textbooks are very difficult to carry around and hold in their hands. The expense of physical textbooks when compared to electronic textbooks was also a recurring topic. Yet, they believe that physical textbooks are not only much more convenient to take notes in, but they also support the more visual type of notetaking (see Fig. 2) that they require, which then assists in their comprehension and recall of the material. In addition, students reported the impression that they were reading more deeply and remembered the information more easily because they avoided distractions afforded by electronic devices, such as the constant connection to the Internet. There was also a sense of accomplishment when it came to finishing physical page.

Figure 2. Example of visual notes that a design student made.

B. Electronic Textbooks

1) Definition

Before answering questions similar to those asked during the physical textbook segment, each focus group was asked to complete an activity in which they defined the term electronic textbook. The five Design student focus groups defined electronic textbooks in the following ways:

1. "A tool for learning without physical barriers. It contains lots of text, with additional elements including pictures, audio, and video."
2. "A gadget that allows us to learn wherever we are."
3. "A portable smart device, which is eco-friendly and able to store varied books with internet support."
4. "A digital content that allows easy access by different media and can be easily modified and shared."
5. "Allows a user to read through electronic devices (examples: computer, tablet, & phone), which provides more interactions and information by images, notes, which is more interesting, attractive, and convenient than traditional printed textbooks."

While the definitions themselves vary, they give insight into what the students view as the most important aspects of electronic textbooks. Students placed emphasis on the portability and diversity of the devices. The components of the aspects of the textbook itself is more limited and described simply as books or text with a few other aspects defined.

During this process, they also highlighted several components as important to their current electronic textbooks such as text, animations, images, video, dictionaries, and infographics. Text was considered especially vital to the electronic textbook as students felt that without text, the textbook loses its main purpose. They also highlighted some ways that electronic textbooks have enriched their learning experience such as facilitating communication, increased mobility, and increased interaction between the reader and the text.

The five Engineering student focus groups defined electronic textbooks in the following ways:

1. "A portable device, which includes all notes or text, video, and pictures into one appliance. It is cheap, environmentally friendly, and convenient when comparing to the physical textbook."
2. "A textbook, which does not print out on paper physically, but can be viewed and edited via electronic device like computer, phone, tablet. It has basic features as physical textbook and advanced features such as video, audio, tests, and animation."
3. "A softcopy that provides us useful content academically."
4. "A textbook in a soft copy version. It's the same as a physical textbook."
5. "A non-physical reading material, displayed by electronic devices. The reading experience depends on the user interface of the software."

Once again, definitions differed between groups, but similarities emerged. The groups believed that the electronic textbook was very similar to its physical counterpart, however many groups defined it as including additional advanced features as well.

Engineering students placed value on the electronic textbook's ability to search for keywords and additional components such as animations, video, and images that help facilitate their learning. They believe that the main purpose of electronic textbooks is to help students revise concepts they've learned in the classroom.

2) Usage

The questions regarding electronic textbooks mostly dealt with student habits regarding electronic textbook reading. Overall, design students reported that they spent significantly less time reading in electronic textbooks. When students did report reading in electronic textbooks frequently, they qualified that they were doing a physical reading from the electronic form. This happened most often when the readings were more than just a few pages. More flexibility was reported when reading with electronic textbooks, yet design students still reported reading most often in the classroom at their desk during the lecture. The reason they reported doing so was so that they may better understand the concepts that the professor is discussing. They reported that the average time they spend with an electronic textbook at times increases to usage during the entire day if they are working on a project. The majority of the time they access electronic textbooks they will use laptops, but if they had access to tablets they would do the reading on that device. They will use the phone if they need to do a short reading and they feel the convenience outweighs the limitations such as discomfort during reading and dizziness. They reported the preference for laptops was to avoid eye fatigue and also that when they wanted to save pages or chapters; there is more storage space in their computers than on their phones. Engineering students reported that they spent on average less than half of their time reading in electronic textbooks at home or while travelling in the afternoon and evenings. The increase in reported reading during traveling was because of the convenience electronic textbooks afford. The students access their electronic textbooks on laptop computers most of the time with only a few tablet owners reporting reading on that device. Phones were considered an extreme option and only used while travelling or when absolutely necessary for revision immediately before an exam.

Investigation into the task requirements of academic reading in an electronic textbook was undertaken as a part of these focus groups. Students were questioned about what types of supporting activities they completed during their electronic textbook reading to help facilitate their comprehension of the material. Design students reported using highlighting tools, music to help them focus, and Microsoft Word or the comment function to take notes. While design students reported taking notes while reading electronic textbooks, they reported taking less notes than when using physical textbooks. Reasons for the limited notetaking were reported as difficulties with the annotation components and the fact that the components do not support the more visual types of notes, which they feel better facilitate their learning than text only notes. Engineering students reported using built in encyclopedia functions, dictionaries, search functions, highlighting, and screen capture functions most often. When they did take notes, they reported to either hand writing them or putting them in a separate Word document because they would not refer back to the textbook later. Several students stated that they do not take any notes when they move to an electronic textbook because of the inconvenience caused by the medium.

Even though students were not explicitly asked about physical and cognitive ergonomics issues related to electronic textbooks, both engineering students and design students brought this subject up. Both groups cited eye fatigue as a major concern associated with the use of electronic textbooks, so students prefer regulating electronic textbook usage to very short readings. One student described the situation succinctly, "If I need to read a long article, for example 20 pages, I would print it out instead of looking at the monitor. But if I only read for just one or two pages, I then will just read it on the monitor." The eye fatigue would, in turn, caused what students described as dizziness or issues reading the text closely. Many students reported skipping lines while reading or reading the content that was only based on the exact concept they need to understand and not complete the full reading. Design students also discussed how they would rather print long readings instead of viewing them online to facilitate their learning, believing that the addition of too many components may destroy their creativity. While engineering students stressed electronic textbooks were easier to carry and allowed for more mobility when completing their readings. They also reported to printing any long electronic readings they may have to complete.

Students also reported several technical issues and other aspects, which influence their interaction with electronic textbooks. Design students repeatedly reported the battery on their mobile devices as negatively impacting their academic reading. They also reported the time it takes to scroll through the text as a hindrance to their reading. Finally, they complained about the small size of the text and described how it made reading more difficult. Both groups of students also discussed how the ease of sharing and downloading electronic textbooks facilitated their learning. Also the usage of electronic textbooks allowed them to avoid the inconvenience of going to the library, identifying the call number, finding the book, waiting in lines, and then carrying it with them, which was a common complaint for both groups of design and engineering students. Accessing the texts online minimized the time it took for students to be able to begin their readings. The ability to quickly and easily go from one text to another was another reported convenience to electronic textbooks. In line with this, students reported that it was easier to identify new resources based on keyword searching. When they were able to identify a core concept they needed to learn, they would type it into the library website or Google to find more resources that referenced that concept. And while students discussed their dislike of reading electronic textbooks on their phones, they reported the positive affect on their time management. Students stated that using electronic textbooks on their smartphones allowed them to read in bed, read when they had spare time while waiting for friends, or read while stuck in unexpected traffic. In addition, students reported that the ability to take digital notes makes them less likely to lose said notes. Some students reported taking pictures of their physical notes to avoid this, while other students only took screenshots of pages or sections of the textbook that they though would be valuable to them later. Students also discussed how cost, mobility, and environmental friendliness made using electronic textbooks more desirable.

While many of these technical advances were reported to have a positive influence on academic reading and resulted in some positive perceptions, students reported many issues. When taking notes, students felt that typing instead of writing made it more difficult to remember and digest the concepts they needed to learn. Engineering students also wished for the ability to draw or write manually in their electronic textbooks, but reported that the current technology that allows these actions are buggy and slow making them unusable. The search functions that students found exceptionally helpful, they also reported as harmful to their reading. Students from both disciplines stated that they missed information when they tried to quickly complete their readings to avoid eye fatigue by searching for and only reading the sentences regarding the required concepts. Both groups of students felt that this negatively impacted their understanding of the material as a whole and put them at a disadvantage. Another major issue that came up with every student regarding electronic textbooks was distractions. Notifications from social media and messaging applications were reported as a major issue, which hindered focus during their academic reading sessions. Students also found that they lost time and focus when searching for keywords they found within their books online; they reported finding themselves playing online games or watching hours of YouTube videos simply by switching to their browser.

C. *Future Electronic Textbooks*

The future of electronic textbooks was investigated in many ways. Overall, design students reported that they would be more likely to use electronic textbooks if they were more interactive. They also desire more features such as accurate text to speech, improved bookmarks that used a sentence or word to mark a place, manipulatable images, and improved responses from the technology when attempting to select or highlight text. Design students also reported a desire for improved text displays, which would reduce eye fatigue, such as e-ink technology or the ability to select the colors and contrast between text and background based on individual preference. Engineering students also agreed that they would be more likely to use electronic textbooks that were more interactive. They believed that this type of electronic textbook would facilitate their learning, speed up their work progress, and make them more efficient students. They wanted less text and more components such as 3D and manipulatable pictures and videos to help illuminate the concepts. The majority of the groups suggested ways of doing this that are not feasible with the current technology available commercially. Frequently both disciplines requested holographs or projection systems for the images so that they would be able to interact with them in what they described as a more interesting or detailed way. These students also placed large emphasis on better annotation tools. They felt that a more natural input for annotation would help facilitate their learning of the materials and if they could write with their finger or a pen and have the information be recorded within the electronic textbook that it would be ideal for their learning experience. Both disciplines thought that electronic textbooks would be improved by shorter blocks of text. Some students even believed that simple summaries of the main concept would assist them during their academic readings.

When students were presented with information regarding the answers from the previous survey, design students agreed that the top five components chosen were appropriate (see Table I). They believed that text was more vital to the learning experience than students in the survey rated it, but agreed that the readings they have to read are diverse and that a lot of it seems unimportant to them, which could have influenced the ranking. Students reported that multimedia ranking first was understandable based on their discipline but thought that the importance of the information from the text should not be subverted. Design students also reported that the findings of the undesirable components from the survey were valid (see Table II).

TABLE I. COMPONENTS DESIRED BY STUDENTS

Rank	Desired Components	
	Design Students	*Engineering Students*
1	Multimedia	Text
2	Bookmarks	Highlighting Tool
3	Highlighting	Multimedia
4	Text	Bookmarks
5	Translation, Dictionary, and Encyclopedia	Annotation

While students liked the idea of speech to text in their electronic textbooks, they eventually decided that the benefits of the tool were not appropriate for electronic textbooks because the note taking required for academic reading requires more thought than afforded by speech to text tools. Only one group felt that link to experts for answers to questions should be included in electronic textbooks. These students felt that this tool could combat the limited amount of time they have with their course tutors. The other four design focus groups did not believe this component was necessary at all. Engineering students thought survey respondents had overestimated the importance of text and underestimated components such as 3D images. The student participants felt that these were the core features that should be included in future electronic textbooks and that they are in line with the traditional conventions that are already in place. They believed that this type of response was because respondents chose components they were more familiar with and could envision. Other than that, students believed the other components chosen as desirable and undesirable were valid. They especially agreed with the inclusion of a time management system as an undesirable component as they felt it would cause added unnecessary pressure to their reading experience.

TABLE II. COMPONENTS UNDESIRED BY STUDENTS

Rank	Undesired Components	
	Design Students	*Engineering Students*
1	Hide Unimportant Aspects	Hide Unimportant Aspects
2	Speech to Text	Time Management System
3	Time Management System	Speech to Text
4	Link to Experts	Text to Speech
5	Text to Speech	Project or Print Annotations

After this general information was gathered, students were asked to complete the final activity in which they were given free rein to create the perfect representation of an electronic textbook for their discipline. As this was without the constraints of current technology, many of the solutions students presented would not be fully functional at this time. Design students produced results that were more visual in nature. The majority of focus groups provided sketches of their visions of future electronic textbooks, keeping notes on functionality and features surrounding the sketch while the other groups provided more descriptions on the functions with sketches supporting those (see Fig. 3).

Their electronic textbooks often took inspiration from applications such as Adobe Illustrator's interface and included the ability to add notes or photos directly inline, shorten forms of the text with emphasis on important concepts rather than of paragraphs, adjustable line spacing and text size, a table of contents, video, audio, adjustable images, bookmarks, the ability to synchronize across devices, translations, a dictionary, and an encyclopedia. They felt that highlighting and annotation tools would no longer hold as much importance future electronic textbook because there would be much less text but still included them. Yet, both groups of students felt they were still vital to the learning experience and included them. They did stipulate that the current rigid structure of these two components were no longer acceptable. Highlighting needed to be more free form and easier for students to accomplish, whereas annotation tools needed to have a better physical input. Typing notes into the annotation tool was considered to be a hindrance in learning the material. Students felt that handwriting better suited their notetaking styles. They reported that this type of notetaking would allow them to draw their own pictures or create lists in bullet point forms to better recall and comprehend the materials. Students also often built in the ability to hide unimportant content automatically by extending the text by clicking on the bullet point text. Many similar components appeared in the engineering future electronic textbook, yet the representation all groups of engineering students chose to convey their textbook was a list form (see Fig. 4). This electronic textbook also relied on less text, but included some discipline specific aspects like interactive equations.

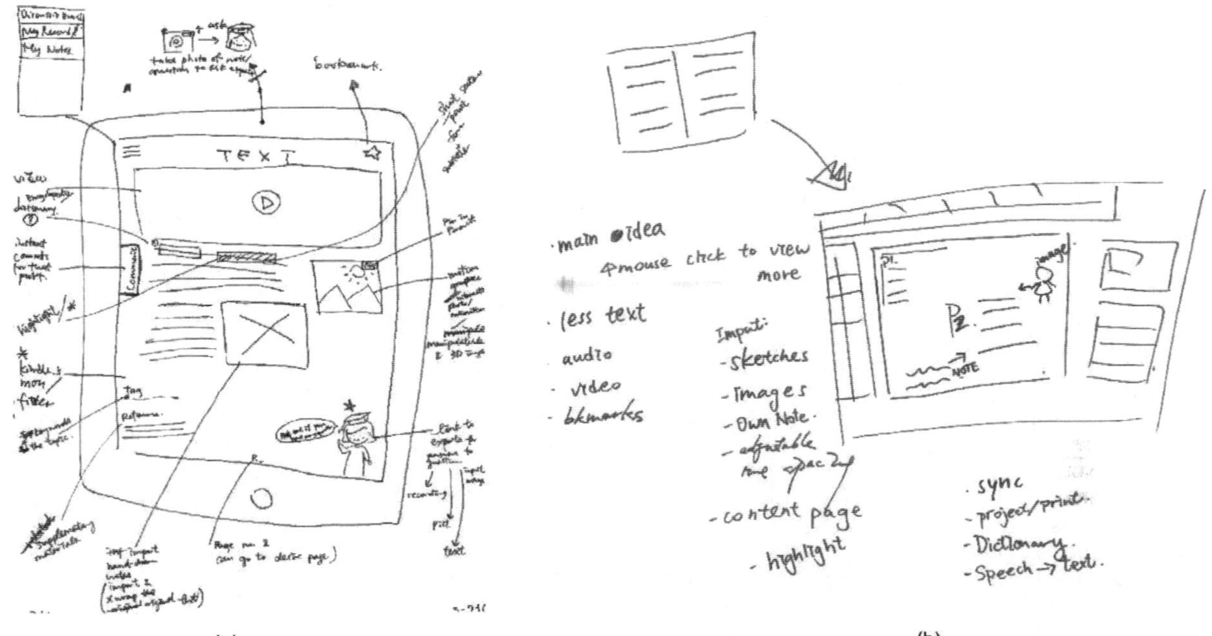

(a) (b)

Figure 3. Two examples of the future electronic textbook by Design Students.

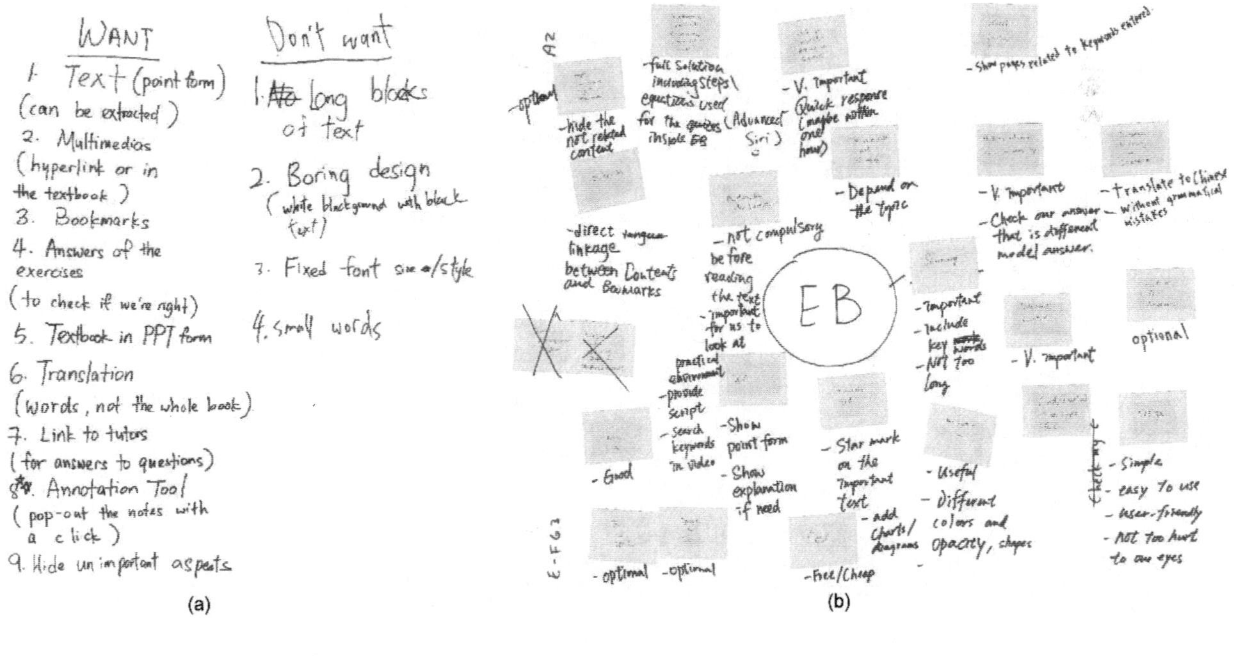

(a) (b)

Figure 4. Two examples of the future electronic textbook by Engineering Students.

IV. Discussion

This section discusses the findings from the focus group sessions as a whole and in relation to past literature.

A. *Student Usage*

Student usage of physical and electronic textbooks differed in both disciplines in all aspects of use. However, both disciplines of students felt that current electronic textbooks did not meet their needs as well as the physical textbooks did. They described usage of electronic textbooks as something that was encouraged by peers, faculty, or necessitated by circumstance. Faculty support of the use of electronic textbooks for their courses has been found to generally increase student usage of electronic textbooks [22]. The mobility offered to students by electronic textbooks do change where and when they do their studies, something that students did describe as a convenience that outweighed many of the limitations of electronic textbooks. In line with previous studies, even with this ease of downloading and mobility, students still reported that they preferred physical textbooks [2]. In addition, even with the increase mobility, students tended to access their electronic textbooks on the more cumbersome technology that is harder to use while mobile, which is in line with previous literature that found that users of stationary computers such as desktops were more likely to have experienced reading with an electronic textbook [22]. Similar to what past research has uncovered about this phenomenon, students reported not wanting to read long blocks of text in an electronic textbook [3] and that feelings of nostalgia [23] make it difficult for them to adapting to the new medium. Nearly all of the focus group participants reported that they would print out long readings, rather than printing them on the screen. If printing of the materials was not a function that was built into the electronic textbook, students would go so far as to find a work around. Students discussed going out of their way to find copies of the textbooks that lacked Digital Rights Management (DRM) restrictions and even to taking screenshots of the pages and later printing them. Printing out pages from electronic textbooks allows for students to continue to experience the four affordances of spatial flexibility, manipulability, tangibility, and tailorability, which students are nostalgic about in regard to print textbooks [12].

Supporting activities also changed for many students. They found themselves taking notes less, several going so far as to report no longer engaging in any supporting activities, and, as past research has found, they were becoming frustrated with built in functions such as bookmarking, highlighting, and annotation tools [11]. Repeatedly, students reported taking notes in the physical form was easier and allowed them to see their notes with the concepts, which later assisted in revising the material. Those that took electronic notes would do so in a Word document so that they may include outside material along with their summaries of important concepts, such as pictures or links to other reference material. They then reported that they would not go back to the textbook where their notes could be taken in context of the larger material.

In addition to the change in supporting activities, the addition of the inherent distractions during reading of electronic textbooks is a serious issue that needs to be addressed. Students stressed that certain components added to enhance electronic textbooks or the simple act of switching to a browser to search for a keyword adds time on to the total reading experience and that past research has informed us will interrupt their deep reading and overall comprehension of the materials [24]. By investigating current use of both types of textbooks, the differences in usage, issues that may arise and understanding the reasoning behind the usage design recommendations, such as shortening blocks of text and finding opportunities to incorporate aspects reminiscent of the four affordances, such as the ability to see notes on the page instead of hidden within an icon, can be made for future electronic textbooks. Also this type of comparative investigation allows for an understanding of technical and ergonomic issues, which emerge from the shift in mediums that can then be designed to avoid.

B. *Future Textbooks*

Student preference for design attributes of electronic textbooks was similar in both disciplines of design and engineering, yet several components differed. Overall, all students agreed that text should be limited to the most important information presented in a shortened paragraph or bullet point form. More information could then be accessed through hovering over the text or similar interaction. Students also felt that creating textbooks that were more interactive would facilitate their learning and allow them to truly understand and engage with the material. Based on student responses, making this type of change would rectify the shift in reading style away from what scholars identify as surface reading back to deep reading [9, 10], which past research has proven necessary for succeeding academically. While these reported changes may make electronic textbooks more appropriate for the type of reading required, previously reported interaction may have been influenced by current ideas of electronic textbooks like the students in the focus groups reported with the past survey results [16]. In addition, student enthusiasm for these components may later wane, but previous studies show that should do little to the effectiveness of the components [25].

Because of the issues associated with students' dislike of long blocks of text and subsequent effect on reading quality, it is recommended that designers incorporate short blocks of text [26] with extended information hidden. The loss of information in long form can be supplemented with components such as multimedia or other engaging components. Although limiting the text may make the information easier to students to digest and read, changes still need to be made regarding the supporting tasks. The common request for electronic textbooks to include a more natural input method for notetaking, which would be closer to handwriting has been already implemented in some e-reading applications such as Evernote (see Fig. 5), this technology is still reported to be cumbersome and not available in many of the applications students use during their academic reading. Based on student feedback during the focus group sessions, more advanced and user friendly versions of this component would be well received and assist in encouraging students to use electronic textbooks. Many of the students also reported desiring a stylus to take notes. While this may make taking notes more reminiscent of taking notes with physical pen and paper, it may create additional complications, which were not previously present in electronic textbooks. Examples of this would be creating a situation where the stylus has to be replaced when a student misplaces them, causing an additional expense or making taking notes when completing academic reading whilst traveling more difficult than it was previously reported.

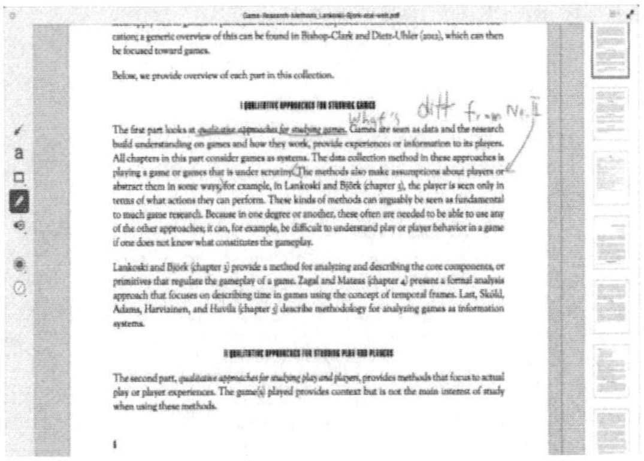

Figure 5. Example of notes taken in an electronic text in Evernote.

C. Comparison of Disciplines

While there were many similarities in responses on the components and format in which future electronic textbooks were presented from both engineering students and design students, there were some fundamental differences. One of these differences was highlighted during the second activity in which it became apparent that while similar requirements may be requested, the ways in which students think and interact with each other and academic materials are different. Design students felt comfortable creating a visual representation of what they thought their perfect discipline specific future electronic textbook was and worked together from the start to create their ultimate proposal, each of them adding to the proposed textbook as they saw fit. This can be associated with the nature of design being undertaken as a team project, especially as taught at this university. On the contrary, engineering students presented their answers in a list form and instead of compromising and discussing opinions during the creation process, waited until after their individual lists were made to try and unify their answers. They also requested that one student write the final list and would only switch designated writers if they felt they did not support the inclusion of a component as strongly as another student. This could be attributed to the often solitary nature of engineering projects, at least in the early stages of work.

When examining the differences in component inclusion, the discipline requirements become apparent. While both groups of students wanted to be able to add their own photos to the text inline or with obvious icons to remind them of their inclusion and have text represented in bullet form, engineering students did not feel that taking their own notes were absolutely necessary in the new textbook and questioned the requirement for inclusion of this tool, though including it later in their final recommendations. When asked about their hesitance surrounding the inclusion of the component, they stated that the information was now in point form and they no longer needed to take notes but could still see value in the inclusion of the component. On the contrary, design students felt no hesitance surrounding the component and still wanted to take their own notes, indicating that this was a requirement based on the interdisciplinary and creative aspects of the design process. Engineering students also requested the component interactive equations to be included in their future textbook, which is consistent with a discipline that requires the use of equations in their work over those that do not, such as design.

Based on the educational requirements of both disciplines of students, it is important to ensure that components change based on the needs of the students and the concepts that the electronic textbook is trying to convey to their readers. Researchers have called for this in the past, requesting that textbooks are coherent and the content is tailored to the reader groups who will be using them [27]. This concept should be extended to the components that will be used in electronic textbooks. Hartley proposed in the past that "changing the way we write textbooks is one way in which we can make a major improvement in the quality of instruction" and proposed that electronic textbooks could do this with different examples for different readers [27]. This already accepted idea, can be extended from the content design and into the technical design. Some educators are already calling for digital textbooks, which bring together different types of content such as multimedia and text to create an electronic textbook that will be more interactive [28], which is echoed in the responses of the student participants in the focus groups.

V. Conclusion and Future Work

Overall, students believe that future electronic textbooks need to be improved to become more interactive to facilitate their learning and help them fully engage with the material. Some examples of changes that both engineering and design students believed would be beneficial to their academic reading process were 3D and manipulatable images, multimedia related to the concepts, and better annotation tools, which allow them to add more than just textual notes related to the topic. Also, students from both disciplines were not averse to changes in textbooks, which are currently outside of commercial technical capabilities such as holographic images. Although, students can agree on these components, when comparing two similar disciplines that share many fundamental characteristics with differences in approaches, it becomes apparent that we need to adopt an approach to textbook design that tailors electronic textbooks to meet discipline specific needs.

From the findings of these focus group sessions, some design criteria can be identified for future electronic textbooks. The future electronic textbooks need to become more interactive, discipline specific, and with less text. Also, discipline specific components are vital, such as interactive equations in engineering textbooks, to better facilitate the understanding of their work and engaging with the material.

While design recommendations such as these have important applications to industry and academia, more research should be conducted to truly verify the practical validity and educational repercussions of the components suggested. The educational perspective should also be investigated to understand the use of electronic textbooks as a teaching aid. This perspective is best investigated on an individual basis because of the changing opinions on appropriate classroom instruction techniques of the individual instructors.

ACKNOWLEDGMENT

The authors would like to thank The Hong Kong Polytechnic University for their support. This research was supported by the Research Grants Council for the Hong Kong PhD Fellowship Scheme (1-904Z).

REFERENCES

[1] K. A. Sheen and Y. Luximon, "Focus Group Study on Student Perception of Electronic Textbooks," ACHI 2016, The Ninth International Conference on Advances in Computer-Human Interactions, 2016, pp. 110-115.

[2] A. Denoyelles, J. Raible, and R. Seilhamer, "Exploring students' e-textbook practices in higher education," Educause Review. [Online]. July 2015. Accessed November 2016. Available from: http://er.educause.edu/articles/2015/7/exploring-students-etextbook-practices-in-higher-education

[3] M. Kropman, H. P. Schoch, and H. Y. Teoh, "An experience in e-learning: using an electronic textbook," Beyond the comfort zone: Proceedings of the 21st ASCILITE Conference, 2004, pp. 512-515.

[4] N. K. Herther, "Ebooks everywhere: the digital transformation of reading," Searcher, vol. 19(6), pp. 22-31, 2011.

[5] B. Shneiderman, Designing the user interface: Strategies for effective human computer-interaction. Boston, MA: Addison Wesley Longman, 1998.

[6] D. A. Kolb, "Learning styles and disciplinary differences," in The modern American college, A. W. Chickering Ed. London: Josey-Bass Limited, pp. 232-255, 1981.

[7] J. Defazio, "Challenges of electronic textbook authoring: writing in the discipline," Communications in Computer and Information Science, vol. 352, pp. 8-14, 2012.

[8] G. Fairbairn and S. Fairbairn, Reading at university. Buckingham, England: Open University Press, 2001.

[9] B. J. Phillips and F. Phillips, "Sink or skim: textbook reading behaviors of introductory accounting students," Issues in Accounting Education, vol. 22(1), pp. 21-44, 2007.

[10] J. Bowden and F. Marton, The university of learning. London: Psychology Press, 2003.

[11] K. O'Hara and A. Sellen, "A comparison of reading paper and on-line documents." Paper presented at the Proceedings of the SIGCHI conference on Human factors in computing systems, Atlanta, GA, 1997, pp. 335-342.

[12] A. Sellen and R. Harper, The myth of the paperless office. Boston, MA: MIT Press, 2003.

[13] R. McFall, "Electronic textbooks that transform how textbooks are used." The Electronic Library, vol. 23(1), pp. 72-81, 2005.

[14] X. Tian and B. Martin, "Value chain adjustments in educational publishing," Publishing Research Quarterly, vol. 29(1), pp. 12-25, 2013.

[15] D. C. Mauk and J. Oakland, American civilization: an introduction. London: Routledge, 2014.

[16] K. A. Sheen and Y. Luximon, "Relationship between academic discipline and user perception of the future of electronic textbooks," Procedia Manufacturing, vol. 3, pp. 5845-5850, 2015.

[17] S. R. Vaughn, J. S. Schumm, and J. M. Sinagub, Focus group interviews in education and psychology. Thousand Oaks: Sage Publications, 1996.

[18] B. G. Glaser and A. L. Strauss, The discovery of grounded theory: strategies for qualitative research. New Brunswick, NJ: Transaction Publishers, 2009.

[19] L. Millward, "Focus groups," in Research Methods in Psychology, G. Breakwell, S. Hammond, and C. Fife-Shaw, Eds. London: Sage Publications, pp. 274-292, 1995.

[20] R. Krueger, Focus Groups: A Practical Guide for Applied Research, 2nd ed. Thousand Oaks: Sage Publications, pp. 36-45, 1994.

[21] D. Stewart and P. Shamdasani, Focus groups Theory and Practice. Beverly Hills, CA: Sage Publications, 1990.

[22] J. R. Miller, A. W. Nutting, and L. Baker-Eveleth, "The determinants of electronic textbook use among college students," American Economist, vol. 58(1), pp. 41-50, 2013.

[23] B. E. Massis, "E-book readers and college students," New Library World, vol. 111(7/8), pp. 347-350, 2010.

[24] S. Aamodt, "Does the brain like e-books? A test of character," Room for Debate: A New York Times Blog. [Online]. March 2016. Accessed November 2016. Available from: http://roomfordebate.blogs.nytimes.com/2009/10/14/does-the-brain-like-e-books/?_r=0

[25] M. Bode, M. Khorami, and D. Visscher "A Case Study of Student and Instructor Reactions to a Calculus E-Book," PRIMUS, vol. 24(2), pp. 160-174, 2014.

[26] P. F. Chong, Y. P. Lim, and S. W. Ling, "On the design preferences for ebooks," IETE Technical Review, vol. 26(3), pp. 213-222, 2009.

[27] J. Hartley, "Textbook design: current status and future directions," International Journal of Educaitonal Research, vol. 14(6), pp. 533-541, 1990.

[28] Z. Miners and T. Z. Freedman, "A Kindle for every student," US News. [Online]. August 2009. Accessed November 2016. Available from: http://www.socialsmarts.com/USNewsandWorldReportKindleAug20092.pdf

DiClas-Grid,

Discussing and Classifying eHealth Interventions

Saskia M. Akkersdijk*, Saskia M. Kelders†,
Louise M. A. Braakman - Jansen ‡ and Julia E. W. C. van Gemert - Pijnen§
Center for eHealth Research & Disease Management
University of Twente,
Enschede, the Netherlands
Email: {s.m.akkersdijk*, s.m.kelders†, l.m.a.braakman-jansen‡, j.vangemert-pijnen§}@utwente.nl

***Abstract*—It is often unclear why one eHealth application is successful and the other is not, because eHealth is usually approached as a black box. Evaluation is often done in the same way for treatment like and non-treatment like application, with a focus on effects and outcomes. This leads to applications being wrongfully put away, because their expected measurements did not performed as well as expected. But in reality, it may not be fair to use these measurements for these applications. Based on discussions around the terms of user and usage, as well as differences found among eHealth application when looking at the literature, two dimensions were selected. These dimensions help discussion needed to make conscious choices during the (re)design and evaluation process of eHealth applications and to opening the black box. These two continuous dimensions are: use-structure and caregiver involvement. Combining them in a grid results into the DiClas-grid. The position on the DiClas-grid influences what a 'user' and 'usage' means in the application, but also has implications for how to best evaluate and (re)design the application. To further help facilitating discussion, six complementing dimensions to the DiClas-grid are discussed. The DiClas-grid is a discussion and classification tool that can help make conscious choices in (re)design and evaluation of applications.**

Keywords*—*eHealth; discussion; classification; evaluation; design.

I. Introduction

In Akkersdijk et al. [1] the grid was introduced. This article elaborates on the grid, names it, and further explores the possibilities of the grid.

In eHealth, there is a diverse range of seemingly successful and unsuccessful applications and interventions. We observe that applications that represent a kind of treatment, for example eMentalHealth interventions [2], tend to be more successful than those that are more supportive of nature, for example those that try to change behavior [3]. It is often unclear why one application is successful, when the other is not. A reason why we cannot always explain the difference in success is because eHealth is often approached as a black box, without knowledge of what happens inside of this box. We search for the effects of the black box and focus on outcomes. Examining eHealth technology from a holistic perspective, in which the technology has value itself, makes it possible to also focus on the mechanisms behind the success. To find these mechanisms it is necessary to open the black box. An important reason why we observe differences in success might be that we evaluate non-treatment-like applications the same way as we evaluate treatment-like interventions, i.e., focused on outcome measures or usage numbers. This results in applications that might be wrongfully put away because their expected measurements did not perform as well as expected, while in reality it may not be 'fair' to use these measurements for these applications. In this paper, we search for a way to give more insight in the black box of the application, by helping with a (re)design and evaluation of that application.

One of the ways evaluation of applications is often done, is by measuring to which extent therapies are followed as intended. One of the ways applications are often evaluated is by measuring to which extent therapies are followed as intended. This measurement of adherence is one of the primary determinants of success in treatments [4], and overall effectiveness of health systems decreases by poor adherence [4]. Although adherence is used as one of the primary determinants of success in therapy, there are also examples in eHealth of applications with a low adherence that are successful. An example of this is QuitNet, a program for smoking cessation [5][6][7]. Adherence to this program is low (23%) [8], but the program can be successful in promoting cessation and preventing relapse [5]. These studies show that it is possible for an eHealth application to have a low adherence but still be successful for a certain group of users.

It is often assumed that a higher exposure and more usage of an application leads to a better outcome. Studies with eMentalHealth interventions often find this high dose – response relationship (also called a usage – outcome relation). An example of this is the study of Bolier et al. [9]. However, this assumption does not hold for all applications. Donkin et al. [10] further explored the usage – outcome relation. The study of Donkin investigates which usage metrics are important in predicting and explaining outcomes for an internet-delivered trail targeting depressive symptoms for those with risk factors for or diagnosis of cardiovascular disease. Their study shows that there is not always a linear dose – response relation, but could be curvilinear (e.g., reaches a saturation point where no further benefit is obtained), or even more complex.

There is a broad range of different eHealth applications and variety in how these applications should be used. These variations can be put on a continuous scale. At one end of the continuum we see applications that require the user to use the intervention in a specific way, for example a fixed order of the modules. These applications are often a (web-based) program of a method, course or intervention. At the other end of the continuum we see applications that leave the usage free, without a strict protocol for each user. Another important factor that varies among different eHealth applications is the involvement of a caregiver. Some eHealth applications are used in close collaboration between patient and caregiver, others with no involvement of a caregiver at all and all variations in between.

Knowing where your application is positioned on these two dimensions can help with (re)designing and evaluating your application. These two dimensions form the DiClas-grid, and applications can be put somewhere on this grid depending on its usage-structure and caregiver involvement. The positioning of an application on the two dimensions influences the term 'user' and 'usage' but also has implications for the way we can (re)design and evaluate the application. The aim of this paper is to present a tool to give more insight in the application, which helps with (re)design and evaluation of that application.

In Section II (The DiClas-Grid), we will take a closer look at the grid, after which we will discuss implications based on the different positions an application can take on the grid in Section III (Implications). We will end this paper with a discussion and conclusion in Section IV (Discussion and Conclusion).

II. THE DICLAS-GRID

In this section, we will take a closer look at the two dimensions of the DiClas-grid (see Figure 1). We will first look at the dimension of use-structure, after which we will look at the dimension of caregiver involvement. Finally, we will describe some eHealth applications and their positions on the DiClas-grid.

The dimension of use-structure has at one end of the continuum applications that force or require the user to use the intervention in a specific way (railroading them). This can be in a specific order, for a specific number of times or lessons, or for a specific duration. These interventions often have a specific end that is known beforehand and are often based on theories about mental health behavior like acceptance and commitment therapy (ACT) or cognitive behavioral therapy (CBT). Because they often find their origin in known theories and therapies, they are often more 'treatment' like and help deliver a kind of short-term care. As discussed in the introduction, 'Living to the full' is a good example for this end of the continuum. The intervention consists of nine lessons, which have to be completed in a specific order in a 12 weeks. Whether participants worked through a lesson in one session or in multiple sessions was up to them [11][12].

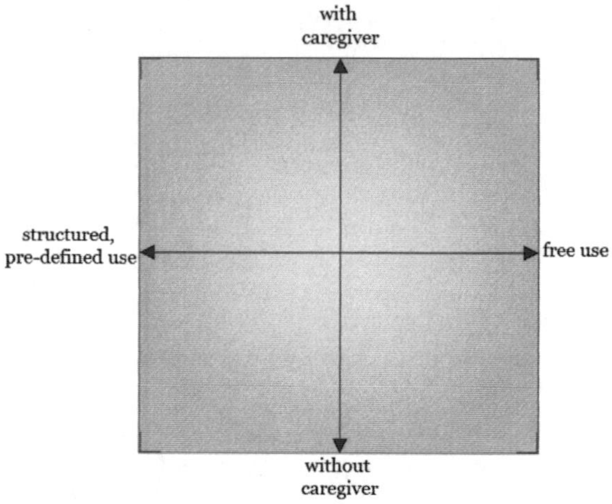

Fig. 1. The DiClas-grid

At the other end are applications that leave the usage free without a strict protocol for each user. There is no specific order or duration for which this application should be used, therefore, they have no specific end. These free-to-use applications often focus more on support and long term care. As discussed in the introduction, 'QuitNet', the application for cessation treatment, is a good example for this end of the continuum. This website offers advice to quit smoking, assistance in setting a quit date, tailored information, assessment of motivation and nicotine dependence, practical counseling (skills training and problem solving), tailored assistants in selecting pharmacotherapies and intra- and extra-treatment social support. How QuitNet is used is completely up to the user [5][6][7].

The vertical dimension represents the amount of caregiver involvement, which varies among eHealth applications. For example, treatment-driven applications often involve caregivers, while lifestyle interventions often can be used autonomously. Research indicates that caregiver involvement is important, but it is not clear what the dosage and frequency of involvement should be [13][14][15][16][17]. In applications that target people with chronic conditions, usually, there is some form of caregiver involvement. However, these applications often struggle to find their fit into daily life, and adherence is often low [18]. Users find it difficult to embed these applications in their own life, while caregivers struggle to embed them into their daily practice [19]. Nonetheless, caregiver involvement is often found to be necessary to ensure adherence and increase effects for web-based interventions for people [20][21].

To illustrate the positioning of an eHealth application on the DiClas-grid, we will now position four applications on the DiClas-grid: 'Living to the full', 'QuitNet', 'Minddistrict' and 'My Health Platform'.

As discussed in the introduction, **'Living to the full'** (LttF) consists of nine lessons, which have to be completed in a specific order in a 12-week period. While there are different versions of this intervention, we will now focus on the version with automated feedback and without involvement of a caregiver [22].

We would place 'Living to the Full' at the bottom left corner on the DiClas-grid (see Figure 2) for the following reasons:

- horizontal dimension: Usage of 'Living to the Full' (such as how it is used, how often) is pre-defined. Exactly when (time) a lesson is completed is left to the user, and it contains some extra options that are optional for the user to complete. Therefore, we would place 'Living to the Full' almost all the way to the left on the horizontal dimension.
- vertical dimension: 'Living to the full' is a standalone program without caregiver support, usage is completely left to the user. Therefore, we would place 'Living to the Full' completely at the bottom of the vertical dimension.

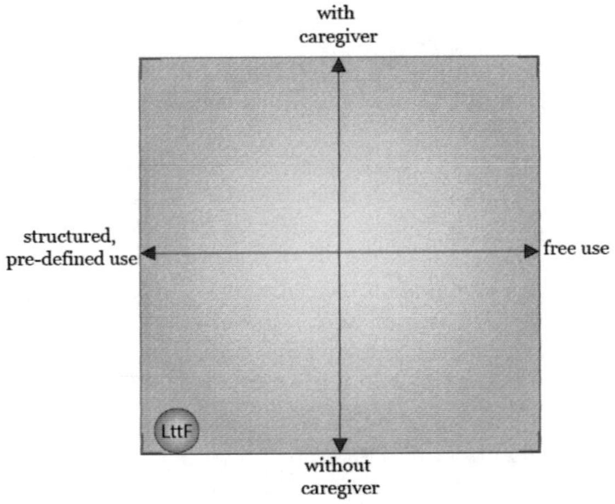

Fig. 2. Positioning of 'Living to the full' on the grid

'QuitNet' (QN) is a website for cessation treatment. There are two versions of QuitNet: a basic version, and an enhanced version, which provides direct access to online cessation counselors and social support systems. Usage frequency of the program is left to the user [5][6][7]. For this example we will discuss both versions. Firstly, we will look at the basic version, **'QuitNet basic'** (QN basic), after which we will look at the enhanced version, **'QuitNet enhanced'** (QN enhanced).

We would place the basic version of QuitNet (QN basic) at the bottom right of the DiClas-grid (see Figure 3 (QN basic)) for the following reasons:

- horizontal dimension: Usage of the basic version of QuitNet (such as how it is used, how often and when it is used) is completely left to the user. Therefore, we would place QuitNet completely at the right side.
- vertical dimension: There is no involvement of a caregiver in the basic version of QuitNet. Therefore, we would place QuitNet completely at the bottom of the vertical dimension.

We would place the enhanced version of QuitNet (QN enhanced) at the right side of the DiClas-grid, at the lower half of the vertical dimension (see Figure 3 (QN enhanced)) for the following reasons:

- horizontal dimension: Usage of the enhanced version of QuitNet is the same as for the basic version (completely left to the user). Therefore, we would place QuitNet completely at the right side.
- vertical dimension: Usage of QuitNet in the enhanced version does provide direct access to online cessation counselors, though usage of this feature is up to the user. Because we have no knowledge about how often this feature is used, we assume that half of the users used this feature and probably not that often. Therefore, we would position QuitNet on the lower half of the caretaker involvement dimension.

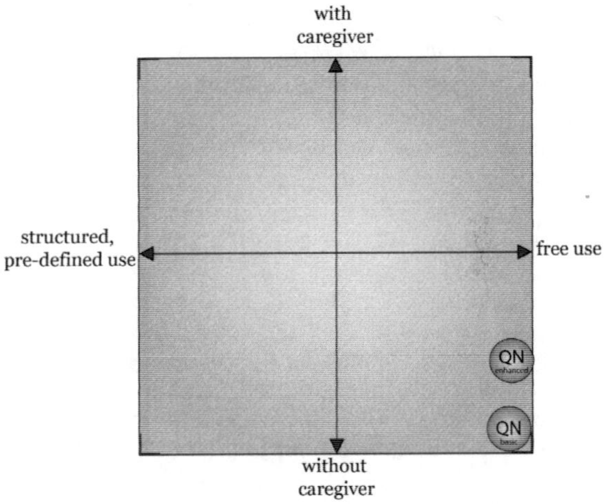

Fig. 3. Positioning of the basic and enhanced version of'QuitNet' on the DiClas-grid

'Minddistrict' [23] provides an online doctor's office for caregivers. It is designed to help deliver personal care tailored to the client, enhance independence in clients, help to go beyond routines, and is always accessible. The platform assists caregivers from triage through blended treatment and relapse prevention. The online platform provides caregivers with an overview of their clients and their progress. The online platform gives the caregiver various tools to do part of the treatment and communication online. Communication between the caregiver and client is enabled through a secure messaging system or by video calling. A triage instrument of questionnaires, either pre-defined or custom-made, can be assigned to clients. Caregivers can give additional psychoeducation through online "modules", consisting of text, video, animation and/or exercises. These modules can be tailored to the client by removing parts of the modules or adding parts from another module. Clients can complete modules on their own or in cooperation with their caregiver. There are many modules to choose from, with subjects ranging from self-help, addiction, chronic pain, eating disorders, depression and anxiety, rehabilitation, ADHD, and everything in between. Finally, there are self-monitoring diaries that can be added to the treatment.

Placement of Minddistrict on the DiClas-grid can be done by looking at the context of both users of the application individually. Placement on the DiClas-grid would be different in a different context.

In the context of the way a client/patient works with it, we would place Minddistrict at the left-hand side, in the upper-side of the vertical dimension on the DiClas-grid (see Figure 4 (MD tech)) for the following reasons:

- horizontal dimension: How many modules a user should complete and the order in which they should be completed is fixed for a patient. How often a client should do a session is agreed with the caregiver. When exactly a session is completed is up to the client. Therefore, we would place Minddistrict (looking at it from the context of the client/patient) just a small bit to the right out on the horizontal dimension.
- vertical dimension: Usage of Minddistrict is prescribed by the caregiver and modules are selected and adjusted by the caregiver. The client can do modules by themselves or in cooperation with their caregiver. Therefore, we would place Minddistrict at the top half of the DiClas-grid. We would not place Minddistrict completely at the top because clients still can do the sessions on their own and not all sessions only in cooperation with their caregiver.

In the context of implementation and the way a caregiver works with it, we would place 'Minddistrict' at the right-hand side, in the top of the vertical dimension on the DiClas-grid (see Figure 4 (MD impl)) for the following reasons:

- horizontal dimension: Which modules and sessions are selected is completely left to the caregiver. Therefore, we would place Minddistrict at the right side of the grid.
- vertical dimension: The client can do modules by themselves or in cooperation with their caregiver. We would not place Minddistrict completely at the top because clients still can do the sessions on their own and not all sessions only in cooperation with their caregiver. This is not changed compared to looking at Minddistrict from the context of the client/patient.

In case of Minddistrict, the caregiver tailors the content of the technology closely to the patient, and caregiver involvement is high. Seen from the perspective of the client/patient as user of the technology Minddistrict is very structured. However, seen from the perspective of the caregiver as user of the technology, and the way Minddistrict is implemented, Minddistrict is free in use.

Placing Minddistrict on the DiClas-grid when taking a different context into account shows that the context of an application is an important factor when placing an application on the DiClas-grid. Changing the context in which you look at the application can change the placement of an application.

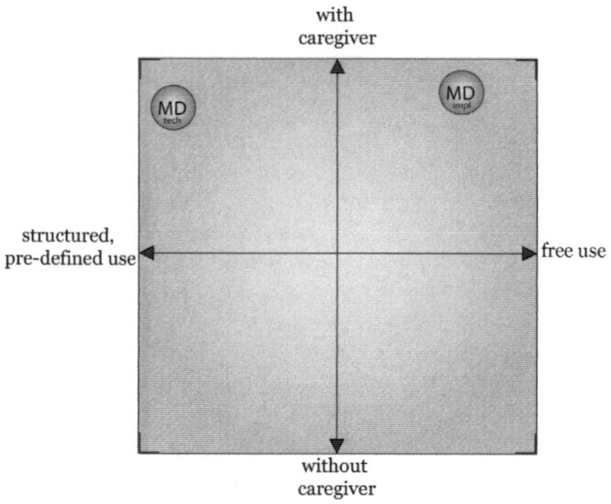

Fig. 4. Positioning of Minddistrict on the DiClas-grid for both contexts(technology and implementation)

'My HealthPlatform' (MHP) is an online platform to support self-care and self-management for people with a chronic illness (e.g., increased cardiovascular risk, COPD, Diabetes mellitus type 2). It is designed to help users keep an overview of and be a director of their own health and lifestyle, alone or in cooperation with a caregiver or expert. In MHP they can monitor their health, find information about their conditions, but also use one of the lifestyle coaches (quit-smoking, nutritional, and exercise coach). While the usage of most of the platform is unstructured, the lifestyle coaches follow a 12 week schedule.

We would place My HealthPlatform at the right-hand side, in the middle of the vertical dimension on the DiClas-grid (see Figure 5) for the following reasons:

- horizontal dimension: Usage of MHP (such as how it is used, how often, and whether or not a coach is used) is left to the user. We would not place MHP completely at the right side, because the coaches do require the user to use them in a specific way and for a predetermined number of weeks.
- vertical dimension: Usage of MHP is mostly left to the user. When MHP is used in cooperation with a caregiver, the caregiver is able to see at home measurements of the user, which provides more insight in the health status of their patient. Because MHP is used with and without caregiver involvement, we would place MHP in the middle of the vertical dimension.

You can position MHP on a different position on the DiClas-grid based on other arguments. In this case, especially the vertical dimension of the DiClas-grid leaves room for discussion. We would like to emphasize that when we would ask multiple people to position the same application on the DiClas-grid we are very likely to end up with as many different positions as we asked people. We would like to argue that this is perfectly fine, because the main purpose of the DiClas-grid is to help you think about certain characteristics of you application and about the implications of the positioning on the DiClas-grid.

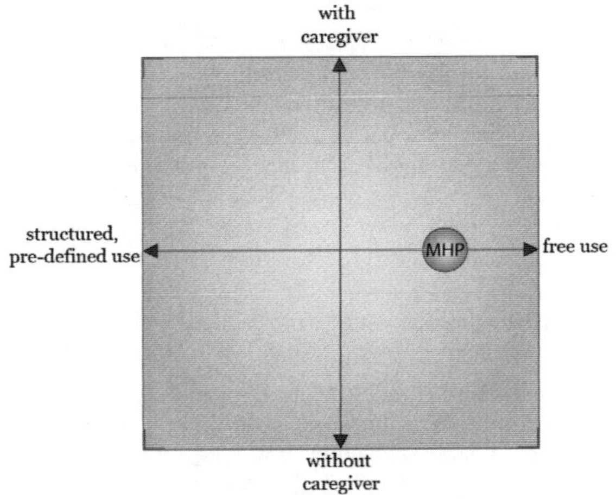

Fig. 5. Positioning of My HealthPlatform on the DiClas-grid

This section consists of examples of how placing an application on the DiClas-grid can work. The examples show that based on another argumentation you can place an application on a different position (as shown in the example of My HealthPlatform), and that the context of an application can play an important role in positioning on the DiClas-grid (as shown in the example of Minddistrict). In the next section, we will talk about some of these implications of the different positions on the DiClas-grid.

III. Implications

Positioning on the DiClas-grid has several implications for the terms 'usage' and 'user' and for the (re)design and evaluation of eHealth applications. In this section, we will discuss some of these implications. We will start with the implications on the terms 'user' and 'usage', after which we will discuss implication for (re)design and evaluation.

A. User

Defining when someone is a user is quite clear when you are dealing with applications that are on the left-hand side of the DiClas-grid. A person that uses the application is a user, and one who does not is not a user. With applications that leave the usage up to the user the way people use the application can vary widely, which leads to a discussion about the term 'user' in this context. We will discuss some questions around the term user, after which we will give our view on the answers.

An important question is: when does a person become a user of the application? This is important because we, for example, use the number of users as an outcome measurement, or we want to know the reach of an application. There are several possible answers to this. We could argue that a person who uses the application is a user, but is there a minimum amount of usage before that person becomes a user, or is 10 seconds enough? And what about someone who does not use the application for a long period of time? Is that person not a user during this period? And could we define certain activities in the application that a person must have done before that person is marked as a user?

For applications that focus on monitoring health or increasing health awareness (mostly positioned on the right-hand side of the DiClas-grid), we can argue that by only becoming aware of such an application a person could potentially be triggered to become more involved in his own health. This means that, in order to have an effect on a person, it does not automatically require that person to use that application. Is this person then a user? We might argue that this person is not a user of the application because he/she did not interact with it. However, the application could still have an effect. In this case the person is not a user in the most common sense of the word, but due to the effect that the application had it balances on the edge of the definition of 'user'.

When we consulted the people who used MHP (Figure 5), it became clear that they had their own view on being a user. There were quite a lot of people who had used the application only a couple of times and therefore, declined to join several studies (interviews, questionnaires, and usability testing) because they did not see themselves as 'users'. In their minds, their definition of a user involves a certain number of reoccurring visits to the application, entering some monitoring data into the system, or participating in the program of a coach. Because they did not meet their own standards of the term user, they thought they could not participate in the study. This example shows that using the system does not equal being a user, at least not for the people who used it. People might have expectations about the intended usage of an application, and it is relevant to communicate the intended use to avoid misunderstanding about the usage.

For evaluation purposes, the definition of what we would call a user can focus on several aspects:

1) The percentage of registered users who see themselves as user, could be a measurement for evaluation of an application. The number of registered users who see themselves as user tells you about their involvement with an application and this in turn can show which role the application holds in their lives and whether the application helps them.
2) You do not always know beforehand who will be a 'user'. When evaluating an application, it is important to define which group of people can be defined as a user, and this group does not always include the people you expected beforehand. For a certain type of evaluation, questionnaire, or interview, a specific group of users might be suitable.
3) An important user that is often forgotten is the caregiver. The caregiver can have his own section in the application where he can see caregiver-specific functionality. He can have his own version of the application, or he can have the same functionality as a patient user. It is important

to realize that the caregiver is a user as well, a user with different needs than a 'patient'. Additionally, both users (the caregiver and the 'patient') affect each other and how they use the application, which means that both types of users should be included in the (re)design and all evaluations.

Finally, a point that we need to take into consideration is that the user, in the sense of usage characteristics, can change over time. That means that when someone starts using an application, their usage can change over time. This change can be caused by a different expectation of the content (they were expecting to find A in the application and were using it that way, but it actually contained B and now they have adapted their usage to B). But is can also be the evolution of the user over time. Knowing that the user can change is important to keep in mind during the design and evaluation process.

B. Usage

The term 'usage' (in the context of an application) can mean a lot of things, such as: How often people return to a website, paths that users follow on a website, how often certain elements on a site are used, etc. This can all be measured by logging user actions with timestamps on a website. Depending on the position on the horizontal use-structure dimension of the DiClas-grid measurements can be used differently and they tell different things about the system/application. Measuring the use of a system or application is useful and insightful for both ends of the spectrum, but evaluating this use differs and the implications/interpretations are different.

For 'railroaded' applications, like 'Living to the full' (see Figure 2), usage measurements can tell you much about the applications. 'Railroaded' applications, positioned at the left-hand side of the DiClas-grid, often are similar or represent a therapy. The user has to follow the structure within the application, do certain actions in a certain order, and use it for a certain amount of time for it to be successful. Therefore, we can define 'normal', or 'ideal' use. We can compare the measured use with the way the application should be used (this can be whether someone completed the application, or the use within an application). Knowing where the occurred usage deviates from the normal or ideal use can help identify problems with the application, or give an explanation why an application does not have the expected results.

Achieving the goal of the application is not completely dependent on the use (the amount and which parts). While a high-dose response relation is often found in eMentalHealth interventions [9], we know that this relation can be far more complex [10]. With applications that leave the user free (right-hand side of the DiClas-grid) the duration of usage is often longer and different situations can be seen than with the use of a 'railroaded' application. For applications positioned at the right-hand side of the DiClas-grid there is no easy definition of 'normal' use, in quantity or in order. This is in contrast with applications positioned at the left-hand side of the DiClas-grid, where the 'correct' following of the structure is essential.

Because there is no prescribed use for applications at the right-hand side of the DiClas-grid, we cannot measure to which extend the measured usage deviates from the optimal use. For example, the measured use of an application positioned at the right-hand side of the DiClas-grid (like 'MHP' in Figure 5) could show users that were dormant for maybe months or years, after which they suddenly used it again. This is unlikely to happen in an application that is 'railroaded'. Because we cannot easily define 'normal use' and it is more likely to vary for different users, adherence, in which we compare the occurring usage with the optimum usage, cannot easily be measured for applications on the right-hand side of the DiClas-grid.

The occurring usage and the use-structure of an application go together. When your application is 'railroaded' and positioned at the left-hand side of the DiClas-grid, all users have a similar usage pattern, while with an application that leaves the use up to the user the occurring usage patterns can vary greatly. It might be good to take this into consideration with evaluation, but also during the design process.

Even though measurements like adherence might not really be suitable for applications that are positioned at the right-hand side of the DiClas-grid, usage measurements can still be very valuable. These measurements can tell you much about the interaction with the application, which parts are used most often, which parts are often used subsequent of each other, after which part users often stop, etc. Knowing more about the interaction with the application is valuable for improving applications, but can also be valuable for finding mechanisms behind application success. Combining usage measurements with use context (what triggered the session) can be used to find a better fit of the content to the context, or improve interaction with the application. By improving the system, and better tuning it to the needs of the users (based on context en measured usage) we can probably increase the effect of applications.

Finally, when we are looking at the usage of an application we should not forget to observe the usage of the application by the caregiver. Caregivers play an important role in the usage of an application by their 'patients' because their usage can be driven by input of said caregiver. When a caregiver does not work with the application as intended or adequately, this will influence the usage of the 'patient' user as well. When the application is meant to be used with a form of caregiver involvement and the caregiver is less involved than the 'patient' expects, the 'patient' user will experience less added value of the application.

C. Implications on (re)Design

With an existing application, the DiClas-grid can be used during the redesign process. Determining the current position of the application on the DiClas-grid based on the characteristics of the application can help you to reflect on your current application by facilitating the thought process about your applications and its characteristic. A first step in the redesign process is to reflect whether it is feasible to reach the objectives of the application from its current position; is it possible to accomplish the goal of the application from this position or is the position of the application on the DiClas-grid not suitable for the goal of the application. The second step is to determine if the current position is the best, or if there are better alternative positions. When the current position and the desired position are known, the next step is to identify their differences. Knowing these differences, it is then possible to determine if the application should be changed and can give a indication about how the application should be changed.

Designing an application is not supposed to be an individual process. It should involve all stakeholders [24], one of the most important being the end users. They have to use it, and can indicate what is most important for them [25]. The DiClas-grid is meant as a discussion and classification tool to help in the design and evaluation process. It can help facilitate the thought process of design choices and their effects to make better conscious decisions.

D. Implications on Evaluation

All implications based on how an application is used (the position of an application on the use-structured on the DiClas-grid) have consequences for the evaluation procedures. 'Railroaded' applications can be evaluated by measuring the usage and comparing it to the optimum usage, in contrast to applications on the right-hand side of the DiClas-grid where usage can vary widely. Because usage can vary so widely, it is also harder to link measured effects to a specific element of the application. Applications on the right-hand side of the DiClas-grid are less feasible to evaluate with an Randomized Controlled Trial (RCT), because they are used for a much longer time, which makes it difficult to keep the circumstances constant. Secondly, because large groups and free use often occur together, it might be harder to find changes in measurements like quality of life for applications on the right-hand side of the DiClas-grid. This does not imply that these changes are unimportant. It would be more suitable to evaluate applications that are positioned on the right-hand side of the DiClas-grid on processes rather than on effects. On the other hand, applications that are positioned on the left-hand side of the DiClas-grid are easier to evaluate on effects, because they have a fixed setting and use-time.

When discussing the implications on the term of user, we discussed that a user (in the sense of usage characteristics) can change over time. This can be true for both sides (left and right) of the DiClas-grid. It is more likely to occur on the right side of the DiClas-grid, because usage there has more freedom. This asks for a process evaluation rather then a effect evaluation. Evaluations more focused on process can include methods like, analysis of logdata [26], machine learning (supervised or un-supervised) [27], Markov modeling [28], [29], Market-basket analysis [30], or time-series analysis [31].

IV. Discussion and Conclusion

There are different ways to classify eHealth that provide an overview of eHealth, such as device driven, based on the medium the technology uses (web-based, mobile apps, etc.), context-of-care driven (eCare, eTherapy, eAppointment, ePrevention, etc.), or actor driven (based on the interaction between the actors of such a system). The DiClas-grid we propose is not meant to replace these classifications, because they provide an overview that our DiClas-grid does not provide. However, our DiClas-grid serves as an extension of these. The different classifications mentioned above serve a different need, while they did not serve our need for a simple way to have some guiding when (re)designing and evaluating eHealth applications. We were looking for a better way to help make a conscious choice in order to find a better fit (in (re)design and evaluation). Positioning of an eHealth application on the DiClas-grid helps to become more aware of implications this has (as discussed in Section III).

Blended care always is a combination of face-to-face and online therapy, and both modalities contribute substantively and procedurally to the treatment process [32]. Therefore, blended care itself always has a high caregiver involvement. When only looking at the technology this is not automatically true, because there can be forms of blended care in which the caregiver plays a small role in the technology.

Based on the position of an application on the DiClas-grid we discussed that the term 'user' can include a different group of people. Those who 'use' an application do not always perceive themselves as a 'user', because they have expectations about the intended usage of an intervention. It is relevant to communicate the intended use to avoid misunderstanding about the usage. The percentage of users who see themselves as a user might be an additional measurement for evaluating an application, because it includes values about involvement with the application. An important group of 'users' that is often forgotten are the caretakers. They often also use the application, and their use or their communication about the application influences the use of the application by their patients.

Adherence is an important measurement for applications positioned on the left-hand side of the DiClas-grid. For these applications, we can define 'normal' or intended use. Because 'normal' or intended use is often a lot harder to define for applications positioned on the right-hand side of the DiClas-grid (there often is no prescribed use) and usage patterns can vary widely, the measurement of adherence might not be suitable. However, usage measurements can be valuable for process evaluations and improvement of the application.

The DiClas-grid can help with the (re)design process by gaining more insight facilitated by the thought process needed for placement of the application on the DiClas-grid. It is important to think about whether the intended (or current) position of the application is suitable for the goal of the application, or whether another position might be better.

Positioning of an application on the DiClas-grid (left versus right, and with or without caregiver involvement) influences which sorts of evaluations are suitable. Evaluations can be focused on process or effects and positioning on the left-hand of the DiClas-grid are more suitable for effects evaluations than positions at the right-hand side of the DiClas-grid.

When using a time series in a process evaluation, the DiClas-grid can also help determining the frequency and number of point measurements. Because, positioning on the DiClas-grid gives an indication about the usage.

We selected the two dimensions of the DiClas-grid based on our needs in the discussions we had around the terms of user and usage, as well as differences we found among eHealth application when looking at the literature. The DiClas-grid is a good way to have a simple tool to compare applications and to facilitate discussions around the (re)design and evaluation process. To complement the DiClas-grid other dimensions can help to facilitate discussions around design choices and evaluation. It is interesting to think about what these dimensions can be. In the following section we will discuss, in random order, some ideas about other dimensions as well as what different positions on these dimensions can implicate.

The first dimension is also a dimension of use, but in a different context: a social context. Whether a user uses an application solitary or the application provides a form of contact or social support can make a difference in the dynamic of the application and how it is used. The social aspect can for example come from others with the same condition/goals, others using the same application, or maybe from their own social network. The occurring dimension can be found in Figure 6 (A). Hardiker and Grant [33] found that social aspects of use was one of the four factors which influences public engagement with eHealth. Placing an eHealth application on this dimension can help in the discussion if, and how to incorporate the social context in the application. It also can help to determine how to evaluate this part of an application. When social context is an important factor in your design, it is important to evaluate this accordingly.

Use context consists of two elements. The first element (see Figure 6 (B)) is whether the application has to be used in private only, or in a more public setting in which sharing of (personal) information plays a role (such as a work environment). Both settings ask for a different approach and different decisions in the design process. At one end of the dimension is an application that is only used in a private setting, while at the other side of the dimension there will be applications used in a public setting, a less comfortable setting and a setting in which sharing (personal) information with others plays an important role. This will influences privacy matters. But also, the willingness to share information [34].

The second element, and third dimension, is how well the application fits in the patient's regular life and schedule (see Figure 6 (C)). If you keep getting reminders to exercise more while you are at work and in a meeting, chances are that you stop using the application. Also data exchange between you and your caregiver sounds great, and if your caregiver gets them without problems even better. But what if your caregiver then has no time to do something with the data you send during a consultation because working with the data does not have a place in the protocols, and he keeps running out of time to actually look at your data? Therefore, an interesting dimension (mostly from an evaluation point of view) might be implementation/fit into daily life, with at one end of the spectrum applications that fit smoothly into daily life and at the other end applications that have an awful fit. It can be argued that all applications should be unobtrusive, because the application is more likely to be used when it is unobtrusive and does not require lots of effort. However, it can also be possible for an application to have a less than perfect fit into daily live especially because this is annoying, motivating the user to change his behavior to the intended goal of the application (where the intended behavior is less obtrusive in daily life). As long as the application supports the changes required to make this change it is unobtrusive, as is also discussed in Laurie and Blandford [35]. Discussing which factor contribute to a perfect fit [36], how to achieve a better fit, and which elements lead to this fit can also be interesting.

A fourth dimension could be a dimension of user target (see Figure 6 (D)): Does the application target a large diverse group, or does it target a small uniform group? Think about an application that targets users with only diabetes versus an application that targets users with a chronic disease. Firstly, when you find that you target a diverse group, it also might be that it is not clear what makes the group you target unique. Therefore, try to think of characteristics of the target group that make this group unique. Secondly, when only targeting a small specific group you can tailor the application more specifically to their needs. Serving a large diverse group, you have to make sure you cover their needs enough to not lose them, or find clever ways to still tailor to their specific needs. When evaluating an application that targets a large diverse group it is possible that the application works fine for a specific group within the larger group. This should be looked at while evaluating.

A fifth dimension is related to the goal of the application: Is the goal of the application disease management, prevention or treatment of an illness? This dimension could be a scale with prevention at one end, and treatment at the other, while disease management is exactly in the middle (see Figure 6 (E)). What kind of consequences does positioning on this dimension have? The content of the application would be different for all of the applications depending on the positioning. Intrinsic levels of motivation of your users probably also would vary, for the simple reason that users of an application that is focused on treatment probably experience problems, and therefore, are more likely to be intrinsically motivated. On the other hand, users of an application focused on prevention probably do not yet experience problems, therefore, there is less (obvious) need to used the application. This influences decisions about content of the application as well as functioning of the system itself, and the service. Positioning of an application further to the right of this scale also influences the evaluation. When measuring effects of an application that is focused on treatment, you look for positive change in specific variables. When dealing with disease management these variables are more far fetched (there probably is not one outcome measurement, but a larger combination). Effects of an application focused on prevention can take a long time to become clear and measurable. Secondly, the duration over which the application should be used differs depending on the position on this dimension [2]. Prevention and lifestyle applications in general have a longer duration, while treatment often is shorter.

A sixth dimension could be: How proactive or reactive is the application? A proactive application is an application that for example gives alerts, send reminders, or emails. A reactive application is an application that only reacts to what the users does with the application (if it even is reactive). The dimension can be seen in Figure 6 (F). Some parts of this dimension are already implemented in applications. However, most eHealth applications are still mostly reactive. Proactive, fine-tuned apps are still work for the future, but have a lot of potential.

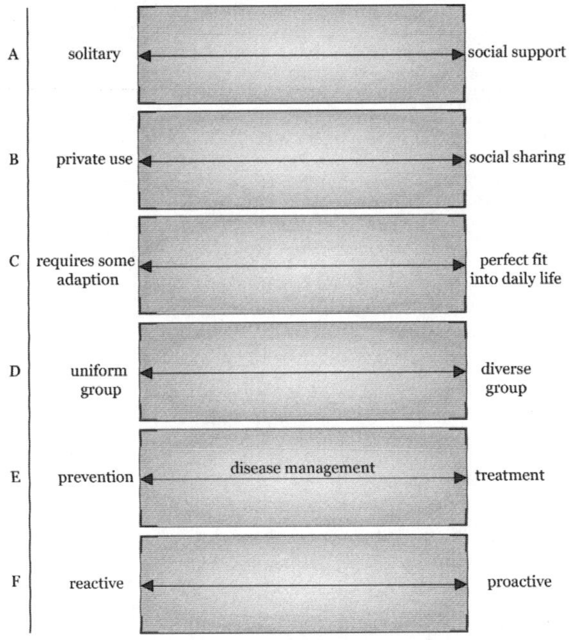

Fig. 6. All dimensions: from top to bottom: A) social context, B) use context, C) fit into life, D) target users, E) goal dimension, F) reactive/proactive

We have discussed six complementing dimensions for the DiClas-grid. These dimensions can help facilitate discussion about other subjects for design and evaluation purposes. They do not replace the DiClas-grid, or a specific dimension of the DiClas-grid, but serve as an addition.

While the six dimensions complement the DiClas-grid, they do not focus on keeping users and their experience. Other important subjects to have a discussion about include e.g., persuasiveness, tailoring of the application, engagement of the user and involvement of the user.

The DiClas-grid is a tool to classify eHealth, to gain insight, facilitate thought processes, and start discussions, and is not meant to be a formal and rigid model. The DiClas-grid combined with the dimensions as discussed can help facilitate discussion and help make conscious choices around many different subjects concerning evaluation and design of eHealth applications.

REFERENCES

[1] S. M. Akkersdijk, S. M. Kelders, J. E. W. C. van Gemert-Pijnen, and L. M. A. Braakman-Jansen, "The Grid , Classification of eHealth Applications Towards a Better (re) Design and Evaluation," in *Eighth International Conference on eHealth, Telemedicine, and Social Medicine, eTelemed 2016*, Venice, 2016, pp. 98–103.

[2] S. M. Kelders, R. N. Kok, H. C. Ossebaard, and J. E. W. C. Van Gemert-Pijnen, "Persuasive system design does matter: a systematic review of adherence to web-based interventions." *Journal of medical Internet research*, vol. 14, no. 6, p. e152, jan 2012, [accessed: 2016-11-30]. [Online]. Available: http://www.pubmedcentral.nih.gov/articlerender.fcgi?artid=3510730&tool=pmcentrez&rendertype=abstract

[3] L. T. Webb, J. Joseph, L. Yardley, and S. Michie, "Using the internet to promote health behavior change: A systematic review and meta-analysis of the impact of theoretical basis, use of behavior change techniques, and mode of delivery on efficacy," *J Med Internet Res*, vol. 12, no. 1, p. e4, Feb 2010, [accessed: 2016-11-30]. [Online]. Available: http://www.jmir.org/2010/1/e4/

[4] E. Sabat, "Adherence to long-term therapies," World Health Organization, Tech. Rep., 2003.

[5] A. Graham, N. Cobb, L. Raymond, S. Sill, and J. Young, "Effectiveness of an internet-based worksite smoking cessation intervention at 12 months," *Journal of Occupational & Environmental Medicine*, vol. 49, no. 8, pp. 821 – 828, 2007.

[6] J. Saul, B. Schillo, S. Everred, M. Luxenberg, A. Kavanaugh, N. Cobb, and A. L, "Impact of a statewide internet-based tobacco cessation intervention," *Journal of Medical Internet Research*, vol. 9, no. 3, 2007.

[7] N. K. Cobb, A. L. Graham, B. C. Bock, G. Papandonatos, and D. B. Abrams, "Initial evaluation of a real-world internet smoking cessation system," *Nicotine & Tobacco Research: Official Journal of the Society for Research on Nicotine and Tobacco*, vol. 7, no. 2, pp. 207 – 216, 2005.

[8] D. S. Kelders, "Understanding adherence to web-based interventions," Ph.D. dissertation, Universiteit Twente, Enschede, 2012, [accessed: 2016-11-30]. [Online]. Available: http://doc.utwente.nl/81967/

[9] M. L. Bolier, M. M. Haverman, P. J. Kramer, D. G. J. Westerhof, P. H. Riper, P. J. A. Walburg, B. Boon, and P. E. Bohlmeijer, "An internet-based intervention to promote mental fitness for mildly depressed adults: randomized controlled trial," *Journal of medical internet research*, vol. 15, no. 9, pp. 1 – 18, 2013, [accessed: 2016-11-30]. [Online]. Available: http://doc.utwente.nl/83748/

[10] L. Donkin, I. B. Hickie, H. CHristensen, S. L. Naismith, B. Neal, N. L. Cockayne, and N. Clozier, "Rethinking the dose-response relationship between usage and outcome in an online intervention for depression: Randomized controlled trial," *Journal of Medical Internet Research*, vol. 15, no. 10, p. e231, 2013.

[11] D. J. van Gemert-Pijnen, D. S. Kelders, and P. E. Bohlmeijer, "Understanding the usage of content in a mental health intervention for depression: An analysis of log data," *Journal of medical internet research*, vol. 16, no. 1, p. e27, 2014, open access, [accessed: 2016-03-08]. [Online]. Available: http://doc.utwente.nl/88658/

[12] M. S. Kelders, T. E. Bohlmeijer, and E. J. Van Gemert-Pijnen, "Participants, usage, and use patterns of a web-based intervention for the prevention of depression within a randomized controlled trial," *J Med Internet Res*, vol. 15, no. 8, p. e172, Aug 2013, [accessed: 2016-11-30]. [Online]. Available: http://www.jmir.org/2013/8/e172/

[13] G. Andersson and P. Cuijpers, "Pros and cons of online cognitive–behavioural therapy," *The British Journal of Psychiatry*, vol. 193, no. 4, pp. 270–271, 2008.

[14] ——, "Internet-based and other computerized psychological treatments for adult depression: A meta-analysis," *Cognitive Behaviour Therapy*, vol. 38, no. 4, pp. 196–205, 2009, [accessed: 2016-11-30]. [Online]. Available: http://dx.doi.org/10.1080/16506070903318960

[15] Z. Hilvert-Bruce, P. J. Rossouw, N. Wong, M. Sunderland, and G. Andrews, "Adherence as a determinant of effectiveness of internet cognitive behavioural therapy for anxiety and depressive disorders," *Behaviour Research and Therapy*, vol. 50, no. 78, pp. 463 – 468, 2012, [accessed: 2016-03-08]. [Online]. Available: http://www.sciencedirect.com/science/article/pii/S0005796712000708

[16] P. Musiat and N. Tarrier, "Collateral outcomes in e-mental health: a systematic review of the evidence for added benefits of computerized cognitive behavior therapy interventions for mental health," *Psychological Medicine*, vol. 44, pp. 3137–3150, 11 2014.

[17] V. Spek, P. Cuijpers, I. Nyklícek, H. Riper, J. Keyzer, and V. Pop, "Internet-based cognitive behaviour therapy for symptoms of depression and anxiety: a meta-analysis," *Psychological Medicine*, vol. 37, pp. 319–328, 3 2007.

[18] M. F. Sieverink, D. S. M. Kelders, D. L. M. Braakman-Jansen, and D. J. E. van Gemert-Pijnen, "The added value of log file analyses of the use of a personal health record for patients with type 2 diabetes mellitus: Preliminary results," *Journal of diabetes science and technology*, 2014, published online before print, [accessed: 2016-11-30]. [Online]. Available: http://doc.utwente.nl/89290/

[19] M. F. Sieverink, D. L. M. Braakman-Jansen, Y. Roelofsen, S. H. Hendriks, R. Sanderman, H. J. Bilo, and D. J. E. van Gemert-Pijnen, "The diffusion of a personal health record for patients with type 2 diabetes mellitus in primary care," *International journal on advances in life sciences*, vol. 6, no. 3&4, pp. 177 – 183, 2014, [accessed: 2016-11-30]. [Online]. Available: http://doc.utwente.nl/93390/

[20] J. D. Piette, A.-M. Rosland, D. Marinec, N. S.and Striplin, S. J. Bernstein, and M. J. Silveira, "Engagement with automated patient monitoring and self-management support calls: Experience with a thousand chronically-ill patients," *Medical Care*, vol. 51, no. 3, pp. 216 – 223, 2013, [accessed: 2016-11-30]. [Online]. Available: http://doi.org/10.1097/MLR.0b013e318277ebf8

[21] C. D. Mohr, P. Cuijpers, and K. Lehman, "Supportive accountability: A model for providing human support to enhance adherence to ehealth interventions," *J Med Internet Res*, vol. 13, no. 1, p. e30, Mar 2011, [accessed: 2016-11-30]. [Online]. Available: http://www.jmir.org/2011/1/e30/

[22] S. M. Kelders, E. T. Bohlmeijer, W. T. Pots, and J. E. van Gemert-Pijnen, "Comparing human and automated support for depression: Fractional factorial randomized controlled trial," *Behaviour Research and Therapy*, vol. 72, pp. 72 – 80, 2015, [accessed: 2016-11-30]. [Online]. Available: http://www.sciencedirect.com/science/article/pii/S0005796715300061

[23] Minddistrict. (2016, November) Minddistrict. [Online]. Available: http://www.minddistrict.com/

[24] D. L. van Velsen, M. M. Wentzel, and D. J. E. van Gemert-Pijnen, "Designing ehealth that matters via a multidisciplinary requirements development approach," *JMIR research protocols*, vol. 2, no. 1, pp. e21 –, 2013, [accessed: 2016-11-30]. [Online]. Available: http://doc.utwente.nl/87481/

[25] M. J. Wentzel, D. L. van Velsen, M. M. van Limburg, N. de Jong, D. J. Karreman, R. Hendrix, and J. E. W. C. van Gemert-Pijnen, "Participatory ehealth development to support nurses in antimicrobial stewardship," *BMC medical informatics and decision making*, vol. 14, no. 1, pp. 45 –, 2014, open access [accessed: 2016-11-30]. [Online]. Available: http://doc.utwente.nl/91558/

[26] F. Sieverink, S. Kelders, S. Akkersdijk, and M. Poel, "Work in Progress : A Protocol for the Collection , Analysis , and Interpretation of Log Data from eHealth Technology," in *Persuasive Technology Fourh International Workshop on Behavior Change Support Systems, BCSS 2016*, no. April, 2016, pp. 56–60.

[27] J. Han, M. Kamber, and J. Pei, *Data mining: concepts and techniques*. Elsevier, 2011.

[28] E. Seneta, "Markov and the birth of chain dependence theory," *International Statistical Review / Revue Internationale de Statistique*, vol. 64, no. 3, pp. 255–263, 1996, [accessed: 2016-11-30]. [Online]. Available: http://www.jstor.org/stable/1403785

[29] J. Borges and M. Levene, "Evaluating variable-length markov chain models for analysis of user web navigation sessions," *IEEE Trans. on Knowl. and Data Eng.*, vol. 19, no. 4, pp. 441–452, Apr. 2007, [accessed: 2016-11-30]. [Online]. Available: http://dx.doi.org/10.1109/TKDE.2007.1012

[30] S. Anand, A. Patrick, J. Hughes, and D. Bell, "Kdd: Techniques and applications a data mining methodology for cross-sales," *Knowledge-Based Systems*, vol. 10, no. 7, pp. 449 – 461, 1998, [accessed: 2016-11-30]. [Online]. Available: http://www.sciencedirect.com/science/article/pii/S0950705198000355

[31] R. B. Penfold and F. Zhang, "Use of interrupted time series analysis in evaluating health care quality improvements," *Academic Pediatrics*, vol. 13, no. 6, Supplement, pp. S38 – S44, 2013, quality Improvement in Pediatric Health Care [accessed: 2016-11-30]. [Online]. Available: http://www.sciencedirect.com/science/article/pii/S1876285913002106

[32] M. J. Wentzel, M. R. van der Vaart, P. E. T. Bohlmeijer, and D. J. E. van Gemert-Pijnen, "Mixing online and face-to-face therapy: how to benefit from blended care in mental healthcare," *JMIR mental health*, vol. 3, no. 1, pp. e9 –, 2016, [accessed: 2016-11-30]. [Online]. Available: http://doc.utwente.nl/99360/

[33] N. R. Hardiker and M. J. Grant, "Factors that influence public engagement with ehealth: A literature review," *International Journal of Medical Informatics*, vol. 80, no. 1, pp. 1 – 12, 2011, [accessed: 2016-11-30]. [Online]. Available: http://www.sciencedirect.com/science/article/pii/S1386505610002091

[34] A. Lentferink, H. Oldenhuis, O. Kulyk, and M. de Groot, "Self-tracking and Persuasive eCoaching in Healthy Lifestyle Interventions: Work-in-progress Scoping Review of Key Components," in *Persuasive Technology Fourh International Workshop on Behavior Change Support Systems, BCSS 2016*, no. April, 2016, pp. 15–35, [accessed: 2016-11-30]. [Online]. Available: ftp://ceur-ws.org/pub/publications/CEUR-WS/Vol-1573.zip

[35] J. Laurie and A. Blandford, "Making time for mindfulness," *International Journal of Medical Informatics*, 2016, [accessed: 2016-11-30]. [Online]. Available: http://dx.doi.org/10.1016/j.ijmedinf.2016.02.010

[36] D. N. Nijland, D. J. E. van Gemert-Pijnen, D. S. M. Kelders, B. J. Brandenburg, and P. E. R. Seydel, "Factors influencing the use of a web-based application for supporting the self-care of patients with type 2 diabetes: a longitudinal study," *Journal of medical internet research*, vol. 13, no. 3, 2011, [accessed: 2016-11-30]. [Online]. Available: http://doc.utwente.nl/85159/

KINECT-Based Auscultation Practice System

Yoshitoshi Murata, Kazuhiro Yoshida
Faculty of Software and Information Science
Iwate Prefectural University
Takizawa, Japan
e-Mail: y-murata@iwate-pu.ac.jp, kyoshida@ipu-office.iwate-pu.ac.jp

Natsuko Miura, Yoshihito Endo
Faculty of Nursing
Iwate Prefectural University
Takizawa, Japan
e-Mail: natsuko@iwate-pu.ac.jp, y-endo@iwate-pu.ac.jp

Abstract— **Students in medical and nursing schools have to practice auscultation. Students usually learn disease sounds, correct points and order for locating a stethoscope on a body in the practice. Humanoid simulators have been widely introduced to practice auscultation. They are effective to learn disease sounds. However, most humanoids cannot detect whether or not a stethoscope is located on a body and which part of body a stethoscope is placed on. And also, since they are too expensive, the number of them is not enough for the number of students in a class. In this paper, we propose a low-cost and high performance system for the practice of auscultation. In this system, students themselves play the role of a patient, instead of a humanoid, and stethoscope locations on the body are measured with KINECT. Also, appropriate disease sounds including normal ones can be assigned at some points on the upper body. Practicing students hear such disease sounds, synchronized with the movement of breathing, through earphones when a stethoscope is placed on the assigned points. Movements of the upper body from breathing can also be detected by KINECT. Experimental results with a prototype showed that our system could perfectly detect stethoscope locations on a body, except for a few lower points, and it could also detect respiratory changes on the front body. The results of a questionnaire for nursing students and practicing nurses showed that our proposed system was useful for them to learn auscultation.**

Keywords-simulator; auscultation; nursing; KINECT.

I. INTRODUCTION

Generally, practicing auscultation is a required subject for students in medical and nursing schools. Students usually learn disease sounds, correct points and order for locating a stethoscope on the body in the practice. We proposed a new auscultation practice system to learn auscultation techniques effectively [1]. Humanoid-type simulators [2][3][4][5] have been widely introduced into medical and nursing schools, and several reports have found that such simulators improve auscultation skills [6]. These humanoid simulators are effective to learn disease sounds; and they enable determining correct stethoscope locations by marking these points on a mannequin. However, it is impossible to detect whether or not a stethoscope is actually placed correctly on a mannequin. Moreover, correct locations vary among patients according to body size. Cardionics provides a hybrid simulator in which a student plays the role of a patient instead of a mannequin to solve such problems [7]. In this hybrid simulator, patches are attached on a body to identify correct attachment points.

In either type of simulators, students have difficulty practicing auscultation by themselves. Teachers have to support students during practice. Moreover, both types of simulators cannot indicate the learning progress for each student using scores.

In our simulator, students themselves are the practice subjects instead of a humanoid model, and the location of a stethoscope can be detected with KINECT, which is a line of motion sensing input devices made by Microsoft [8]. The correct locations are normalized with respect to the positions of both shoulder joints and both hip joints for each student playing the role of patient. Therefore, our proposed simulator can both show correct locations on a body and can detect whether or not a stethoscope is placed on correct points without patches regardless of the change in body size.

In addition, most existing simulators cannot simulate the timing of breathing or the synchronized forward and backward movements of the upper body. However, our simulator can detect these forward and backward movements of the front body and provide exhalation and inhalation sounds synchronized with those movements.

We have developed a prototype system and evaluated it experimentally. The results showed that our system could perfectly detect stethoscope placement on a body at seven of ten points. Our system could not always detect a stethoscope because three lower points were easily shadowed from KINECT by the T-shirt worn by a student acting as a patient. Also, our system could detect changes of the front body in breathing.

The aforementioned mean that students could learn auscultation practice by themselves using our system and that our system could indicate the learning progress for each student through their increases in score during the KINECT simulation.

Moreover, we found that all students could set up our system by themselves. The results of a questionnaire for nursing students and practicing nurses showed that our proposed system was useful for them to learn auscultation.

After introducing related works in Section II, we describe the concepts and features of our system in Section III. The key technologies of our simulator and the evaluation results are described in Section IV. An application for a teacher to lecture about auscultation was developed. It is introduced in Section V. The questionnaires for students who set up and used our system and for practicing nurses are shown in Section VI. The key points are summarized in Section VII.

II. Related Works

Many kinds of patient simulators have been developed and provided as medical and nursing training tools [2][3][4][5][7][9]. Because we propose a new type of simulator for practicing auscultation, we first discuss existing auscultation simulators, which are divided into three groups: the humanoid model type, the virtual reality type and the hybrid type.

A. Humanoid model type

Kyoto Kagaku Co., Ltd. provides the Lung Sound Auscultation Trainer (LSAT) [2], shown in Figure 1, for respiratory auscultation. There are several small speakers inside a mannequin. Disease sounds are recorded from real patients. This simulator also works for cardiac auscultation by changing from respiratory sounds to cardiac sounds. Sakamoto Model Corporation provides the Sakamoto auscultation simulator [4]. This simulator also works for both respiratory and cardiac auscultation. Sakamoto provides a transparent cover for this simulator, as shown in Figure 2, to illustrate correct stethoscope locations.

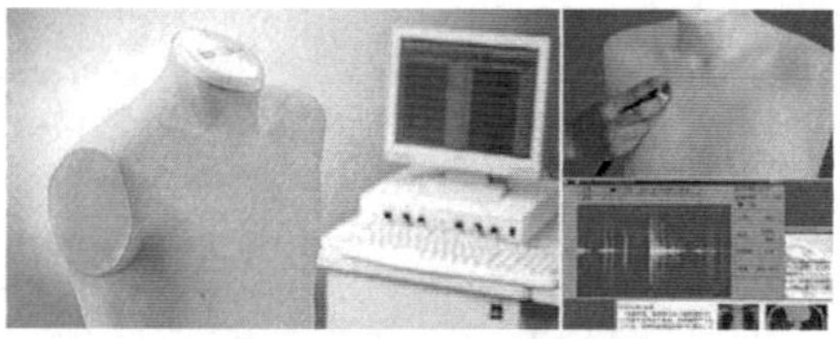

Figure 1. Lung Sound Auscultation Trainer (LSAT) by Kyoko Kagaku

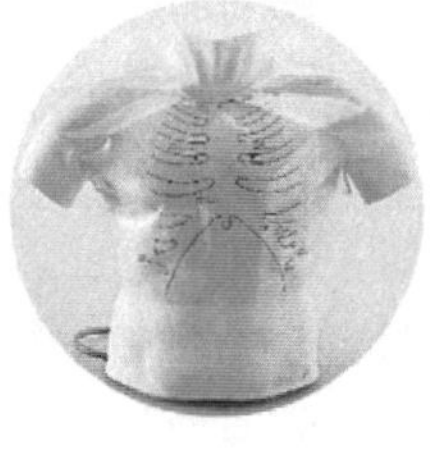

Figure 2. Transparent chest cover, by Sakamoto Model, to illustrate correct stethoscope locations

Although the above two simulators are focused on the upper body, they simulate disease sounds, not the motion of the upper body. And also, they cannot detect where a stethoscope is placed on. Each price is too expensive to buy a few of them.

On the other hand, the SimMan® 3G [4] by Laerdal is an advanced patient simulator that can simulate the characteristics of a real patient, including the blood pressure, heart beat, chest motion, and so on. It is too expensive, however, for a general nursing school to buy.

B. Virtual reality type

Zadow, et al. developed the SimMed system for medical education [9]. By using an interactive multi-touch tabletop to display a simulated patient, as shown in Figure 3, they have created an immersive environment that supports a large variety of learning scenarios. The simulated patient can show skin changes and be animated to show realistic bodily and facial movements. By its nature, the setup allows scenarios to be repeated easily and to be changed and configured dynamically.

SimMed is substantially lower in cost than a full-scale humanoid simulator. It has many functions, however, and is still too expensive for most nursing schools. Moreover, while students can touch the virtual patient on a display, they cannot physically feel the motion of the virtual patient.

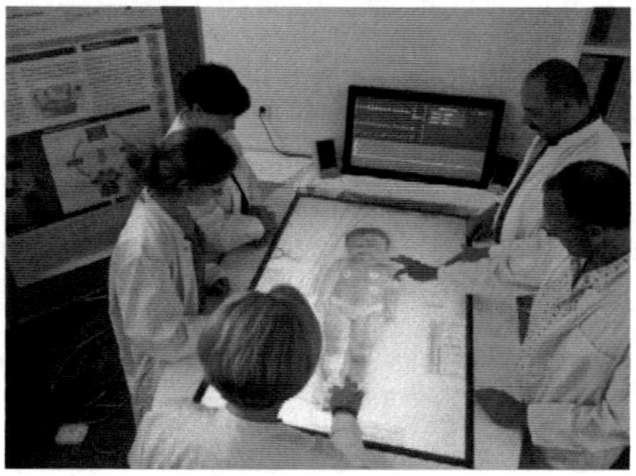

Figure 3. The SimMed system

C. Hybrid type

Cardionics provides the SimScope WiFi Hybrid Simulator™ [8]. As shown in Figure 4, a specific stethoscope has a Bluetooth earphone mounted to hear disease sounds from a PC, and patches are attached on a body to decide the locations for auscultation practices. Therefore, a person who attaches patches on a body has to know the correct locations for each practice. Students have difficulty practicing auscultation by themselves.

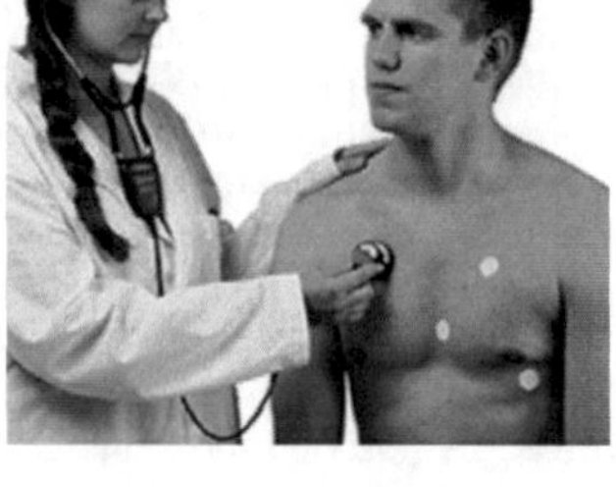

Figure 4. SimScope WiFi™ The Hybrid Simulator$_{TM}$ by Cardionics

III. System Concept

As a matter of course, the introduction cost is adequate for the number of students in a nursing school. Among the nursing skills that students have to learn are the recognition of different sounds between different kinds of diseases and the knowledge about placing correct points and order for locating a stethoscope on a body. In the case of respiratory auscultation, students have to listen to respiratory sounds for more than one cycle. Moreover, the learning progress for every student is consolidated on a cloud server. Therefore, an auscultation practice system requires the following issues:

- Low system cost.
- Simulating real disease sounds at different points on the body.
- Showing correct points for locating a stethoscope on an operation display.
- Judging whether or not a stethoscope is located on shown points.
- Judging whether or not a stethoscope is fixed on a body for more than one respiratory cycle.
- Logs that students practice are stored on a cloud server.

Our practice system comprises a cloud server, terminal equipment, and a PC for a teacher. The terminal equipment and PC for a teacher are connected to the cloud server through the Internet as shown in Figure 5. With the terminal equipment in our system, shown in Figure 6, the students themselves can act as patients instead of having mannequins act as patients—much like in Cardionics's hybrid simulator—to achieve low cost. The stethoscope locations and forward and backward movement of a body during breathing are measured with KINECT. The students can hear disease sounds generated by a PC through earphones. The sound volume for each point is different for locating a stethoscope, as with a real patient. Therefore, a specific stethoscope and patches are not needed. These also enable low system cost.

Log data for a student practicing with terminal equipment are sent to the cloud server and stored. Such stored data are managed in the cloud server. A teacher can access the cloud server and can look at a list of learning progress reports for each student.

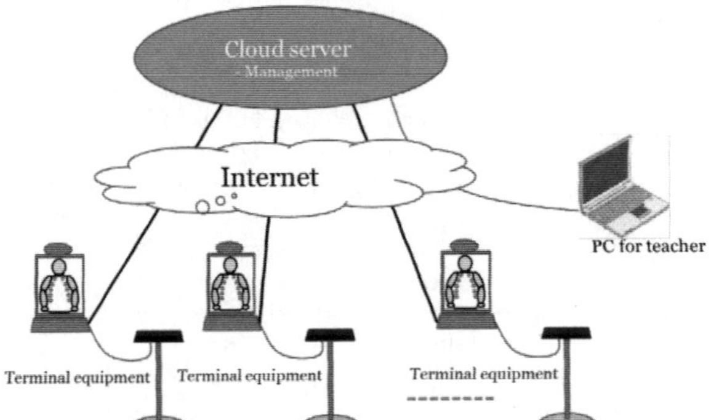

Figure 5. System configuration of our proposed system

Figure 6. Terminal equipment in our proposed system

IV. DETECTION TECHNOLOGIES

The following capabilities are necessary to implement our auscultation practice tool:

- Tracing a stethoscope.
- Detecting a stethoscope's location.
- Detecting whether or not a stethoscope is placed properly on a body.
- Automatically adjusting correct points for locating a stethoscope on a body according to body size.

Also, detecting the inhalation and expiration during respiration is desired.

Because KINECT automatically generates position data for a stethoscope while it is traced, we examined each of these four issues except the second one.

A. *Tracing a stethoscope*

Three candidate tracing methods are shape tracing, color tracing and the AR marker. Since a stethoscope is held by hand, the shape of a stethoscope as viewed through a video camera changes over time. And, size of AR marker on a stethoscope would be too small to detect it with KINECT. Therefore, we chose the color tracing, rather than shape tracing or AR marker. The process of color tracing is shown schematically in Figure 7. Video data from the BGR (Blue, Green, and Red) 32 output of KINECT is converted to HSV (Hue, Saturation, and Value) color data. First, the traced object, i.e., a stethoscope, is pointed, and its hue histogram is generated and stored. Then, masking data are generated for each frame from the HSV data, the minimum saturation, and the maximum and minimum brightness.

The video data for practice is also converted to HSV data, and target areas are separated with the masking data. Noise, including the hue data of the target area, is reduced by a median filter. The output data from the median filter is then traced by the cvCamShift function of Open CV [10]. Since cvCamShift sometimes outputs incorrect data, the hue histogram for the output data is repeatedly compared with the pre-stored hue histogram. When these histograms are equivalent, the color tracing is successful.

We used an experiment to select a target color from among seven choices: "red", "green", "light blue", "yellow", "yellow-green", "pink", and "orange". We used KINECT v1 and examined whether it could detect only a target color. We show the resulting data for the top three colors in Figure 8. "Yellow-green" had the best performance, with no portions having the target color except the target. "Light blue" also performed well, but since the color of a stethoscope tube is "light blue", that was detected. In the case of "yellow", dots of the same color appeared in the bottom-left region of the image. The other colors exhibited more dots of the same color or had a smaller target size. Hence, we chose "yellow-green" as the target color for our experiments.

The aforementioned processing enables obtaining the x_0 and y_0 axis of a stethoscope on a video image window of KINECT. The size of the video image window and the size of the 3D window in KINECT are the same. Therefore, the length z_0 between KINECT and the location (x_0, y_0) can be obtained to assign (x_0, y_0) to the 3D window as shown in Figure 9. This means that the location (x_0, y_0, z_0) can always be obtained.

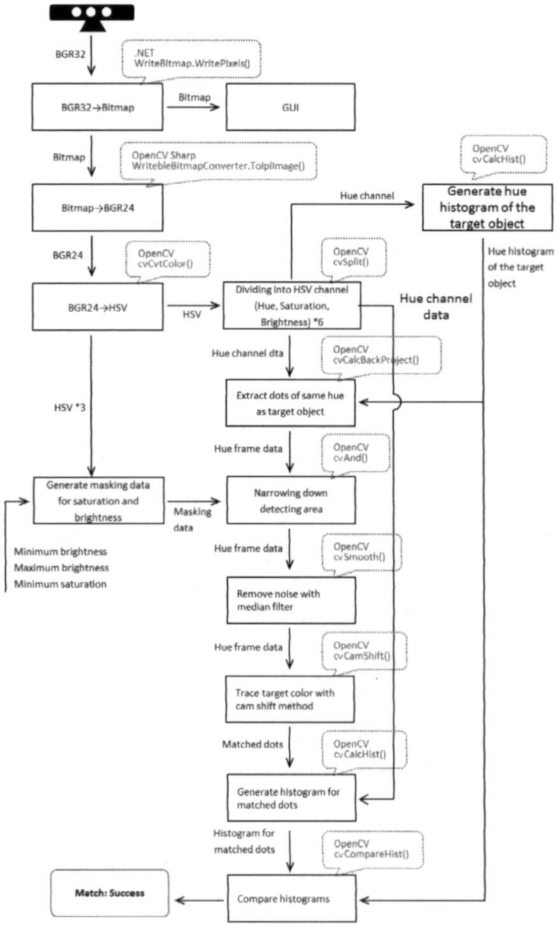

Figure 7. Process of color tracing

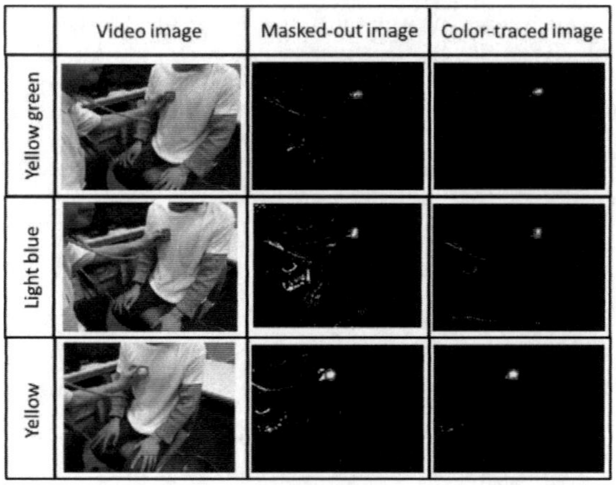

Figure 8. Experiment for deciding a target color

The KINECT camera should be tilted horizontally because the z0 is the shortest length between the flat surface of a KINECT camera and the flat surface going through the measuring point parallel to the surface of the KINECT camera. Also, we recommend the KINECT camera be positioned at about shoulder height.

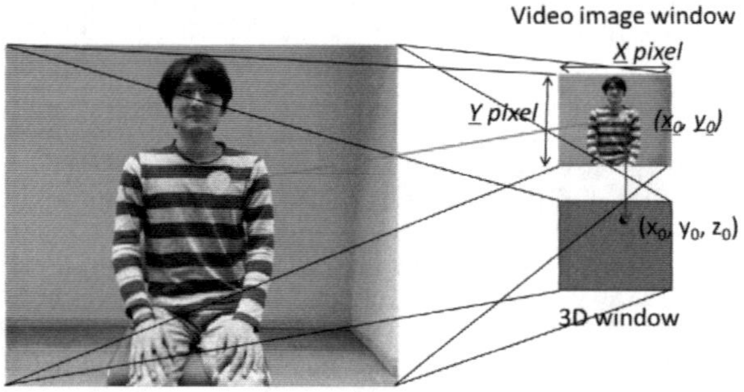

Figure 9. Relationship between video image window and 3D window in KINECT

B. Detecting whether a stethoscope is placed on a body

At this time, we think determining whether or not a stethoscope is placed exactly on a body is unnecessary, so we do not use any sensors on the stethoscope. Instead, we estimate the stethoscope is located on a body, where a stethoscope is placed and fixed within distance S from a body surface for T seconds, as shown in Figure 10. In this figure, L_{bs} is the length between the shoulder and a KINECT, and L_d is the threshold length to determine a stethoscope placed on the body. We used S = 10 cm, and T = 0.3 second to achieve balance between certainty and fast recognition. If stricter detection were required, we could change these parameters. Before measuring length L_{st} between a stethoscope and a KINECT, we determine whether or not the stethoscope is within the outline of a body by using a pre-installed program on a KINECT to get an outline.

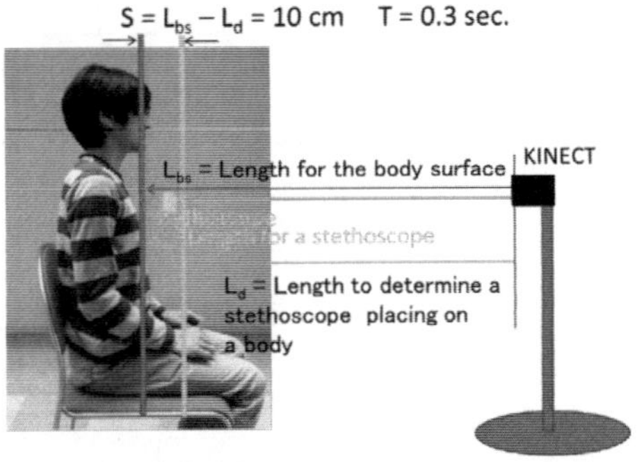

Figure 10. Scheme to estimate whether a stethoscope is placed on the body

We experimentally examined whether or not this method was useful. We started to develop a prototype system by the KINECT v1. However, as described in the next sub-section, the KINECT v1 could not detect the respiration well. Therefore, we redevelop the system by the KINECT v2. The data for the front body were measured by the KINECT v1, and the data for the back body were measure by the KINECT v2. A student acting as a patient wore a white T-shirt with dots marking correct stethoscope locations; a student acting as a nurse placed a stethoscope on these marked dots. We marked 10 dots on a front of T-shirt and 12 dots on the back as shown in Figure 11. These points are decided on the auscultation practice manual. The participants were five male students and five female students for the front body, and three male students and two female students for the back body. Since we think that there is not difference between male and female for the back body, we reduce the number of patients. The experimental results are listed in Table I.

The system sometimes missed when the stethoscope was placed at certain points (points 8, 9, 10) of the front body. As seen from Figure 6, a stethoscope placed on one of these points would sometimes be shadowed from KINECT, especially for women, because these points were below the breasts. The other hand, there are not big sagging on a T-shirt in the back body as shown in Figure 12. Therefore, the system detected a stethoscope was placed on a T-shirt perfectly for every participant.

These are possible solutions to improve detection rate for the front body:

- Switch from a T-shirt to clothing with a tighter fit.
- Attach some dimensional markers to the stethoscope.

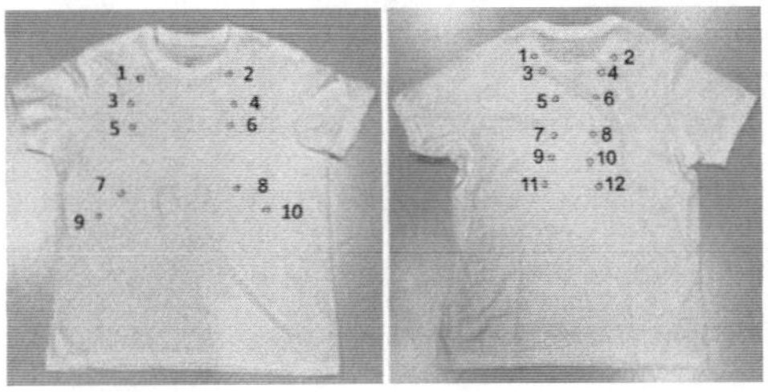

(a) Front view (b) Back view

Figure 11. T-shirt with dots marking stethoscope locations

Figure 12. Scene of experiment for the back body

TABLE I. COUNT OF DETECTING A STETHOSCOPE PLACED ON A BODY

(1) Front body

Point #	1	2	3	4	5
Male (5)	5	5	5	5	5
Female (5)	5	5	5	5	5
Point #	7	8	9	10	11
Male (5)	5	5	5	5	3
Female (5)	5	5	4	3	2

* Measured by KINECT v1

(2) Back body

Point #	1	2	3	4	5	6
Male (3)	3	3	3	3	3	3
Female (2)	2	2	2	2	2	2
Point #	7	8	9	10	11	12
Male (3)	3	3	3	3	3	3
Female(2)	2	2	2	2	2	2

* Measured by KINECT v2

C. Detecting the inhalation and expiration in respiration

Burba et al. used chest motion to detect breathing [11]. They found that detecting the motion of the front body was possible, but they did not describe detecting respiration from the back body. However, practicing auscultation makes it desirable to detect the inhalation and expiration from the back body if feasible.

In our previous study, we found a KINECT v1 could detect the inhalation and expiration from the front body. However, mistakes sometimes occurred in detecting the inhalation and expiration for the front body, and the output data for some measurement points had extraordinary values.

In this section, first, we describe the difference between KINECT v1 and v2 in detecting respiration for the front body. Then, detecting respiration for the back body is described.

1) Difference between KINECT v1 and v2 in detecting respiration for the front body

Since there are two main types of breathing: chest respiration and abdominal respiration, we select six points for measuring movement of the chest and abdomen in this experiment, as shown in Figure 13. Since the stethoscope location is not always fixed, we do not consider it in this experiment. The upper three measuring points are the inner junctions of five vertical lines equally dividing the space between both shoulders into four regions and a horizontal line halfway between the height of the center of the spine and the average height of both shoulders. The lower three measuring points are the inner junctions of the above-mentioned vertical lines and a horizontal line through the hip center. We adopt the same direction of more than 4 points as the resultant direction.

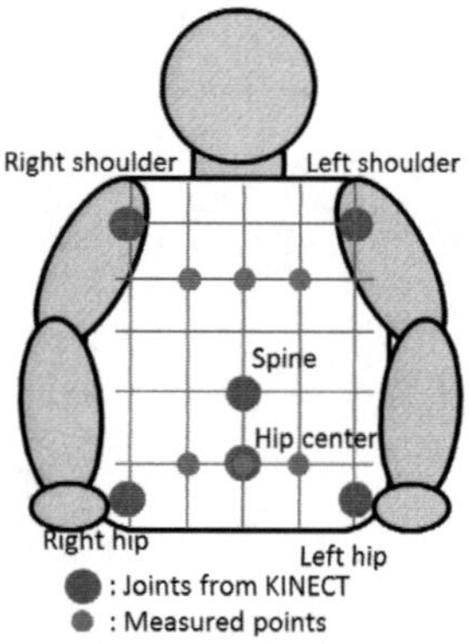

Figure 13. Measuring points for breathing motion

We designed our system to detect the changes from expiration to inhalation and vice versa as quickly as possible. Since there were small but rapid changes in the output data, we used a moving average of 30 samples, with a sampling period of 10 ms. We then have specified that when the sampling data continued to increase 3 times, the breathing mode was expiration; and when the sampling data continued to decrease 3 times, the breathing mode was inhalation.

We measured the number of inhalation and expiration in breathing and their periods with our proposed system, and compared the results with other data obtained by participants keying the up and down arrow keys on a keyboard.

The experimental results were shown in Table II. We measured them for two KINECTs that are K-1 and K-2. Both of them were the same model. At first, we measured for many participants with K-1. Our system counted more and more breaths than did keying for most participants. Data shown in Table II were some of them. In addition, the output data for some measuring points had extraordinary values, as shown in Figure 14 (2). We changed K-1 to K-2 to clear the reason of this problem. Our system with K-2 could count more accurately than that with K-1. However, our system counted fewer breaths than did keying for participant D. And, the extraordinary values were sometimes measured with K-2, too. As shown in Figure 14 (1), our system detected changes in breathing with a delay of about 1 sec. relative to keying when breathings were correctly detected. The measured delay is bigger than we expected.

TABLE II. NUMBER OF BREATHS IN KINECT V1

	Participant	Keying		Proposed system	
		Inhalation	Expiration	Inhalation	Expiration
K-1	A	12	12	16	16
	B	10	10	39	39
	C	14	14	15	15
K-2	C	11	11	11	11
	D	10	10	7	7
	E	15	15	15	15

We think this false detection is derived from the aforementioned extraordinary output data. Also, we estimate the extraordinary output would be derived from a sagging T-shirt and the depth measuring algorithm, which is the random pattern scheme [8]. The random pattern algorithm would be affected from fluctuations of sagging T-shirts as movement from breathing.

We re-programmed the proposed system to KINECT v2; its depth measuring algorithm is the time of flight (TOF) [8]. We experimentally examined the false-detection rate and detection delay for the re-programmed system. The experimental results were shown in Table III and Table IV. The participants were eight male students and four female students. They did not wear our prepared small white T-shirt, but their own clothes. One false detection occurred in about ten breaths for three participants, even though they wore a relatively loose shirt or sweater.

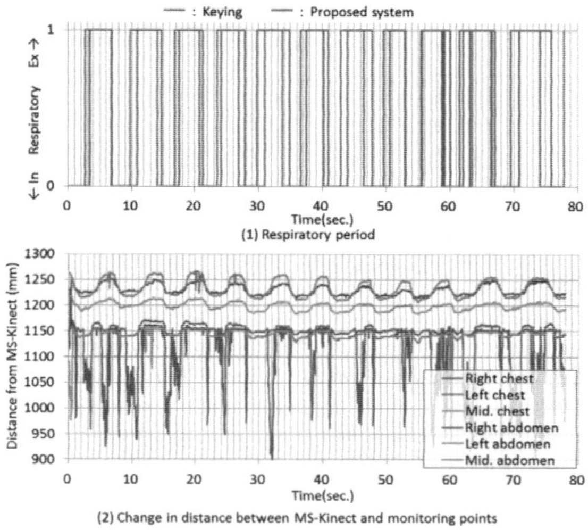

Figure 14. (1) Respiratory period, and (2) change in distance between KINECT v1 and measuring points

Table III. Number of Breaths and detection delay in KINECT v2

No.	Participants	Sex	Keying		Proposed system		Delay time (sec.)		Angle to KINECT Clothes
			Inhalation	Expiration	Inhalation	Expiration	Inhalation	Expiration	
1	A	M	9	9	9	9	0.31	0.70	-Right in front -Y-shirt
2	B	M	10	10	10	10	0.36	0.41	-Left 40° -T-shirt
3	B	M	15	15	15	15	0.26	0.43	-Left 40° -T-shirt
4	B	M	21	21	21	21	0.36	0.40	-Left 40° -Naked
5	C	F	9	9	10	10	0.17	0.37	-Left 40° -Shirt
6	D	F	10	10	10	10	0.71	0.65	-Left 40° -Shirt
7	E	F	10	10	10	10	0.62	0.95	-Left 40° -Shirt
8	F	M	10	10	10	10	0.36	0.48	-Left 40° -T-shirt
9	G	M	10	10	10	10	0.28	0.77	-Left 40° -T-shirt
10	H	M	10	10	9	9	0.80	1.21	-Left 40° -T-shirt
11	I	M	10	10	10	10	0.95	0.95	-Left 40° -T-shirt
12	J	F	10	10	9	9	1.47	0.50	-Left 40° -Sweater

TABLE IV. AVERAGE NUMBER OF BREATHS AND DETECTION DELAY IN TABLE III

		Male	Female
False-Detection	Inhalation	0.01	0.05
	Expiration	0.01	0.05
Detection delay (sec)	Inhalation	0.46	0.74
	Expiration	0.67	0.62

Table IV showed the average of the false detection rate and detection delay. We think the false detection rate was low enough for the auscultation practicing system. However, some measured detection delays were longer than 1 second. We have to improve the detection scheme for inhalation and expiration so that shorter detection delays occur and a high detection rate is possible. The other hand, since the proposed system can start to measure movement of upper body just after recognized a student playing a patient, it is possible to avoid this problem by detecting respiration cycle before a stethoscope placing on a body.

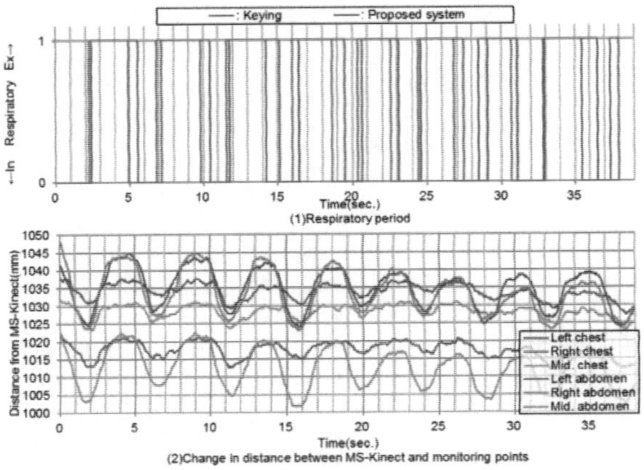

Figure 15. (1) Respiratory period, and (2) change in distance between KINECT v2 and measuring points

As for the high detection rate, KINECT v2 did not find any extraordinary values for every participant, as shown in Figure 15. Because Figure 15 (2) is an example of abdominal respiration, the measured values for every abdominal measurement point varied widely and clearly. No extraordinary values are evident in this figure.

We could not clarify the reason for the extraordinary outputs found by KINECT v1 from these experiments. However, we will develop our system using KINECT v2 in the future.

2) Detecting respiration for the back body

We tried to utilize the previously described program to detect the inhalation and expiration during respiration for the back body. We noticed that the detection results for the back body were opposite to those for the front body in pre-experiment. When a participant breathed in, our system recognized it as expiration. Also, we noticed that the motion of the back body was different from that of the front body. Hence, we measured the motions of both shoulders in addition to the previously described six points for 13 participants. Five of them were female.

There were many participants who beyond our expectation. The measured data values for only two of the thirteen participants varied clearly and widely enough to detect the inhalation and expiration in both the chest and abdomen. Also, we could not get enough data to detect them for other participants. Most clearly changed data corresponded to the breathing shown in Figure 16, and less than optimal data are shown in Figure 17 and Figure 18.

We noted the following issues from these measured data:

- All measured data on the back chest changed to less than the data on the abdomen.
- In the case of abdominal respiration, the shoulders almost never move up and down.

Measurement points on which data clearly changed depended on the person. For example, for both Participant A in Figure 16 and B in Figure 17, the shoulders moved up and down for chest respiration. The other hand, the shoulders of Participant C in Figure 18 did not move.

However, we do not think that this issue is serious problem. The reason we would like to detect motion of upper body derived from respiration is that we think a student playing a nurse feels unnatural when sounds of expiration or inhalation do not synchronize with movement of the upper body. When there is not any movement of a back body, there is not any synchronization between respiration sounds and motion of back body. Therefore, a student playing a nurse would not care for the replay timing of respiration sounds, when a back body did not move. In prototype system, when the system detects motion of back body, respiration sounds are replayed to synchronize with movement of a back body: when the system cannot detect them, respiration sounds are automatically replayed with respiration cycle measured for a front body.

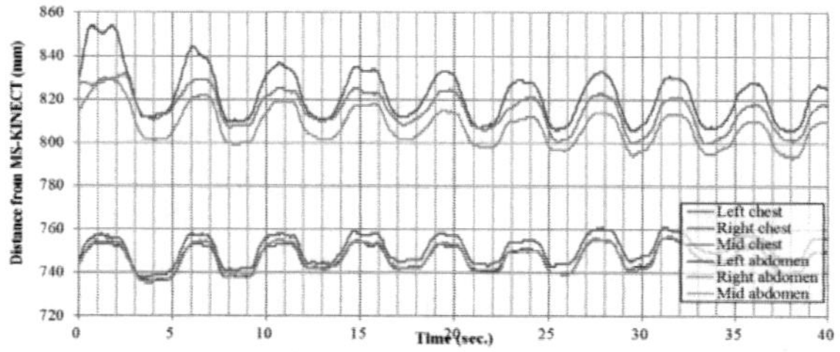

(1) Chest respiration (Back body)

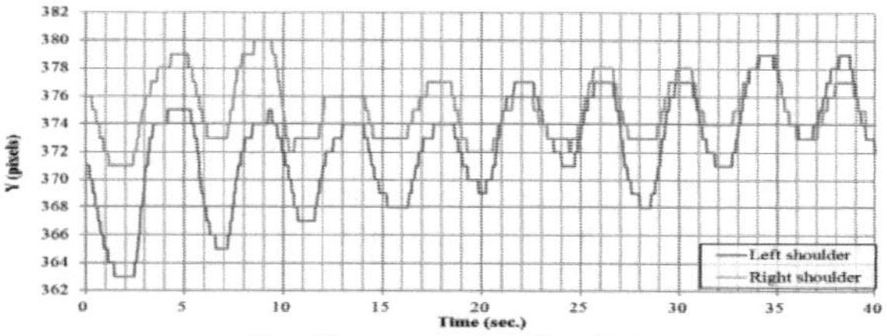

(2) Chest respiration (Shoulder)

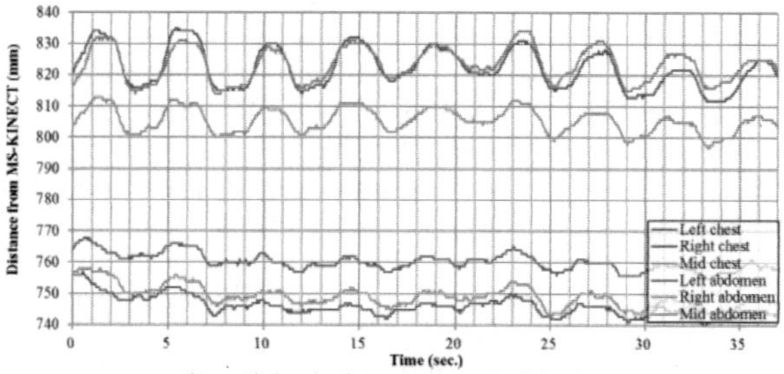

(3) Abdominal respiration (Back body)

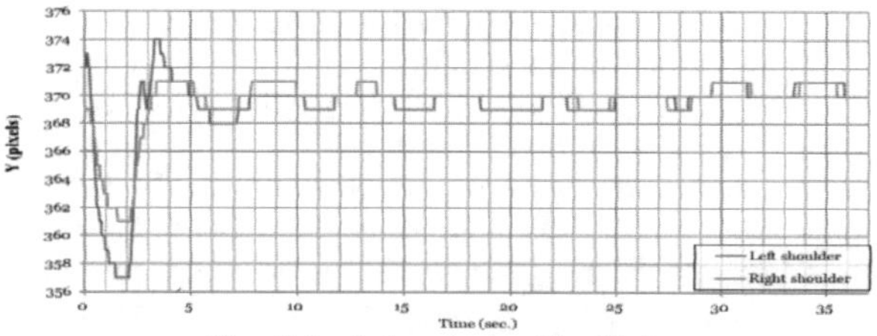

(4) Abdominal respiration (Shoulder)

Figure 16. Example of clearly motion in the back body
Participant A, Sex: Male, Cloth: Prepared T-shirt

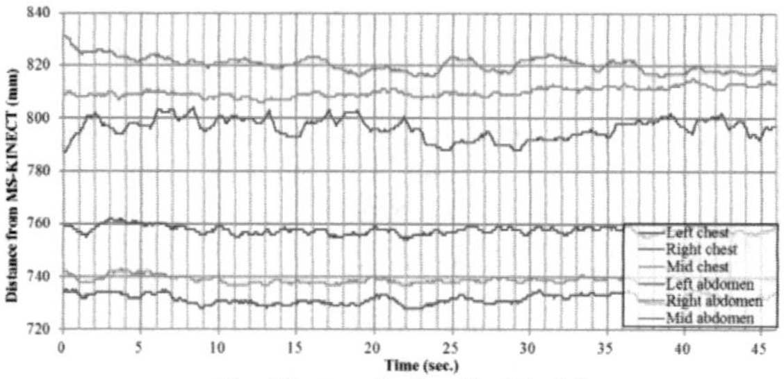

(1) Chest respiration (Back body)

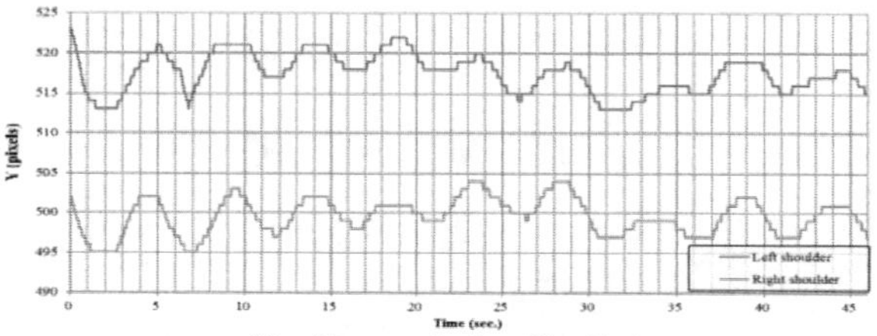

(2) Chest respiration (Shoulder)

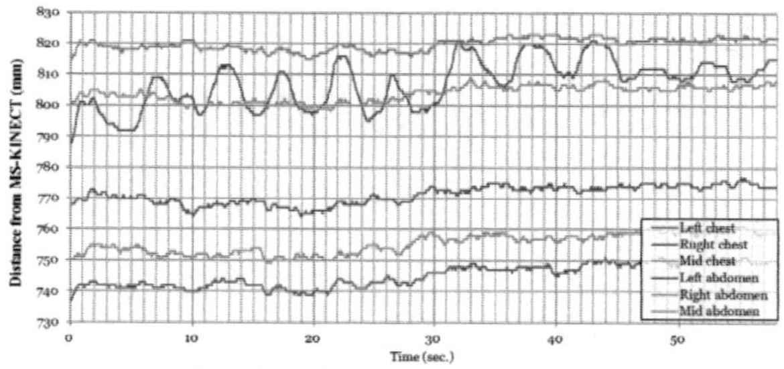

(3) Abdominal respiration (Back body)

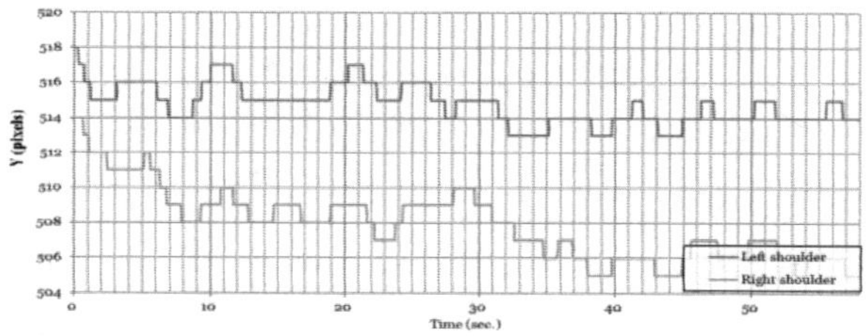

(4) Abdominal respiration (Shoulder)

Figure 17. Example of unclearly motion in the back body (1)
Participant B, Sex: Female, Cloth: Her own loose shirt

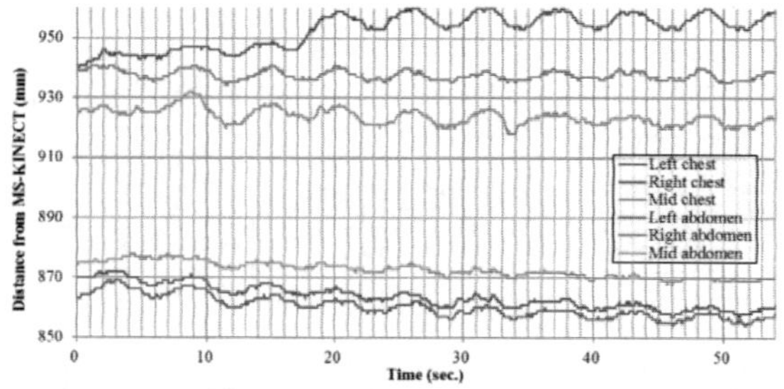

(1) Chest respiration (Back body)

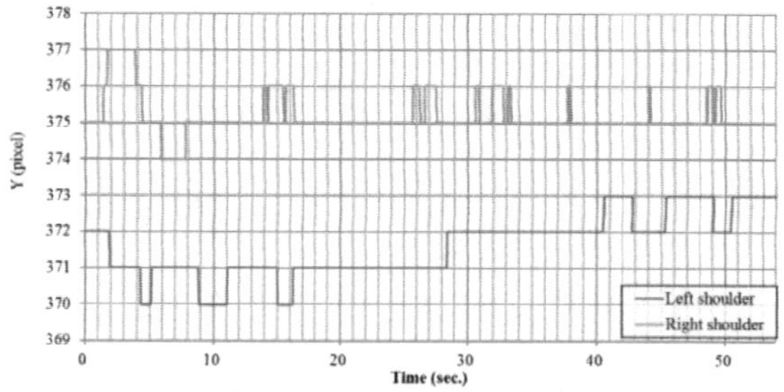

(2) Chest respiration (shoulder)

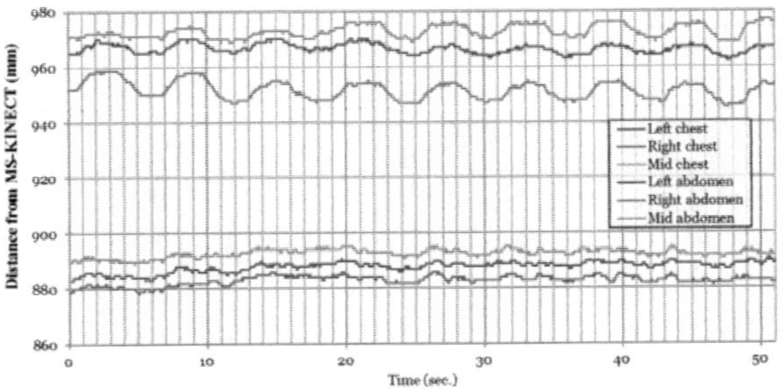

(3) Abdominal respiration (Back body)

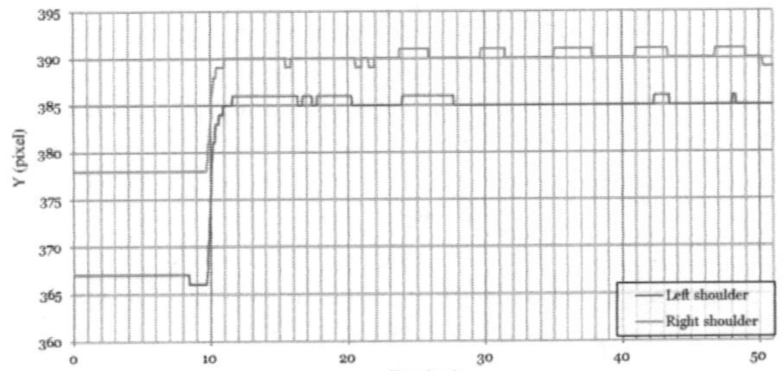

(4) Abdominal respiration (Shoulder)

Figure 18. Example of unclearly motion in the back body (2)
Participant C, Sex: Male, Cloth: Prepared T- shirt

D. *Automatically adjusting according to body size*

The correct points for placing a stethoscope depend on the size of the body; they differ a little between men and women. Also, their X and Y values vary according to the distance between a student playing a patient and a KINECT. Therefore, we estimate correct positions for placing a stethoscope with respect to the positions of both shoulder joints and both hip joints.

Two sets of correct position data and the positions of both shoulders and both hips are measured and stored as standard man and woman data. Since the origin of the output of KINECT's depth camera is at the upper left, it is difficult to compare body sizes among people. Therefore, for each person we reset the origin to the junction of a horizontal line connecting the average right and left hip heights and a vertical line passing through the midpoint between the right and left hip heights. Here, we assume that a person is sitting upright, and that the human body is a little unsymmetrical. The result of resetting the origin for each person is illustrated in Figure 19.

The estimated left-side (X_{LE}, Y_{LE}) and right-side (X_{RE}, Y_{RE}) locations are calculated with the following equations by using the above stored standard locations and the measured positions of both shoulders and both hips:

$$X_{LE} = X_{LS} * X_{mls}/X_{sls} \quad , \qquad (1)$$

$$Y_{LE} = Y_{LS} * Y_{mls}/Y_{sls} \quad , \qquad (2)$$

$$X_{RE} = - X_{RS} * X_{mrs}/X_{srs} \quad , \qquad (3)$$

$$Y_{RE} = Y_{RS} * Y_{mrs}/Y_{srs} \quad , \qquad (4)$$

where
(X_{LS}, Y_{LS}) is the left-side standard location,
(X_{sls}, Y_{sls}) is the standard left shoulder position,
(X_{srs}, Y_{srs}) is the standard right shoulder positon,
(X_{pls}, Y_{pls}) is the measured left shoulder position of the patient, and
(X_{prs}, Y_{prs}) is the measured right shoulder positon of the patient.

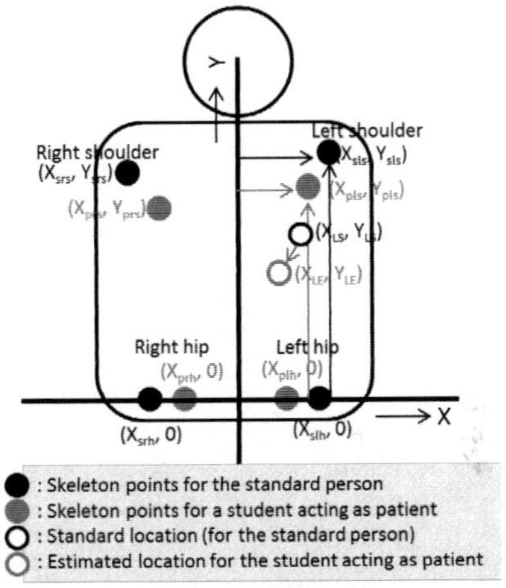

Figure 19. Illustration of adjusting locations for body size

We measured the ten points marked on a T-shirt and skeleton data for three men and three women to validate the proposed automatically adjusting algorithm. Each man wore the same T-shirt like that in Figure 11.

Since the T-shirt was relatively small, it should have closely fit each person, and we predicted the marked points should have adjusted to each person correctly. After selecting one man and one woman each as the standard, we estimated correct points a stethoscope placing by using the above equations and the data for the standard person.

TABLE V. COMPARISON BETWEEN MEASURED AND ESTIMATED LOCATIONS

Location		Male-1		Male-2		Female-1		Female-2	
		X	Y	X	Y	X	Y	X	Y
2	measured	28	136	46	157	53	167	52	157
	estimated	33	142	42	154	35	164	36	155
4	measured	28	118	43	140	55	144	53	133
	estimated	32	127	40	137	45	140	46	131
6	measured	27	94	44	118	52	123	50	107
	estimated	31	107	39	116	44	120	45	113
8	measured	30	51	43	78	55	68	50	59
	estimated	33	72	41	77	56	68	57	64
10	measured	40	36	49	59	62	52	65	54
	estimated	38	57	48	62	63	55	65	52

Since there was not big difference between left-side data and right-side data, we showed only the left-side data in Table V. The location relationships between measured and estimated correct points for a female and male participant are shown in Figure 19. In these figures, shoulder points and correct points a stethoscope placing for a standard person, and shoulder point and estimated and measured correct points a stethoscope placing for other participants are presented.

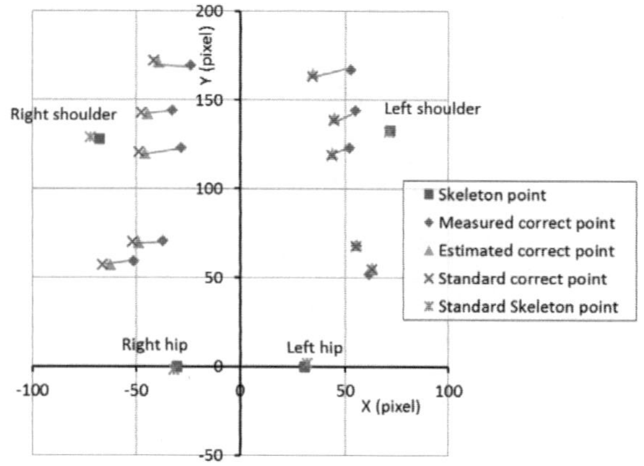

(1) Female-1

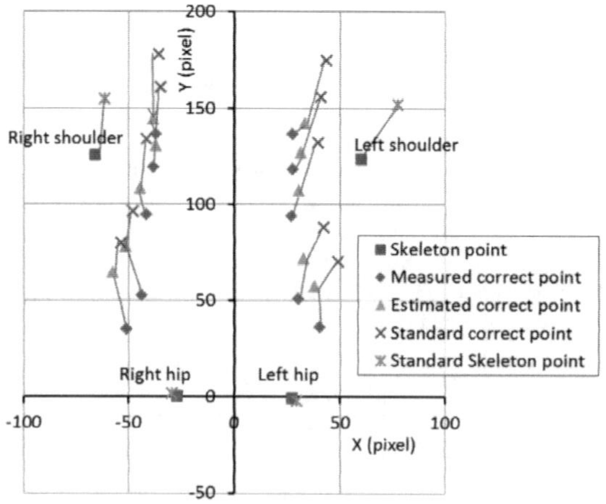

(2) Male-1

Figure 20. Example of measured and estimated correct points

Data unit in Figure 20 was not the millimeter, but the pixel. In these figures, 10 pixels corresponded roughly to 4 cm. Differences between estimated and measured positions depend on participants and measuring points. Maximum difference was about 8 cm. In case of Figure 20 (1), shoulder and hip positions of Female-1 were the same as those of a standard participant. Therefore, estimated correct positions of Female-1 were the same as those of a standard participant. However, locations of measured points were different from location of estimated ones in Y axis. The reason for this difference must be that each person did not wear the T-shirt symmetrically in the horizontal direction. In case of Figure 20 (2), differences between estimated and measured Y values were bigger for lower points, because the T-shirt was less elastic in the vertical direction.

We think the above experiment does not appropriate to evaluate the proposed estimation scheme for adjusting locations to different body sizes after experimenting. We try to devise an appropriate scheme to evaluate this technology.

V. Application

As we described already, every necessary technology to implement a practical auscultation practice system has been achieved. Therefore, we developed a simple application for a teacher to explain practicing auscultation for the lungs.

Our co-authors in the nursing course showed the following functions:

- Dividing the upper body: The lungs and the abdomen were divided, and the right and left lungs were divided up and down at the pit of the stomach level.
- Assigning sounds: An appropriate sound was assigned to each divided area for teaching content.

We developed two programs:

- Sound assigning program: This program assigns a sound to each area, dividing the upper body into five areas.
- Auscultation teaching program: This program detects on which area a stethoscope should be placed, and it replays a sound assigned to that area.

Figure 21 (1) shows the operation window to assign a sound file to each divided area. The front upper body was divided into five areas. The upper three points are the location on both shoulders and the bottom of the throat that was obtained from the Microsoft skeleton program. The pit of the stomach was also obtained from the Microsoft skeleton program. The abdomen is the isosceles right triangle whose top is the pit of the stomach. First, a teacher selects an area. Then, he or she selects a sound by clicking the right radio buttons according to the taught content. Disease sounds are managed using the operation window shown in Figure 21 (2). Because the disease sounds were different for the expiration and inhalation, both of them were prepared. Prepared sound files were assigned to expiration and inhalation for each disease.

Figure 22 shows an example window of the developed teaching program. When a stethoscope is placed on various areas, the area on which the stethoscope is placed changes yellow.

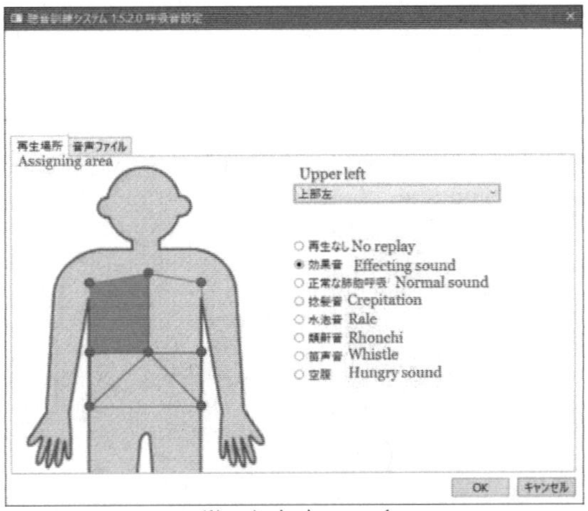

(1) Assigning sound

(2) Management of sounds

Figure 21. Operation window to assign a sound to divided body area

Figure 22. Operation window of the auscultation teaching program

VI. EVALUATION

We think our system is useful for not only students in medical and nursing schools but also for practicing nurses to continue their learning. Hence, we asked nursing students and practicing nurses to test whether or not our system was useful for them. Unfortunately, we did not ask medical students. For this questionnaire, we used the first prototype for which a student heard one sound when a stethoscope was placed on anyplace of the upper body, not the second prototype written in Section V.

A. *Nursing course student*

We asked six nursing students who took physical assessment including practicing auscultation. The questions asked of them and the results are as follows;

Q1: Could you set up this system by yourself?
Yes: 6, No: 0

Q2: Did you hear both a normal sound and a disease sound when you placed a stethoscope on the body?
Yes: 5, No: 1

Q3: Did you hear sounds prior to placing a stethoscope on the body?
Yes: 5, No: 1

Q4: Which simulator is more useful to practice auscultation, the existing mannequin or our system?
Equal: 2, Our system: 4, Mannequin: 0

Q5: Would you like to use this system to practice auscultation?
Yes: 6, No: 0

Their comments were as follows:

(1) Disease sounds can be learned in addition to learning communication with patients.
(2) Imagining a practical patient is easy.
(3) Learning is possible while practicing auscultation naturally.
(4) I'm very interested in this system, because it could utilize a wide range of applications.
(5) I feel the timing of the replayed sounds was a little different from that in real situations.
(6) I feel the replayed sounds were a little different from those in real life, because I could not hear the heartbeats.

On the basis of the answers for Q3 and comments for (5), we redesigned the replay timing of sounds so that they could be heard naturally.

B. *Practicing nurse*

We asked the following questions to 50 practicing nurses. Their age composition is shown in Figure 23. Most of them had several years of experience.

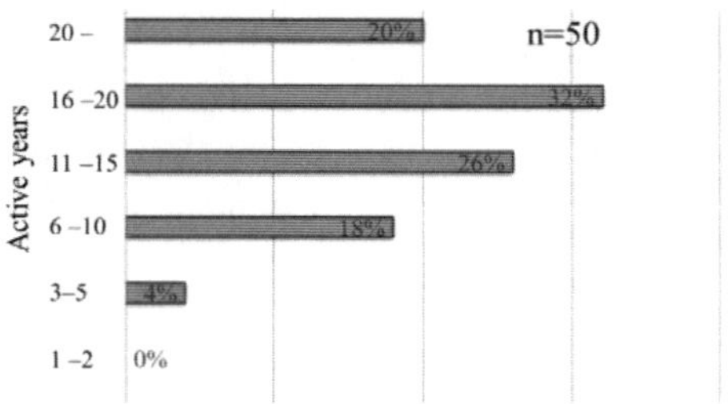

Figure 23. Age composition of questionnaires to active nurses

Q1: Are you interested in this system?

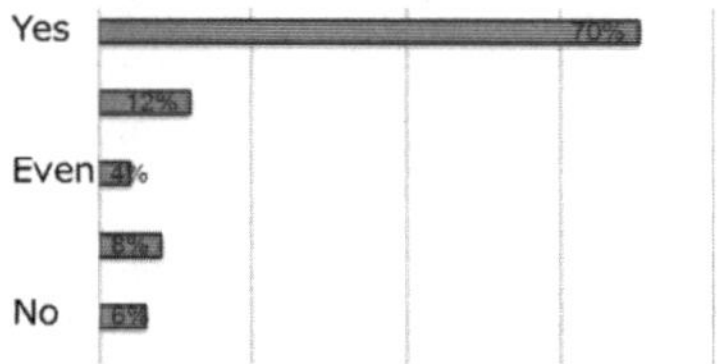

Q2: Does our system imitate real clinical situations?

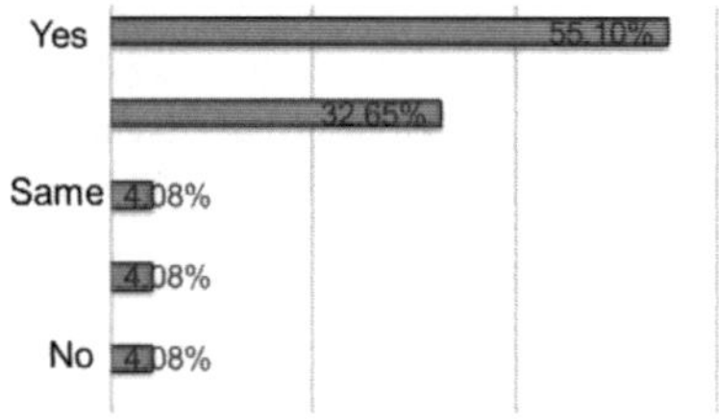

Q3: Is it useful for a student to learn auscultation?

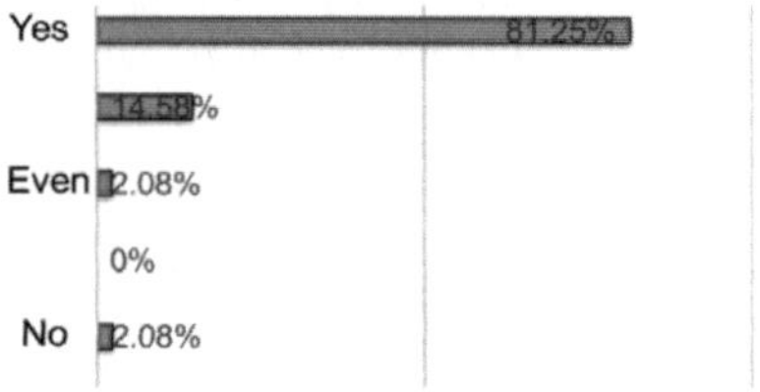

Q4: Is it useful for a practicing nurse to learn auscultation?

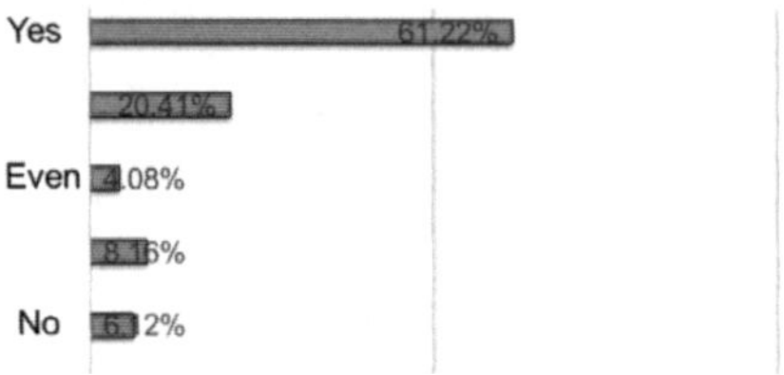

Most of them evaluated our system highly for both nursing students and practicing nurses to learn auscultation. Also, most of them thought our system imitates real clinical situations.

VII. CONCLUSION

We proposed a new auscultation practice system for medical and nursing students. In this system, students themselves play the role of a patient instead of a humanoid model, and the locations for stethoscope placement on the body are measured with KINECT. Therefore, this practice system has low cost. In addition, the system can judge whether or not stethoscope locations are correct. Practicing students hear disease sounds, synchronized with the movement of breathing, through earphones or a speaker.

We developed a prototype system and evaluated it experimentally. The results showed that our system could perfectly detect stethoscope placement on a body, except for three lower points, and that it could detect respiratory changes. Also, it perfectly detected the inhalation and expiration during respiration for the front body. However, the detection delay for respiratory changes was slightly larger than expected. We found that most people breathed without any movement on their back body.

We asked nursing students and practicing nurses to test whether or not our system was useful for them. The results of a questionnaire showed that our system was useful for them to learn auscultation. However, some of them felt the replay timing of sounds was a little different from that in real situations.

We think that it is possible to solve or avoid such problems; and plan to provide production service in a future.

REFERENCES

[1] Y. Murata, N. Miura, and Y. Endo, "Proposal for A KINECT-Based Auscultation Practice System," in Proc. eTELEMED, 2016, pp. 86-91.

[2] Patient simulators for nurse & nursing care training, Kyoto Kagaku Co., Ltd., https://www.kyotokagaku.com/products/list02.html#cate_head01 [retrieved: December, 2016]

[3] Lung Sound Auscultation Trainer "LSAT", Kyoto Kagaku Co., Ltd., https://www.kyotokagaku.com/products/detail02/m81-s.html [retrieved: December, 2016]

[4] Sakamoto Model Corporation, Sakamoto auscultation simulator, http://www.sakamoto-model.com/product/emergency/m164/ [retrieved: December, 2016]

[5] Laerdal, SimMan 3, http://www.laerdal.com/us/SimMan3G [retrieved: December, 2016].

[6] J. Butter, W. C. McGaghie, E. R. Cohen, M. E. Kaye, and D. B. Wayne, "Simulation-Based Mastery Learning Improves Cardiac Auscultation Skills in Medical Students," Springer, Journal of General Internal Medicine, Volume 25, Issue 8, 2010, pp. 780-785.

[7] SimScope WiFi (The Hybrid simulator), Cardionics http://www.cardionics.com/simscope [retrieved: December, 2016].

[8] Meet Kinect for Windows, https://developer.microsoft.com/en-us/windows/kinect, [retrieved: March, 2016]

[9] U. von Zadow, S. Buron, T. Harms, F. Behringer, K. Sostmann, R. Dachselt, "SimMed: Combining Simulation and Interactive Tabletops for Medical Education," in Proc. CHI '13, Proceedings of the SIGCHI Conference on Human Factors in Computing Systems, April 2013, pp. 1409-1478.

[10] Open CV, http://opencv.org/ [retrieved: December, 2016]

[11] N. Burba, M. Bolas, D. M. Krum, and E. A. Suma, "Unobtrusive Measurement of Subtle Nonverbal Behaviors with the Microsoft Kinect," IEEE, Virtual Reality Short Papers and Posters (VRW), pp. 1-4, 2012.

Continuous Noninvasive Arterial Blood Pressure Monitor with Active Sensor Architecture

Viacheslav Antsiperov
Department of Physical and Quantum Electronics
Moscow Institute of Physics and Technology
Dolgoprudny, Moscow Region, Russia
e-mail: antciperov@cplire.ru

Gennady Mansurov
Laboratory of Diagnostic Systems
Kotel'nikov Institute of REE of RAS
Moscow, Russia
e-mail: gmansurov@cplire.ru

Abstract—**The paper presents the latest results of developing a new method of noninvasive continuous blood pressure monitoring. This method is based on the principle of pulse wave compensation. It is shown that sensors for such a measurement should be not only smart, but also active. In this connection, the concept of smart sensors is expanded to the concept of active sensors. The technical design of the active sensor for noninvasive pressure measurement is described. The results of active sensor calibration and testing are under discussion. The last section of the report is devoted to the development of software for active sensor control - its intellectual stuffing. In this section is described and justified a new principle of active measurement of quasi-periodic processes - pulse wave compensation based on prediction patterns. The progress achieved in the research and the ways for further investigation are outlined in the conclusion.**

Keywords - smart and active sensors; compensation method; arterial blood pressure monitoring.

I. INTRODUCTION

Advancements in semiconductor industry substantiate a dramatic growth of rate of innovations in a wide range of human activities - from computer technology and communications to home appliances and transport. This acceleration is due mainly to delivering smaller, cheaper, more reliable solutions able to integrate critical functions such as analog circuits, and embedded microprocessors in a single device. New packaging technologies and methodologies have also improved the cost and reliability of the assembly of packages with small footprints. For example, chip-scale packages have reduced semiconductor package sizes to the dimensions of the die inside. Other innovations, such as the use of a known good die or foldable printed circuit boards, have also provided considerable miniaturization in the overall space occupied by the electronic portion of the modern devices.

Medical technology is a broad field where the above innovations play a crucial role in sustaining health. The development of medical devices and equipment has made significant contributions to improving the health of people all around the world. From "small" innovations like healthcare gadgets that help a person to monitor and manage serious health conditions like asthma, heart problems (see Fig. 1), diabetes, etc., to larger, more complex technologies like MRI machines, artificial organs, and robotic prosthetic limbs, technology has undoubtedly made an incredible impact on medicine.

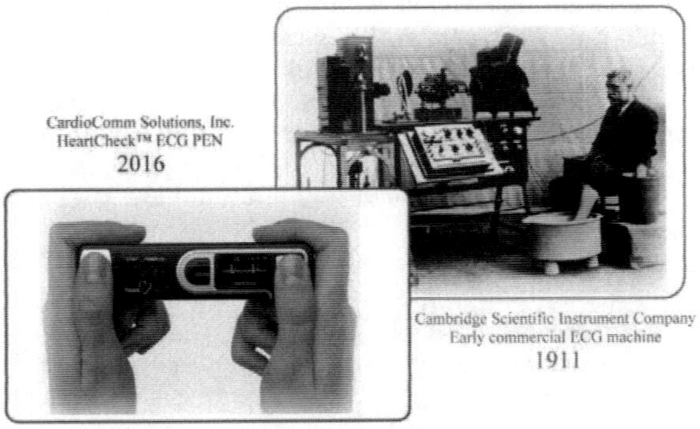

Figure 1. An example illustrating the incredible progress in medical technology: the ECG monitor evolution from the first commercial huge size machine (1911), consuming a lot of electrical energy, to a pocket-sized healthcare gadget (2016), powered by a compact battery.

It should be noted that the development of low-cost, low power, multi-functional medical devices, that are small and can communicate in wireless manner over short distances, like ECG monitor shown in Fig. 1, is only the visible tip of the modern medical technology iceberg. Its main core is that medical devices are getting smarter. Embedded microprocessors, SOC (System On a Chip), SBC (Single Board Computers), integral PDAs (Personal digital assistants) and advanced sensor technology are all common elements of today's medical devices. The computing power of microprocessors and high performance peripherals provide the ability to support several simultaneous processes: the sensor management, signal processing, data collection and analysis, wired and wireless communications and user interface. This gives unlimited possibilities in the design of medical devices. On the other hand, it leads to myriad ways of combining the components in developing well-defined purpose devices, and, accordingly, it requires to investigate a lot of possible solutions and to find the optimal one.

In this paper our experience in the development of a modern high-tech medical device - arterial blood pressure

(ABP) monitor is presented and the results obtained in its implementation, including the main result - the elaboration of the basic architectural components concept, named an Active Sensor by analogy with the Smart Sensor (see Fig. 2) are discussed. Initially, these results have been briefly considered at the Second International Conference on Smart Portable, Wearable, Implantable and Disability-oriented Devices and Systems (SPWID 2016) [1]. In this article they will be discussed in detail.

The structure of the paper is as follows. Section II provides a detailed discussion of the Smart Sensor and the Active Sensor concepts. It also gives strong arguments in favor of the Active Sensor concept for developing medical devices. Section III gives a short overview of existing non-invasive ABP monitoring methods. Section IV introduces a new method of ABP monitoring proposed by the authors – *pressure compensation method.* Section V describes in detail the structure and design of the sensor developed. In Section VI the results obtained in real experiments with active ABP monitoring are reported. In Section VII some ways to overcome the shortcomings associated with the incomplete compensation of time–varying blood pressure caused by a simplified structure of the PID (proportional–integral–derivative) feedback controller are scheduled. Finally, the Conclusion briefly summarizes the problems discussed in the article.

II. FROM SMART SENSORS TO ACTIVE SENSORS

One of the main trends in developing modern devices and instruments in medicine as well as in a rather broader field – the development of industrial control and monitoring systems – is a full implementation of smart sensors [2].

The basic architectural components of the Smart Sensor are shown in Fig. 2 (a more detailed description can be found in IEEE 1451 Standards, created by the IEEE Instrumentation and Measurement Society's TC-9 Technical Committee on Sensor Technology). The Smart Sensor consists of four main components: the *Sensing Unit* (SU) that collects real-world analog data, the *Signal Conditioning* circuitry and the *Analog to Digital Converter* (ADC) destined to condition and convert data to the digital domain, and, finally, *microprocessor* (μP), which by means of Applications installed performs the main tasks of the Sensor. Task processing can be supported by *Memory* and *Communication Unit*; some results can be displayed by *Local User Interface* (IEEE 1451 general model of a Smart Sensor).

The most important component of the Smart Sensor architecture is certainly a microprocessor (μP). As mentioned above, the computing power of microprocessors provides both the ability to support multiple simultaneous processes and the processing of data associated with each of them. Data processing can be very complicated: nonlinear, adaptive to data variability, incorporating artificial intelligence (AI) elements, etc. However, it should be noted that due to the unidirectional data flow from SU to the microprocessor, the latter does not affect the conditions of data measurement – it can only manipulate in some way the data measured.

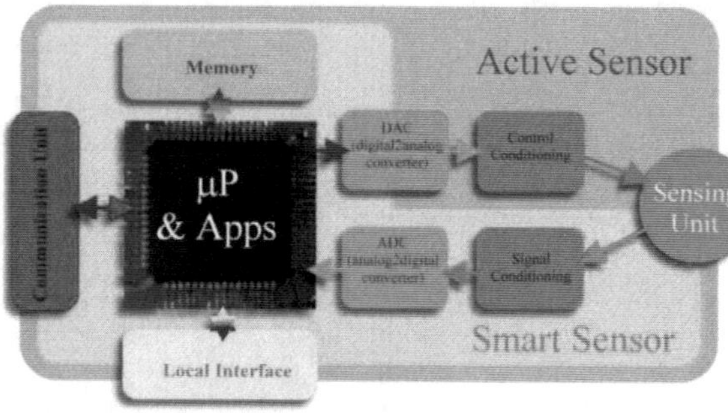

Figure 2. The principal components architectures of Smart Sensor (green) [2] and Active Sensor (blue) [1]. The latter extends the circuit of Smart Sensor by adding the Sensing Unit control sub-circuit.

The above fact is of little importance in artificial systems, whose behavior is stationary or is forecasted at least statistically. The situation is different when somebody deals with living systems, e.g., with a human. As noted in [3], due to highly non-linear, non-equilibrium nature of living systems even the slightest differences in the physiological state of the bio-system used and in the conditions obtained in a experiment can, in consequence of deterministic chaos, assume singular importance.

We came face to face with this situation and its inherent problems when a few years ago began to develop a smart medical device – arterial blood pressure (ABP) monitor. At that time we thought that supplying the classic tonometer by some tracking system, followed by block forming pressure dynamics prediction and flexible active sensor controller based on the prediction, we will be able to monitor ABP continuously in time in absolute units – in mmHg. It is possible because such a device would have to implement the proposed earlier *compensation method* for measuring the pressure. Of course, to compensate the variable pressure in real time that monitor had to include powerful computational tools. We believe, that the use of embedded microcontroller with smart applications can produce much better results than a standard digital tonometer can. So, the use of a built-in microcontroller (μC) and its periphery gives the monitor characteristic features of smart sensors, as shown in Fig. 2. But it turned out, that for the realization of this idea we had to go even one-step further – to introduce a SU control feedback loop, which depends on data observed. Thus, the idea of an Active Sensor was created, whose components architecture expands the architecture of the Smart Sensor.

The basic architectural components of the Active Sensor are shown in Fig. 2. The Active Sensor circuit extends the circuit of the Smart Sensor by adding the Sensing Unit control sub-circuit. In contrast to the data measurement path, the control sub-circuit is directed in the opposite way - from the microprocessor to SU, although composed essentially of the same, but performing the inverse functions components - *Digital to Analog Converter* (DAC) and *Control Conditioning* circuitry. Thus, if in the Smart Sensor SU is simply, in unidirectional way connected to the microprocessor, in the Active Sensor it is included in a loop circuit consisting of the data measuring and controlling sub-circuits.

Before moving on to the practical implementation of the above rather abstract concept of an Active Sensor, it is appropriate to review briefly the set of ideas and approaches existing today in the field of non-invasive arterial blood pressure monitoring.

III. ABP Monitoring Approaches

One of the first (1876) noninvasive devices suitable for measuring human arterial blood pressure was a plethysmograph invented by Etienne Jules Marey. Its functioning was based on the recording of pulsed changes in the limb volume caused by ABP pulsations [4]. Upon measurements, the patient's forearm was placed into plethysmograph, which was filled with warm water. A change in the blood content in the forearm arteries resulted in water displacement from the plethysmograph, which in turn moved the float and stylus that recorded the plethysmogram, as shown in Fig. 3.

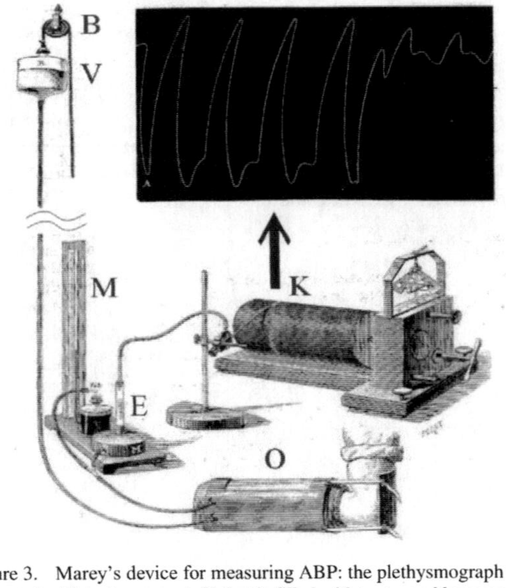

Figure 3. Marey's device for measuring ABP: the plethysmograph O is connected with both reservoir V that is filled with water and hung up on a block B and a mercurial manometer M; the volume change is written in sooty tape K by a stylus connected with float E [4].

In the Marey's experiment, volume changes were recorded at constant pressure in the plethysmograph O, which was set by the height level of water reservoir V and reflected in monometer M connected to O. In his experiments, Marey obtained many interesting results; he found the dependence of the plethysmogram amplitude on the plethysmograph pressure [4]. In principle, on the bases of these data, as it became known later, Marey could have determined the two main parameters of ABP, both systolic and diastolic pressure. However, due to difficulty of the implementation and ambiguity of interpretation of the result, Marey's plethysmographic method was not used for a long time (about 100 years). At the end of the 19th century, more attention was paid to methods of measuring ABP that are based on sphygmomanometry, monitoring of changes in the character of the heart pulses (tense pulse, soft pulse, etc.), which is manifested in micromovements of arterial walls. The main task of sphygmomanometer devices was to find such an external pressure on an artery that would balance the counterpressure of the blood with the bloodstream ceasing and the pulse, correspondingly, disappearing. The most successful sphygmomanometer out of the first devices was the device of Samuel von Bashe, created by him in 1881. With its help he determined systolic pressure, although only approximately. Gustav Gartner in 1899 improved this device and referred to it as a tonometer.

The search for a noninvasive way to measure ABP culminated in two discoveries that made successes in the cardiological studies in the 20th century possible: description of Riva-Rocci's sphygmomanometer in 1896 and discovery of Korotkoff's sounds in 1905. Riva-Rocci's method is compression of a brachial artery using a special cuff that relates to a mercurial manometer. Air is pumped into the cuff using a bulb until the pulse in a radial artery disappears and, then, it is slowly released. ABP is determined by manometer indications, when the pulse appears. Thus, the device makes it possible to assess systolic pressure. Korotkoff's discovery of the particularities of sound effects upon decompression of a brachial artery in 1905 underlies an alternative method for determining systolic and diastolic ABP. It is believed that Korotkoff's sounds relate to impairments in the laminar flow of blood in arteries upon their crossclamping. This impairment is accompanied by shock phenomena, and their sound makes it possible to hear the pulse within the range between systolic and diastolic ABP.

Technological advances in the field of microelectronics and computer engineering made it possible in the second half of the 20th century to improve Riva Rocci's and Korotkoff's classic method for measuring ABP [5]. The improvements were mainly connected with the intention to maximally exclude the effect of the human factor on measurements. Thus, it was found that the most convenient estimate of the pulse for automatic procedures of determination of parameters of ABP is the amplitude of pressure oscillations in an occlusion cuff, a fact that was discovered by Marey a hundred years before! In 1976 the first bedside automatic instrument for measuring ABP (Dinamap825), which employed a modified method of Marey's, was produced. The transition from a plethysmograph to a typical occlusion cuff combining a device for creation of external compression and a pulsation sensor, which is not very accurate, but acceptable for measuring ABP, gave birth to a new oscillometric technique. Today, a portion of automatic and semiautomatic AP meters that are based on the oscillometric technique constitute approximately 80% of all the devices. Manufacturers intensely compete in the field of study of algorithms for processing measured data to increase the accuracy and reliability of the results obtained. Recent devices have an increased intelligence level (including fuzzy logic models), and they could be fully related to smart sensors (see Fig. 2).

The methods mentioned assume a single measurement of the ABP parameters. It is often sufficient for general assessment of the state of the cardiovascular system. However, there are many situations that require continuous beat-to-beat ABP monitoring. At the worst, it is implemented invasively directly inside one of the peripheral arteries upon introduction of a catheter into it. However, since invasive methods relate to patient discomfort and there is a risk of complications upon introduction of a catheter, noninvasive methods are preferable for routine measurements.

Jan Peňáz was among the first to put forward a method of continuous non-invasive blood pressure monitoring [6]. His method described in 1973 was aimed at reducing the risks of arterial cauterization. The Peňáz method employed the idea of "vascular unloading", based on the assumption that in the "unloaded state" the pressure inside the blood vessel is equal to the outside pressure.

The basic element of the device proposed by Peňáz is a small finger cuff (Fig. 4 (A)) that has an infra-red (IR) light source on one side and a light receiver on the opposite side. The blood volume of the finger is estimated via the absorbance of IR light. The signal obtained by such a plethysmograph is further used in a feedback loop to control the pressure in the cuff. The pressure is controlled in such a way that blood volume in the finger is kept constant in time and equal to the volume, which corresponds to the unloaded vessels state defined during the calibration process. In this case the oscillations of the controlling pressure are approximately equal to pressure in the arteries. Later some formulae were proposed for recalculating pressure from finger vessels to brachial arteries, which made it possible to verify the method with respect to classical procedures.

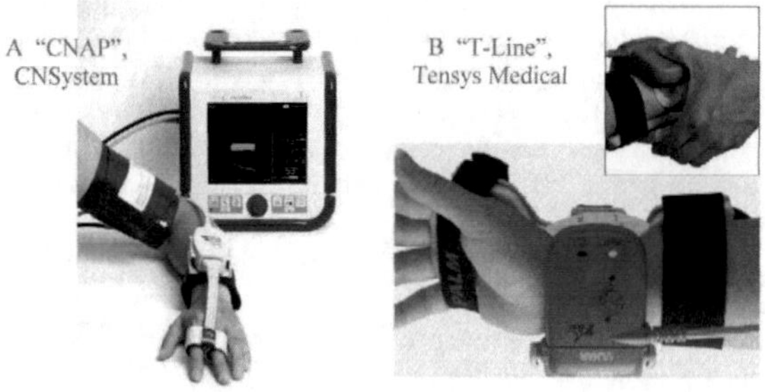

Figure 4. Modern approaches to ABP monitoring systems.

Another technique, which provides continuous non-invasive blood pressure monitoring is arterial tonometry [7]. Like the Peňáz method arterial tonometry is based on pulse oscillation estimates, but here the principal of arterial unloading is different. In this case the cuff is placed on the wrist, so the sensor is over the radial artery (Fig. 4 (B)). The sensor presses the artery to the radial bone until it is flattened enough but not occluded. At this intermediate position arterial wall tension becomes parallel to the tonometer sensing surface and arterial pressure is then the remaining stress (perpendicular to the surface) measured by the sensor. The pressure needed to flatten but not occlude the artery is known as the "proper hold-down pressure" and is calculated by a complicated algorithm, which includes the preliminary estimate of systolic, diastolic and pulse pressures over a range of "hold-down pressures".

Currently, the devices employing these methods include the CNAP^{TM} (Peňáz's approach, Fig. 4 (A)) and T-line from Tensys Medical (arterial tonometry approach, Fig. 4 (B)).

The methods of continuous non-invasive blood pressure monitoring have both advantages and disadvantages [5]. We believe that the main drawback of these non-invasive methods is the following: irrespective of the method of vascular unloading the control of the unloading is exercised using integral parameters (blood vessels filling, overall sensor force, sensor displacement, etc.). It enables monitoring an average pulse wave of ABP but does not guarantee the details of the pulse form. We proposed a new method of arterial blood pressure monitoring that is aimed at local unloading of arterial walls by compensating local pressure.

IV. Compensation Based Approach to ABP Monitoring

When analyzing well–known methods of non–invasive continuous measurement of blood pressure, we conclude that the best results of monitoring non–stationary dynamics of blood pressure are achieved by the so–called compensation methods or methods like them.

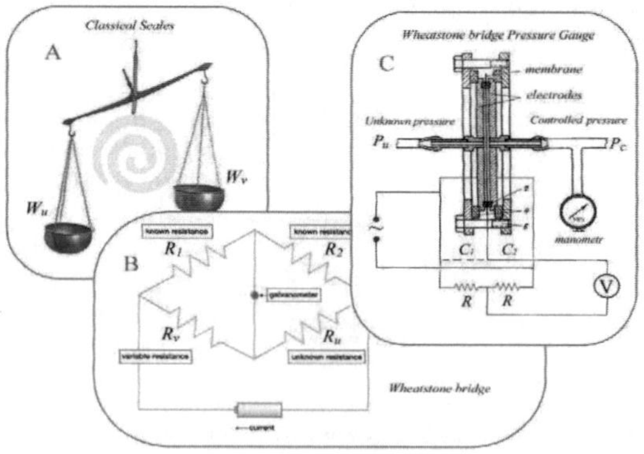

Figure 5. Measuring physical quantities using the compensation method: (A) measuring object weights W_u using a balance scale; (B)a bridge circuit of measurement of unknown resistance R_u; (C) measuring pressure P_u in aggressive media by controlling the undeformed state of a membrane.

Compensation methods are applied for measurement of various physical quantities and are based on the compensation of an unknown measured value by controlled counter value and nullification of their difference. The simplest example of the compensation method is the use of balance scales on which unknown mass W_u is measured using a set of weights W_v, Fig. 5 (A). A scale null indicator is a vertical position of a beam or connected arrow pointer.

The compensation methods are high-precision, applied for mechanical and electrical measurements, and can be implemented in the form of bridge and half-bridge circuits (Figures. 5 (B), 5 (C)). In some cases, to increase measurement precision at deviations from zero that are less than the discreteness of a controlled physical quantity, a calibrated scale of a null detector can be used. Note that compensation methods are usually used to measure the static variables – constant unknown resistance R_u in bridge circuit – Fig. 5 (B) and the steady pressure P_u in aggressive environment – Fig. 5 (C).

We consider the compensation method as the fundamental basis to measure the varying blood pressure. The application of this method for measuring the non–static dynamic quantity has become possible for two reasons. First, the fact that blood pressure changes are not so fast, its rhythm is of the order of one beat per second, and its spectrum fits into the range of a few tens of Hz. Second, there are relatively cheap high-performance microcontrollers (ATMEL, MICROCHIP, STMicroelectronics, etc.) available at present, for which the ABP dynamics represents almost quasi-static changes.

V. The Structure and Design of the Sensor

As noted above, our method of measuring ABP is aimed at local, without cuffs, unloading of arterial walls by means of compensating intra–arterial pressure by controlled pressure. Our method of measurement is like the method of bridge measurement of unknown pressure P_u in Fig. 5 (C), but it has dynamic, "adaptive" features as in the Peňáz method [6]. The appearance of the device developed for continuous ABP monitoring and its typical use for measuring intra–arterial pressure is presented in Fig. 6.

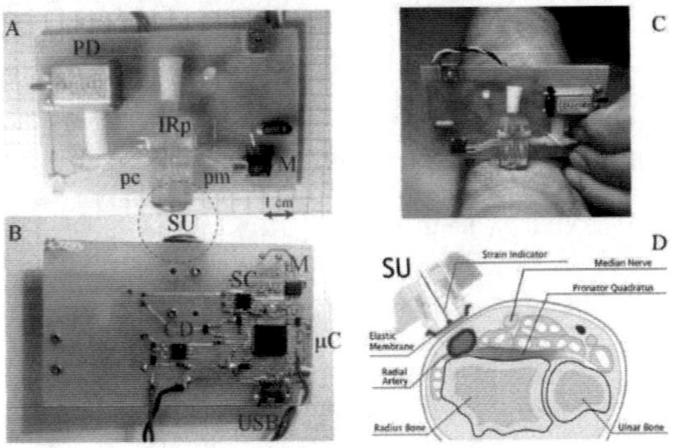

Figure 6. Active sensor for continuous blood pressure monitoring with a fluid-filled SU camera and an aperture covered by an elastic membrane; its typical arangement on the patient's wrist.

Note that the device developed (Figures 6 (A), 6 (B)) conforms exactly to the concept of an Active Sensor (see Section II, Fig. 2). The main role here is played by the programmable microcontroller µC (STM32L152RBT), that has the communication I/O (USB) on one side, and on the other side it has *pm* (pressure monitor – data measuring sub-circuit) and *pc* (pressure control – SU controlling sub-circuit) interfaces to the measuring unit SU (sensor unit).

The data measuring sub-circuit *pm* contains membrane displacement indicator SC (signal conditioning) and pressure gauge with instrument amplifier M. The SU controlling sub-circuit *pc* includes the pump PD and the chip for pump control CD (control driver). Both sub-circuits – *pm* and *pc* – are tubes coming from the fluid-filled cavity of the measuring unit SU. SU is a camera with an aperture covered by an elastic membrane (of red color in Fig. 6(D)) and containing a thin rod with one end attached to the membrane, the other end partially overlaps IR-radiation flux inside optoelectronic infrared pair IRp. Thus, this rod gives us a way to measure the membrane deformation as it serves as an indicator of displacement.

When the pressures in the SU chamber and directly behind the membrane differ, the latter is deformed with respect to one side or another and the rod is shifted in the direction of the deformation. This displacement leads to a change in the shuttering of the radiation flow IRp, which is then recorded by the displacement indicator SC. If the SU pressure is chosen so that the membrane returns into a flat, undeformed state, then the pressure out of SU can be measured due to the equality of the pressures on both sides of the membrane. If SU is established directly above the radial artery as it shown in Figures 6 (C), 6 (D), then we measure the arterial pressure under the assumption that it is equal to the external pressure. This constitutes the idea of the compensation method that we proposed.

Our last improvement of the ABP monitor presents a somewhat different technological SU implementation. Here a fluid, filling SU camera is substituted by the air. We had to make this step as the previous construction of the ABP monitor had to be supplied by sealed SU camera to prevent fluid leakage. To prevent the permeability, it requires to supply the monitor by additional control tools and additional mechanical components, which could bring about a considerable complication of device setup. Thus, it was decided to develop a monitor with a chamber filled with air, because small air leakage being acceptable in this case.

Moreover, it turns out that a substitute of working liquid by air opens new ways of SU construction implementation – a controlled air leakage could be used as a toll controlling pressure in the SU chamber. It turned out that in this case it can be done without the elastic membrane covering the aperture! The skin surface within SU aperture could act as a membrane.

The appearance of the ABP monitor with air filled SU chamber and its general structure are shown in Fig. 7. Besides the main chamber with compensating air pressure P over aperture (similar in construction to fluid filled SU chamber) there are channels diverting air with their chambers and pressure P_{out} gauges. These channels are located near the aperture on the surface of the SU contacting with wrist. Thus, here like in the previous version of ABP monitor is possible to pick out two sub-circuits – pm (data measuring sub-circuit) and pc (controlling sub-circuit). The pm sub-circuit by means of an outer pump controls the compensating pressure P in a main chamber of SU. The pc sub-circuit measures the difference $\Delta P = P - P_{out}$, based on which the control is carried out. The control is aimed at providing such meanings of ΔP, that ensure minimum air flow from aperture to the channels diverting air. This implies the existence of the thinnest channel between the SU surface and the surface of wrist. In other words, the aim of the control is to keep the dynamics of the compensating pressure in such a way that the surface of the skin would be separated from the SU, but would not move away from it, as shown in Fig. 7 (C). In general, construction described reminds a sensor for classical tonometry forward in [8], but the advantage of our ABP monitor is that it enables to make direct pressure measurement and it does not have mobile mechanical parts.

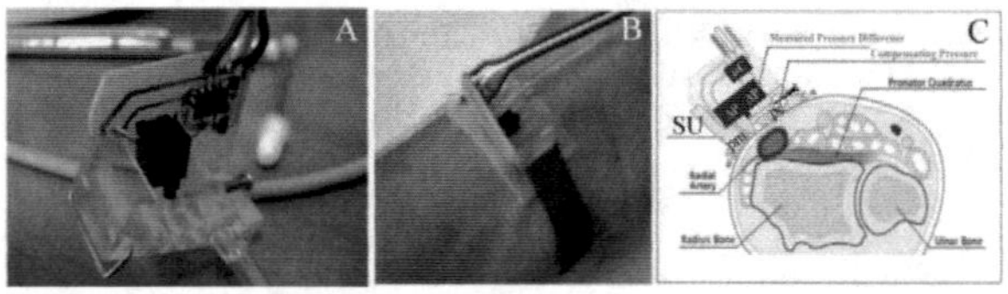

Figure 7. Active sensor for continuous blood pressure monitoring with an air-filled SU camera and without a membrane; its typical arangement on the patient's wrist.

The new construction of ABP monitor undoubtedly fully corresponds to the concept of the Active Sensor. Moreover, it uses the same microcontroller and applications exercising data processing and control were modified slightly. The first experiments gave the same results, which were obtained by the authors earlier and will be described in the next section.

VI. Experiments with the Sensor Developed

The qualitative description of the functioning of the ABP monitor given above could be illustrated by quantitative results obtained by authors in the experiments with the device. Fig. 8 (A) gives the dependence of the SU membrane deformation displacement (in units of the ADC of the displacement indicator SC) on a uniform increase in the pressure inside SU at constant external pressure (the pressure difference at the membrane P is the indications of M in mmHg that are counted from the atmospheric pressure). It follows from the plots that in small (±5 mmHg) range of pressure difference on both sides of the membrane the signal of the displacement indicator SC is proportional to it. Fig. 8 (B) presents similar dependences of the displacement of the membrane of SU and pressure P inside SU (that is counted from the atmospheric pressure) upon the compensation of uniformly increasing external pressure, with the pressure control *pc* being on. The compensation pressure P is created by a pump, when the voltage with pulse-width modulation, which in turn is formed by the microcircuit CD, is applied to its driver PD.

Without specifying the details of the mechanism of formation, note that the resulting P is proportional to the control signal y, which is directed by the microcontroller μC to CD in response to the membrane displacement ε that is measured by SC (and preceding stored values). The control algorithm was initially chosen as the simplest PID controller [9], with its control signal y_j at the time moment (reading) j being formed according to the classic recurrence formula:

$$y_j = y_{j-1} + k_I \varepsilon_j + k_p (\varepsilon_j - \varepsilon_{j-1}) + k_d (\varepsilon_j - 2\varepsilon_{j-1} + \varepsilon_{j-2}), \quad (1)$$

where y_{j-1} is the preceding signal, ε_j, ε_{j-1}, ε_{j-2} are the current and preceding displacements of the membrane from the flat position, and k_I, k_p, k_d are the PID integral, proportional and derivative coefficients, that are selected empirically. It follows from the plots in Fig. 8 (B), that using PID controlling (1) it is possible for a sufficiently wide range of changes of the external pressure (10–170 mmHg) to keep the membrane of SU near the equilibrium position.

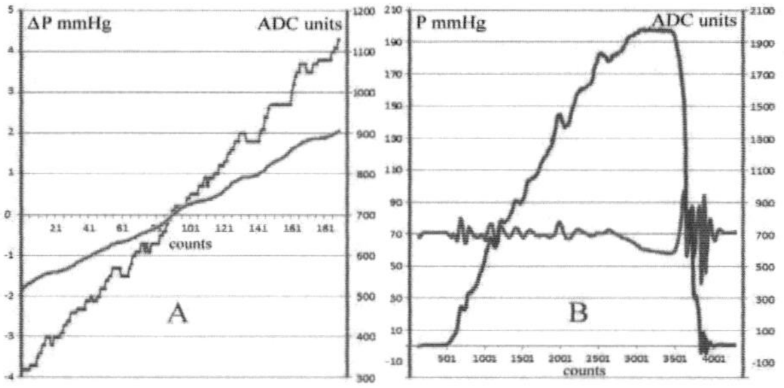

Figure 8. The dynamics of the SU membrane displacement (ADC units) in response to changes of the pressure difference inside and outside the SU.

Based on the calibration data numerous experiments on monitoring blood pressure were carried out. The results of one of the measurements are given in Fig. 9. The lower part of the figure graph shows that the use of the PID control makes it possible to hold the membrane close to undistorted (flat) state for all the time of blood pressure measurement. The quality of such regulation can be estimated by the value of uncompensated difference in pressures equivalent (proportionate) to deviations of the membrane from the flat position (a kind of the membrane jitter).

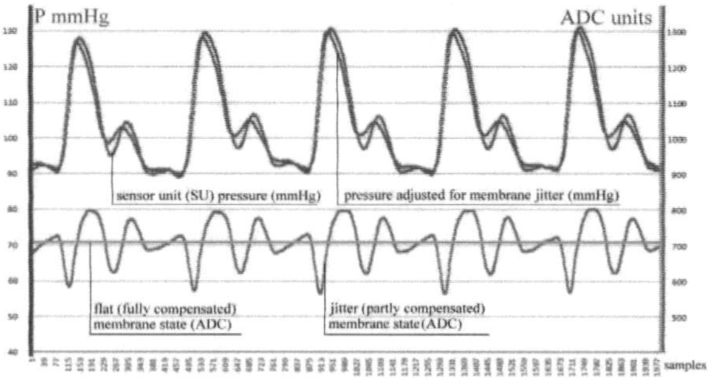

Figure 9. The dynamics of the SU membrane displacement (ADC units) while compensating the time-varying outside blood pressure.

To illustrate the compensation Fig. 9 (top) shows a graph of pressure P inside the chamber SU (the data from pressure gauge M), as well as the same data corrected by the membrane jitter values. In comparison with P the uncompensated difference in pressures ΔP is not large. Fig. 9 also shows that P, on average, coincides with the estimate of ABP, but at the moments of a rapid change in the pulse waveform the controller cannot trace these changes.

The overall impression of the first results of monitoring blood pressure with the help of the above–described active sensor is ambivalent. On the one hand, the sensor does its task – it can be used not only to measure the main parameters of blood pressure – systolic and diastolic blood pressure (range of ABP), but also to observe the varying pulse wave, and not through indirect measurements but by measuring the pressure itself (in mmHg). On the other hand, as it could be initially assumed, a simple PID controller does not provide full compensation, which leads to the distortion of the pulse wave components. Practically, the compensation algorithm based on PID control proved to be very difficult in changing its settings.

VII. Pulse Pressure Compensation Control on the Basis of Pulse Wave Patterns

These shortcomings associated with the incomplete compensation of time–varying blood pressure have a simple explanation.

Firstly, the PID controller used is the special case of linear regulators class and it is well known that linear methods for treating biological signals, especially, linear adaptation methods are used in limited ranges of their changes and only under strictly controlled (for example, laboratory) conditions. This is due to highly non-linear non–equilibrium nature of living systems [3]. Even a small change in physiological state can lead to considerable changes in the result.

Secondly, PID control takes into account only local characteristics of the signals, as it is customary in the theory of dynamic systems. In the case of dynamic systems, such a regulation is natural, as such systems are deterministic. However, living systems and their subsystems are known to be poorly described by models of dynamic systems, even if there is a freedom of choice of corresponding differential equations coefficients. Living systems are much more consistent with models of stochastic, non–deterministic systems.

Thirdly, in solving technical problems low order of PID control is preferred due to its ease of implementation. When somebody deals with a complex biomedical signal, particularly, an ABP signal, low order of control is a drawback.

These facts indicate that the pulse wave of blood pressure is much more like a wideband pulse signal than the sum of harmonic components. In view of the foregoing it can be assumed that a random point process is a closer mathematical model of blood pressure pulsating. A point process is an increasing sequence of time moments (points) of certain homogeneous events, such as the arrival pulse wave, with a random length of time intervals between them. Excluding the changes in the pulse wave pattern, and treating them as a stream of homogenous sequence of events, it would be legitimate to use theoretical methods dealing with the problems associated with the heart rate.

One such well-developed theory today is the theory of radar signal processing. A major problem of this theory is determining the unknown arrival time of an electromagnetic pulse emitted by the radar and reflected from a target. The method of matched filtering should be noted as one of the most effective methods for solving this problem. In this method the filter response is matched with the radar pulse, so that the maximum signal of the filter output is observed at the time, when the reflected pulse arrives. Essentially, the matched filter generates a covariance of sent and received pulses, and it is well known that if both pulses have the same, up to a constant factor, shape, the maximum of the covariance will be achieved when they coincide. The corresponding displacement between pulses will estimate the time of arrival.

The principle of the matched filtering could be interpreted in a different way, more suitable for our tasks. Let us assume at some moment, the time T between the ABP pulse wave is known along with the shape of the current pulse, further considering the shape of the next pulse to be the same as the current one, the future of signal can be easily predicted.

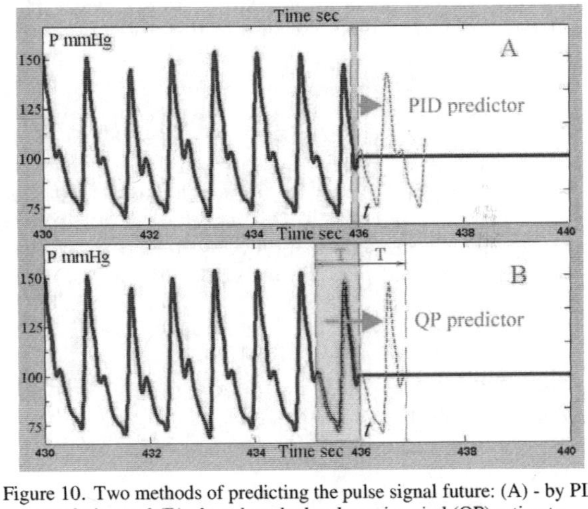

Figure 10. Two methods of predicting the pulse signal future: (A) - by PID regulation and (B) - based on the local quasi-period (QP) estimate.

For this purpose, it should take an existing fragment of the signal of duration T, immediately preceding the current time and move it to the time T into the future (see Fig. 10). From the theory of the matched filter, it follows that the expected ABP pulse will coincide with the prediction. This idea is illustrated in Fig. 10, which also shows a comparison of the proposed signal prediction with the prediction carried out by PID. As stated above, the PID controller makes a prediction based on current, local estimates of the fundamental frequency, the amplitude of the main component and amplitudes of neighboring harmonics.

VIII. Conclusion and Future Work

Summing up the results of the investigation it should be noted that for compensatory ABP measurement specifically the current pattern of the pulse wave efficiently predicts the expected signal. This pattern must be dirigible enough to change significantly with changes in the state of the object measured as well as changes in the conditions of its active measurement.

For this reason, realized by the classic regulators including the PID controllers, patterns in the small parametric models of ABP waves are of little use for active blood pressure measurement. An idea of forming the adequate patterns in the task is given in the above qualitative reasoning. It is based on the property of quasi-periodicity of ABP signal.

This property lies in the fact that high variability of the period of heart contractions occurs only at long time intervals, whereas at short time intervals of several seconds or a few heart beats its changes are generally small and fall within a couple of percent.

Therefore, estimating the current period, more precisely the quasi-period T and getting a pulse wave pattern as a signal fragment of T duration immediately preceding the current time moment gives us a good pattern of the pulse wave. Thus, the task of building a pattern for current pulse of the signal is reduced to the task of effective evaluation of its current quasi-period.

The results of developing a new method for monitoring ABP using active sensors and their control based on the pulse wave patterns that have been discussed in this study have demonstrated their high potential in numerous experiments that have been performed by us to test and improve the device developed. This makes it possible for the authors to voice the hope of good prospects of further development up to its adoption in healthcare practice in the very near future.

Acknowledgment

This work was supported by RFBR (Russian Foundation for Basic Research) grant 14-07-00496-a.

References

[1] V. Antsiperov, G. Mansurov, and B.Bonch-Bruevich, "Development of an Active Sensor for Non-invasive Arterial Blood Pressure Monitoring," Proc. of the 2-d International Conference on Smart Portable, Wearable, Implantable and Disability-oriented Devices and Systems (SPWID 2016), 2016 - Valencia, Spain, pp. 1–5, ISBN: 978-1-61208-480-0. Available: http://www.thinkmind.org/index.php?view=article&articleid=spwid_2016_1_10_80004/ 2016.11.12.

[2] M. Chaudhari and S. Dharavath, "Study of Smart Sensors and their Applications," Intern. J. of Adv. Research in Computer and Comm. Engineering, vol. 3(1), Jan. 2014, pp. 5031–5034.

[3] W. Klonowski, "From conformons to human brains: an informal overview of nonlinear dynamics and its applications in biomedicine," Nonlinear Biomedical Physics, vol. 1, article 5, 2007, DOI: 10.1186/1753-4631-1-5. Available: https://nonlinearbiomedphys.biomedcentral.com/articles/10.1186/1753-4631-1-5 / 2016.11.12.

[4] E.-J. Marey, "Pression et vitesse du sang," in Physiologie Experimentale, Masson, Paris, 1876.

[5] E. Chung, G. Chen, B. Alexander, and M. Cannesson, "Non-invasive continuous blood pressure monitoring: a review of current applications," in Front Med, vol. 7(1), 2013, pp. 91-101.

[6] J. Peňáz, "Photoelectric measurement of blood pressure, volume and flow in the finger," Digest of the 10th International Conference on Medical and Biological Engineering, Dresden, 1973, p.104.

[7] G. Drzewiecki, J. Melbin, and A. Noordergraaf, "Arterial tonometry: review and analysis," in J Biomech., vol. 16(2), 1983 , pp. 141-152.

[8] G. L. Pressman and P. M. Newgard, "A Transducer for the Continuous External Measurement of Arterial Blood Pressure," in IEEE Transactions on Bio-medical Electronics, vol. 10, no. 2, 1963, pp. 73-81.

[9] K. Astrom and T. Hagglund, "PID Controllers: Theory, Design and Tuning," 2nd Edition, Triangle Park, North Carolina: Instrument Society of America, 1995.

Web Accessibility Recommendations for the Design of Tourism Websites for People with Autism Spectrum Disorders

Antonina Dattolo
SASWEB Lab, DIMA
Università di Udine,
via Santa Chiara, 1
34170 Gorizia, Italy
email: antonina.dattolo@uniud.it

Flaminia L. Luccio
DAIS
Università Ca' Foscari Venezia,
via Torino 155,
30170 Venezia, Italy
email: luccio@unive.it

Elisa Pirone
Università Ca' Foscari Venezia,
Venezia, Italy
email: elisapirone06@gmail.com

***Abstract*—The universal Web design represents an ambitious and open challenge for the current research on the Web. Key aspects are Web accessibility and Web usability by people with the widest possible range of abilities, operating within the widest possible range of situations. Universal design is adaptive for the users, and provides personalised answers to different users. A recent study shows an estimation of the global burden of people with Autism Spectrum Disorders. In the last 24 years, the incidence of autism has a stable prevalence of 7.6 per 1000 or one in 132 persons. They represent a significant number of people. People with Autism Spectrum Disorders are usually solitary and visual thinkers and they could take advantage by the use of the Web. This paper discusses of tourism, website and people with Autism Spectrum Disorders. The aim is to define a set of recommendations for the design of tourist websites for people with Autism Spectrum Disorders, and to present a case study articulated in two tourist, autistic-friendly, websites. The first website considers the area of Rieti, central Italy; it has been validated through expert reviews, and several trials with many autistic, verbal users of a specialised centre for neurological and physical disabilities. The second website contains as a tourist destination the area around Mestre, close to Venice, Italy. In this case, the website has been validated on a single, non-verbal autistic user.**

Keywords–Autistic-friendly website; Accessibility; Usability; Tourism; Autism Spectrum Disorders.

I. INTRODUCTION

In this paper, we consider the problem of developing accessible and usable tourism websites dedicated to people with Autism Spectrum Disorders (ASD), and we extend some preliminary results presented in [1].

Web accessibility and usability represent an open challenge for the current research on the Web. They aim to make Web content more accessible and usable to a wider range of people with disabilities, including blindness and low vision, deafness and hearing loss, learning disabilities, cognitive limitations, limited movement, speech disabilities, photo sensitivity and combinations of these [2].

Usually, Web developers apply general standards, following the guidelines provided by the World Wide Web Consortium (W3C) [3][4], without considering the importance of a universal Web access. However, every country supports the application of the Web Content Accessibility Guidelines (WCAG) 2.0 [2], although adopts a different Web accessibility legislation. WCAG 2.0 are recommended, but become mandatory only in specific cases, such as, for example, in Italy, for the Web sites of public Institutions (see the so-called "Legge Stanca", literally Stanca Act [5]).

In this paper, we are interested in the development of tourism, accessible and usable, websites dedicated to people with ASD. We thus first consider the relation between autism and tourism, and show the existing solutions for tour proposals to these specific users; then, we present the state of art of existing guidelines for the development of websites for people with cognitive disabilities, and extend them with new features specialised for people with ASD. To the best of our knowledge, there are no existing websites that are usable and accessible for people with ASD, and let them freely navigate inside them and take independent decisions (as, e.g., the choice of a tour). We thus propose an interesting case study of two touristic websites for autistic users and we show some experimental results on a set of 9 verbal users, and 1 non-verbal user.

The paper is organised as follows: Section II presents ta brief description of the ASD; it is followed by Section III, dedicated to the state of art in autism and tourism. Section IV introduces the themes of usability and accessibility, which are deepen in Section V, where the discussion is specialised for users with cognitive disabilities; guidelines and related work are presented. Section VI revisits the recommendations for people with ASD. Section VII proposes our case study, focused on accessible and usable tourism, autistic-friendly websites. Finally, Section VIII describes the experimentation made with a group of autistic verbal and non-verbal users and the related results. Section IX ends the paper with a look to future work.

II. AUTISM SPECTRUM DISORDERS

The American Psychiatric Association defines the ASD as neuro-developmental disorders with persistent impairments in social communication and social interaction, and restricted, repetitive patterns of behaviour, interests, or activities [6]. This general definition puts together people with a wide and different set of features and behaviours; so, it is difficult create multi-purpose environments, but some general guidelines may be followed. We have focused our attention on ASD user profiles with the following features.

Typically, the impairments in social communication are related to *language delays*, however, people with ASD often present good visual abilities. Thus, modern speech therapies are usually combined with Augmentative and Alternative Communication (AAC) techniques, i.e., powerful methods that combine different visual components in order to create syntactically and semantically correct sentences [7][8]. Among these AAC techniques a standard approach is to use the Picture Exchange Communication System (PECS), which promotes interactions among users with the exchanging of images [9]. People with ASD are usually visual thinkers, i.e., think in pictures and express their concepts by visualising sequences of pictures of the same object [10]. They have often *limited attention*, i.e., limited time in which they might be able to concentrate on a particular task, thus defining limited, self-contained tasks is very important. They might also suffer of *Sensory Processing Disorders* (SPD), and may thus feel distressed and confused in particular situations. SPD are neurological disorders that cause difficulties with processing information from the five senses (vision, auditory, touch, olfaction, and taste), as well as from the sense of movement (vestibular system), and/or the positional sense (proprioception) [11].

III. AUTISM AND TOURISM: RELATED WORK

The UNTWO (World Tourism Organization) [12] recommends in several publications cited in [13], among the priorities for a global tourism, the accessibility of tourism websites. The numbers are not negligible. In a recent report in [14], ENAT (European Network for Accessible Tourism) indicates, only in Europe, a request for accessible tourism of 350 millions of people with different disabilities; more specifically, the global incidence of people with ASD is of 7.6 per 1000 (one in 132 persons) [15].

For people with ASD, some issues that have to be considered are: social inclusion, e.g., in community activities, emotional well being, e.g., happiness, interpersonal relations, e.g., friendship, and physical well-being, e.g., leisure [16]. All this, may be achieved with a well planned tourism trip, being it an experience that stimulates all the above domains. However, this is not trivial since all the different activities related to a trip, i.e., transportation, accommodation, etc. have to be adjusted to meet special needs.

The first issue that has to be considered is the preparation phase since it is well known that autistic people get stressed in unknown, unexpected situations, thus everything has to be previously planned, and the user has to be prepared to tackle each of the trip steps.

In [16][17] the authors identify some of the issues that a travel agency or whoever organises the trip should consider while planning it for an autistic user:

- *Survey phase*: A questionnaire should be provided in order to identify, which are the main physical, sensory, or language problems the user has, and which are his/her main interests. According to this, the agency can suggest suitable tourism destinations or trips. E.g., a person that gets stressed in very crowded, loud or noisy places should avoid amusement parks, crowded exhibitions, etc.

- *Trip planning*: At this point a trip may be planned by taking into consideration all the provided information. The first thing is transportation, so how to move from a home place to the destination. This can be achieved in different ways, what is important is that all the steps are previously explained to the user. If the transportation is by air the airline company has to be contacted. Some airports offer some pre-planned visits of the airport [17], the airport of Dublin also offers some very interesting on-line material, which visually describes the different parts of the airport, following a logical sequence of actions, which a traveller is supposed to do. It also provides interesting tips on how to face all the different situations while being at the airport and also some general rules to follow during a trip [18].

- *Accommodation*: What seems to be more suitable for autistic users, are small size accommodations, such as small hotels and motels, or preferably the renting of a condo where the user can "feel at home". In the USA, there are some hotels that already provide "autism-friendly accommodations" with special rooms, staff, and meals [16].

- *Tours*: What concerns the tourism activities, is obviously based on the preference of the user. Thus, following the guidelines of the survey is a good starting point. Here, we have to distinguish between one day trips, which obviously have to be planned in the surrounding area, avoiding too much stress for the trip, and longer trips of two or more days. There are obviously different options, depending on the preferences and also considering the sensory problems, which autistic users might have. Possible options are sightseeing in scenic areas since autistic users often like silent places and love taking pictures, visiting historical and cultural sites, museums, aquariums etc. Some museums are "Autism-Friendly Museums", and are prepared to host these special tourists. An example is the Royal air force museum in London, which offers an autism friendly trail that can be downloaded from the site, and has won an Autism Access Award [19], or the Metropolitan museum of art in New York, which offers e.g., a social narrative (PDF) about visiting the Museum with tips, and a Sensory Friendly Map of the museum with the list of quiet and less crowded areas [20].

Obviously, general considerations are that first one should prepare the autistic user to very simple day trips, then these trips might be planned for a longer time. Daily activities should not be too intense, locations should be visited in times of the day during which places are not too crowded, i.e., stressful situations should be avoided as much as possible. A good idea is also to prepare the user for the changes of routine by illustrating the trip using brochures, photos, videos, etc.. A last consideration is that, often, things might not go as expected, thus the trip should be planned with flexibility in order to include last minute changes of planes, i.e., alternative destination, activities and so on.

The site Autistic Globetrotting is an example of a site developed for people with ASD (see [21]). It has the aim of encouraging families with people with ASD to travel around the world; it provides interesting hints for travelling, e.g., how to let a person with ASD pack his/her luggage, interesting destinations and hotels, and so on. It also states that the benefits of travelling outweigh the problems one will face. For instance, travelling provides a hand on experience on some topics, which are studied in books at school (also people people without ASD would benefit from it). Mathematics for calculating money exchange or tips, art by visiting museum and galleries, geography by physically moving from a city to another, literature by visiting homes of famous authors, and so on. It increases flexibility not only in the daily routine but also in the dietary restrictions by encouraging the experience of tasting different food, in the interaction with other people by enhancing social and language skills, in raising family bondings, etc.

IV. Website usability and accessibility

Access to information and communication technologies is a basic human right as recognised in the United Nations Convention on the Rights of Persons with Disabilities (CRPD) [22]. Disabilities might be either physical (visual, auditory, etc.), or cognitive and neurological, and they might be temporary or boundless.

The W3C [2][3] has proposed different guidelines to produce accessible websites; in 2012, a W3C group, named Web Accessibility Initiative (WAI), publishes a draft containing some principles of Web accessibility for people with cognitive or neuronal disabilities [23]; it marks the birth of a task force group, named Cognitive and Learning Disabilities Accessibility Task Force (COGA). Recently, in August 2016, COGA, together the Protocols and Formats Working Group, and the Web Content Accessibility Guidelines Working Group, have published a list of general guidelines useful for most people with Cognitive and Learning Disabilities [4].

However, specific features have to be considered while dealing with people with specific disabilities, such as ASD. As we have previously mentioned, typical problems that people with ASD face are limited attention, sensory hypersensitiveness, different way of learning and reacting to things, problems related to limited text comprehension (e.g., not understanding figurative language, etc.). For such a reason standard usable and accessible websites might not be suitable for such users.

Web usability and accessibility are closely related and their goals, approaches, and guidelines overlap significantly [24].

Usability is all about designing an easy to use website that appeals to as many people as possible. websites should be intuitively usable. Usability is a *quality attribute* that assesses how easy user interfaces are to use; it is is defined in [25] by five quality components:

- Learnability: How easy is it for users to accomplish basic tasks the first time they encounter the design?
- Efficiency: Once users have learned the design, how quickly can they perform tasks?
- Memorability: When users return to the design after a period of not using it, how easily can they reestablish proficiency?
- Errors: How many errors do users make, how severe are these errors, and how easily can they recover from the errors?
- Satisfaction: How pleasant is it to use the design?

For Web developers, a key aspect of usability is following a User-Centred Design (UCD) process to create positive and emotional user experiences [26].

Accessibility is about ensuring an equivalent user experience for people with disabilities. For the Web, accessibility means that people with disabilities can perceive, understand, navigate, and interact with websites and tools, and that they can contribute equally without barriers.

Usable accessibility combines usability and accessibility to develop positive user experiences for people with disabilities. User-centred design processes include both techniques for including users throughout design and evaluation, and using guidelines for design and evaluation. UCD helps making informed decisions about accessible design. Thus UCD is necessary to improve accessibility in websites and Web tools.

The goal of Web accessibility is to make the Web work well for people, specifically people with disabilities. While technical standards are an essential tool for meeting that goal, marking off a checklist is not the end goal. People with disabilities effectively interacting with and contributing to the Web is the end goal [24].

We note that accessibility $\neq$ usability. A website may comply with standards, may pass all the automated accessibility checks, may appear to be accessible, however, it is not necessarily usable. Web pages can be verified accessible by focus groups, and still be inaccessible to a third party. Individual users may have cognitive, technical, or other barriers.

V. Usability and accessibility for people with cognitive disabilities: general guidelines and related work

The improvement of the quality of life for autistic people is a big and open challenge. Several studies recognise that the computer technologies are assuming a significant role for supporting the people with ASD [27][28][29]:

- They arouse the interest of these typology of users, that take major advantage from the interaction through devices with touch screens, such as tablets.
- They help them to develop their abilities and simplify some aspects of their lives.
- They represent a significant help for their families and therapists.
- The same research supports their effectiveness for people with ASD.

The computer technologies offer a large set of possibilities, such as websites, Web apps, affective computing, virtual reality, robotics, multitouch interfaces. In this paper, we mainly focus our attention on websites, and more in general, Web-based applications, since they play a major role in content display both for online and offline use, becoming more accessible with time, and giving the possibility to develop tools that easily meet users needs. Web sites, Web applications, and multitouch interfaces to the Web technologies assume a strong role on the basis of the following considerations [1][30]:

- They are within the reach of all.
- They do not need of specific pre-expertise in their use.
- They do not have elevated costs.
- They play a major role in content display both for online and offline use.
- They are becoming more accessible with time, cancelling the gaps among people, without discriminations.

- They may offer easy and fast customisation.
- They evolve in the time with the new Web standards, tackling problems that are typically linked to more traditional monolithic applications.

An interesting reading [31] is provided by Jamie Knight, autistic developer and senior accessibility specialist: he discusses on cognitive accessibility and provide some practical guidelines.

However, in the last ten years, some works [1][4][27][29][32][33] have been dedicated to define a formal set of guidelines for designing usable and accessible user interface for people with cognitive disabilities:

- In [32], the authors propose a review of the previous literacy and extract a list of 64 recommendations for people with cognitive disabilities, including ASD; only four of them have a frequency of citation, in the revised works, at least 50%. They are:
  1) Use pictures, icons and symbols along with text (75%).
  2) Use clear and simple text (70%).
  3) Consistent navigation and design on every page (60%).
  4) Use headings, titles and prompts (50%).

  The four recommendations can be regrouped in three macro-areas: language (items 1, 2), structure and navigation (item 3), and graphical layout (item 4).
- The recent report [4], cited in Section IV, identify detailed techniques that should enable content to be usable by people with cognitive and learning disabilities. They may be grouped in the following main macro-areas:
  1) Provide a clear structure.
  2) Be predictable.
  3) Use a clear writing style.
  4) Provide rapid and direct feedback.
  5) Include help meaning.
  6) Use a clear design.

  The six recommendations can be regrouped in same three macro-areas: language (items 3, 4), structure and navigation (items 1, 2, 5), and graphical layout (item 6). The macro-area no. 5 suggests the importance to include short tooltips on all icons, jargon; to introduce charts and graphs; to use symbols and images to show meaning [4].
- In [29], the authors analyse the previous literacy between 2005 and 2015 and identify 28 guidelines, specific for people with ASD; then they distribute them in the ten categories (deeply discussed in the cited paper [29]): four of them represent critical interface design aspects, according to the number of extracted recommendations and number of works from which the authors extracted the recommendations:
  1) Visual and textual vocabulary.
  2) Customisation.
  3) Engagement.
  4) Redundant representation.

  The 4 recommendations can be regrouped in the same three macro-areas: language (item 1), structure and navigation (items 3, 4), and graphical layout (item 2).

The Web Content Accessibility Guidelines, WCAG 1.0 and WCAG 2.0 [2] represent the international standard reference model in this research domain. However, different practical and operative guidelines have also been proposed [31][34], together to some systematic studies [27][29][32][33].

Given that most of them consider similar aspects, we have analysed and summarised all of them in the following three macro-areas:

a) Graphical layout: Users with ASD may easily be distracted by secondary contents, thus webpages should be very simple and should not contain information, images, frames that create distractions. The page layout should be consistent throughout the website. Images and white space should be copiously used in order to focus the user attention and to simplify the concepts absorption. Background sounds, moving text, blinking images and horizontal scrolling should be avoided. Attention must be payed to fonts and colours: words should be easily readable, thus should usually be written in plain Sans-serif fonts (e.g., Verdana) of at least 12 points. To emphasise words, bold should be used. The choice of the right color is very important, people with ASD may avoid the navigation in sites with particular predominant colours, e.g., red. Foreground and background colours should have sufficient contrast but not too much, some ASD users find e.g., that black on white, is too visually stimulating.

b) Structure and Navigation: The website should have a simple and logical navigation structure, links should be easy-to-access and to find, and few options should be given in order to avoid the user confusion. The navigation inside the site should be limited by three clicks. Each page should contain the navigation information and navigation buttons at the top and the bottom of the page.

c) Language: The language should be simple and precise so that it does not create ambiguity, secondary and irrelevant information should be avoided. The text should be short and self-contained. The words should refer to things that "can be seen", acronyms and abbreviations, together with non-literal text, and jargon should not be used since people with ASD literally interpret the text content.

VI. SPECIALISING THE GUIDELINES FOR AUTISM

The three macro-areas proposed in previous Section V define a set of general recommendations; in this section we specialise them, integrating new features, specific for the ASD characteristics:

Graphical layout: Limit the text to very few, simple sentences, and add many images in a PECS-like style in order to describe concepts and actions through sequences of images. This is the most important feature, i.e., the copious use of images throughout the site, in order to transmit all messages. Repeat concepts, and in the homepage write a sentence that let the user feel it navigates in its "own" site. Write sentences in bold, of big size and uppercase.

Structure and Navigation: Use simple and sketchy symbolic pictures. If the site is directed to a group of young users, add, when possible, some simple games to involve the user, and also to check his/her level of attention.

Language: Use simple and minimal sentences, and illustrate concepts through images, and not though the written text.

To summarise, the main feature that we think should be added is the use of many figures to explain situations, illustrate actions, etc. This choice is based on an Augmentative and Alternative Communication (AAC) approach that is widely used to improve standard communication. Moreover, we enforce the involvement of the user on the site navigation by adding sentences that personalise the site, and by adding games that increase his/her curiosity. As we will describe in Section VIII, this choice has been proved to be winning in our testing phase.

VII. Our case study: *ScopriRieti* and *ScopriMestre*

There are sites that provide tourism destinations for people with ASD, or that give some tips on how to develop an autistic-friendly website, however, we could not find sites whose aim was the independent (or almost-independent) planning of a trip by a user with ASD.

In this section we propose two tourism websites developed for a target group of young people/adolescents. The first site, called ScopriRieti (literally, "Discover Rieti") [35], has first been proposed in [36]. It is an autistic-friendly site, that the user may use to simply search information, or, what is more interesting, to independently choose one among different possible destinations of a one-day tourism trip in the neighbourhood of Rieti, an Italian town in the north of Rome. This site has been developed for verbal young/adolescents people with ASD. The second site, called ScopriMestre (literally, "Discover Mestre") [37], has recently been developed for a tourism tour around Mestre (a city close to Venice), and it is a simpler version developed for non-verbal young users with ASD.

While developing both sites, we made sure that they met all the accessibility and usability standards we have presented in Section V and Section VI. We first developed the ScopriRieti website in collaboration with therapists and operators of the centre Nemo in Rieti, which hosts different users with neurological and physical disabilities. Following the same guidelines we then developed a similar, simplified version, the ScopriMestre website.

A. *The website structure of ScopriRieti*

We will first introduce in detail the ScopriRieti website, the ScopriMestre website will have a similar structure, in the next section we will point out the main differences.

Graphical layout. The page layout is essential and simple in order to be enjoyable and comprehensible. The background is white and it has just some simple bars in a flexible colour (blue and orange), in order to avoid too much contrast on the colour, and thus prevent visual discomfort. The header of the website contains typical standards such as the logo, the primary navigation menu and the search bar, as shown in Figure 1.

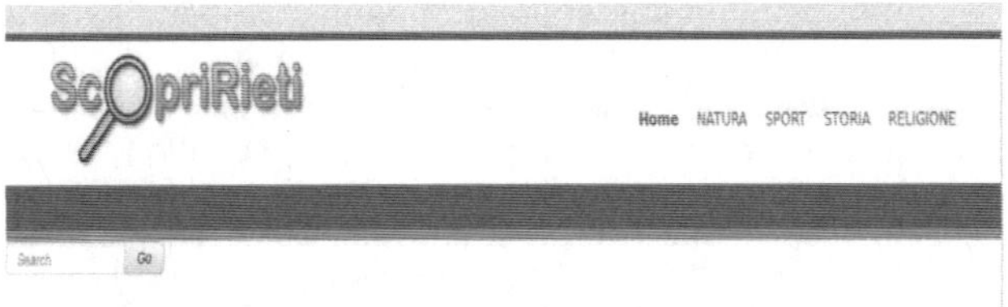

Figure 1. The header bar of our website. The menu contains four main items: nature, sport, history, and religion.

We have inserted many pictures and photos in order to increase the full comprehension of the site content, which is simple, repetitive and synthetic. We have also followed the guidelines and the suggestions proposed by the therapists, and we have included travel itineraries that well suited the users. In the homepage we have written a sentence "Sei un ragazzo del Centro Nemo? Questo sito è stato realizzato proprio per te" (Translation: "Are you a girl/guy of the Nemo center? This site has been developed for you") that would let the user feel happy to navigate inside his "own" site.

We have used simple fonts, big size and uppercase for letters in order to facilitate the reading (see Figure 2. Translation: Discover several new places of Rieti and organise a tour with your friends or your family. What do you like?).

Figure 2. The fonts.

Language. The language used is simple, sentences are minimal and do not contain acronyms or abbreviations (see, e.g., Figure 2). Note that, the site is in Italian since it had to be accessible by Italian disabled users. We have preferred to illustrate all the tourism experience with a sequence of pictures, rather than with some text (see the entire home page in Figure 3). This prepares the user with ASD to a sequence of practical experiences and actions, which he/she has already visualized and pre-processed, and thus reassures him/her.

Figure 3. The homepage.

Structure and navigation. The site is organised in at most four levels. From the logo, the user can go back to the homepage. In some pages, there is a back command. We have not included pop-ups in Javascript, background sounds, moving images in order to avoid sensory annoyance.

The user may navigate from the homepage to the secondary pages by clicking (a) on the images or (b) on the green smiles below, or from the main menu in the header (see Figure 3).

We have discussed with the psychologists, and opted for the use of simple and sketchy symbolic pictures. The idea is that if a user with ASD sees a picture with a bike inside a bike trail (see Figure 4) s/he will conclude s/he can practice this sport.

Figure 4. A bike trail tour.

In the homepage, the user faces a decision "Cosa ti piace?" ("What do you like?"), and has to choose one of the four itineraries, each of one has the same structure and contains: a simple question, an image that represents the general content of the page, links to pictures, videos, four images that link to subsections, which include information on where it is located, how we can go there, what can we do there, and what to bring (see Figure 4). All these links can be used to prepare the user to the trip by visualising in advance a map of the location, the way the trip will be done (by car, by bus, with relatives, etc), how to behave, and what to see. The last link, a backpack, contains instructions on how pack it and what to bring on that specific itinerary (see Figure 5).

Figure 5. What to bring in the backpack.

Some sections contain some games, in which we can check the level of attention of the user. There are questions as: "Have you seen this image during the itinerary?" and so on (see Figure 6).

Finally, at the end of most of the pages we can find a print command ("stampa la pagina") and a back command ("torna indietro"), see Figure 7.

Itineraries. We have chosen four possible itineraries, for a one day tourism trip in the neighbourhood of Rieti and the options are: nature, sport, history, and religion. These topics, well fit these users since people with ASD like to explore quiet and relaxing places (nature and religion), love to move around (sport), are very good at memorising dates and images, and often like to take pictures (history). As we will mention in Section VIII, all these itineraries have all been very much appreciated by the users of the experimentation.

All these itineraries can be done in one day, having as a starting position the centre Nemo, and given that some are urban tours we have pointed out known places inside town to facilitate the comprehension of the destination location.

Figure 6. A partial screenshot of the game, Sì, No (literally Yes, No).

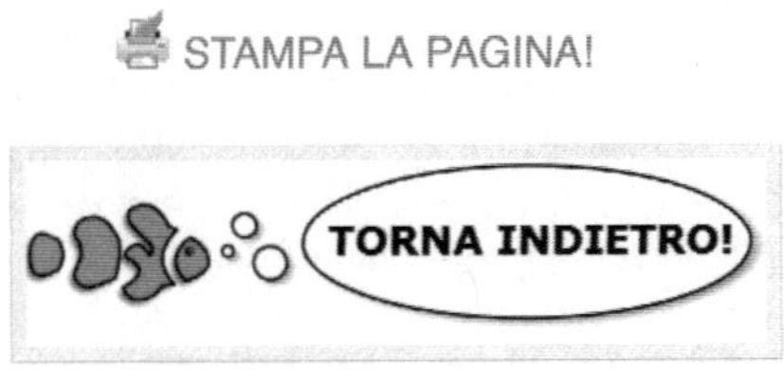

Figure 7. A print and a back command.

B. *The website structure of ScopriMestre*

The structure of this new website is very similar. We have chosen the same graphical layout, fonts and colors, structure and navigation. However, given that our experiments had to be run on non-verbal users, we added as a general requirement additional simplicity in the site presentation. In particular, the main differences between the two sites are:

- *Language*. Given that the users are non-verbal we have decided to use very few short sentences and many pictures, i.e., a simplified language compared to the one of the ScopriRieti website.
- *Itineraries*. We have chosen four possible itineraries, for a one day tourism trip in the neighbourhood of Mestre and the options are: nature, sport, parks, and culture (see Figure 8).
  Given the strongest disability of the new users (they are non-verbal) we have decided to replace more complicated itineraries such as religion and history with parks and culture. In particular, for the itinerary in the nature we have chosen a one day trip along the Sile river, close to Treviso, for sport a trip along the cycling path of Forte Gazzera in Mestre, for parks, a one day spent in the San Giuliano park of Mestre, the biggest of Europe, finally, for culture we have planned a trip to Venice.

Figure 8. The homepage of the ScopriMestre website.

- *Page content and navigation*. We have simplified the explanation of the itineraries and page content, see e.g., Figure 9 where we have used two simple sentences and a single image.

Figure 9. A simple page of the ScopriMestre website.

We have also omitted the part on games, since the games require at least some basic verbal language and some non trivial-level of content comprehension. Finally, we have chosen videos animated by the cry or movment of animals or by activities on water, which are usually liked by people with ASD.

VIII. Experimental results

In this section we describe the methodology we have used and we then illustrate our experimental results first on the ScopriRieti site [35], and then on the ScopriMestre site [37].

A. Methodology.

The development of the both websites has followed different steps:

1) Collection of medical material: The first phase was the search for the medical material on the specific disability, and the study of the characteristics that define these specific users. This is important both to well meet the accessibility requirements, and for the choice of the appropriate content for the site.

2) Literature survey: In the second phase, we have collected all the information about usable and accessible websites for users with ASD, and we have proposed some new interesting features that the site should include (e.g., pictures, games, etc.).

3) Interviews: In the third phase, for the ScopriRieti site, one of our group members (Elisa) has interviewed a group of specialised therapists and professionals of the centre Nemo of Rieti, Italy. This centre hosts people with different neurological and physical disabilities. Elisa has collected information about the different disabilities and about the limits and expectations one could meet while developing the site. She has also participated to different meetings in order to analyse the methodology used by the professionals for approaching these users. What we have realised in this phase is that the site had to be developed for users, which were not "too much serious", given that it is not easy to include users with big behavioural problems in tourism activities outdoor. Thus, we have developed our site for users with non-serious or mild disabilities, that had some elementary ability with the use of a computer and with reading. In the case of the ScopriMestre site, we have discussed with a specialised operator what to change to the previous structure, so that the site could be enjoyable by non-verbal users with ASD and with non-serious disabilities. We have thus decided to mostly simplify the language shortening many sentences, and we have proposed very simple itineraries, which are very well known by all the people living in Mestre (and probably by all the autistic users that will navigate on the site).

4) Website development: The forth phase was the development of the website applying the four categories of guidelines (discussed in Subsection V and in Section VI), and the hints and suggestions collected in the two previous phases.

5) Preliminary test on a single user: The fifth phase was a preliminary test on a single user. We wanted to ensure that the site had been appropriately developed and was comprehensible. We have collected the impressions both of the user and the therapist that was following the meeting. We have thus accordingly adjusted and improved the site.

6) Test on a group of users: In the sixth phase, for the ScopriRieti site, we have then presented this new version to a wider set of users, which were first instructed, and were then left free to navigate and explore the site. We have then collected their impressions. In that case of the ScopriMestre site, for lack of time, we have tested the final site only on the same single user. However, although this site has still to be tested on more users (this is left as future work), as we will see in Section VIII-B, the results on a single user were still very interesting. In this case, given that the user was non-verbal, we have collected the impressions of two distinct operators that separately, and during different sessions, followed the test on the user.

7) Assessment questionnaire: In order to evaluate the perceived accessibility of the Web site, in the last phase we collected the impressions from the relatives of the users by an assessment questionnaire.

B. Outcomes of the tests and of the assessment questionnaire for the ScopriRieti site.

We first present the results for the ScopriRieti and then for the ScopriMestre site.

We have tested the ScopriRieti site with a set of 9 users (1 female and 8 males) with non-serious or mild disabilities, with different backgrounds and general expertise, with some computer skills, and with some interest on this touring activity.

1) Disabilities: Our 9 users (whose names have been omitted for privacy reasons) had the following disabilities:

- U1 (19 years old), U2 (18 years old), and U3 (18 years old)
  ASD and medium mental retardation;
- U4 (19 years old)
  ASD and mild mental retardation;
- U5 (14 years old), and U6 (15 years old)
  Asperger syndrome;
- U7 (22 years old) ASD and psychosis;
- U8 (15 years old)
  medium mental retardation
- U9 (17 years old)
  mild mental retardation.

We have excluded from this group users with serious mental disabilities.

2) Computer skills: All the users had some basics skills on how to use the computer. Depending on the skills, we have left the users to either autonomously navigate, or we have partially helped them. Some have been able to type the name of the site, others have found it already opened. We made sure that could navigate without external distractions.

3) Test results: We have tested the Web site with one user at a time with the goals of completing three simply tasks:

- Task 1: Autonomously navigate in the Web site.
- Task 2: Autonomously choose a typology of itinerary.
- Task 3: Complete the test.

In the following, we describe the test results for each user:

- User U1 has shown very good computer skills; she was able to autonomously navigate inside the site; she has chosen the historical itinerary, and has navigated inside it in a non sequential way, being intensely involved. She has spent a lot of time looking at pictures and videos.
- User U2 was able to autonomously navigate; he has shown interest for the religious itinerary, and, in general, for the preparation of the backpack in all the different itineraries.
- User U3 was almost independent in the navigation phase. He chose the historical itinerary, and got so involved by looking and photos and videos and completed the test saying "I want to go there!". He really liked the proposed games.
- User U4 has navigated inside the sport section; he has intensely observed pictures and videos, and has autonomously discussed, which places he had already explored, and which were new.
- User U5 has shown very good computer abilities. He has chosen the historical itinerary, and has explored it following the sequential sequence, observing all the pictures, and enthusiastically playing the proposed games.
- User U6 had more difficulties on the use of the computer; thus he had found the homepage open. He has chosen the naturalistic itinerary, which he had already visited with the school. He has intensively observed all the pictures, and declared that he wants to go back there with his family.
- User U7 has chosen the religious itinerary; he was very curious about all the churches and saints (he did not know about), thus asking many questions during the navigation and observation of the photos.
- User U8 has chosen the naturalistic itinerary. He was not very skilled in the use of the computer, thus followed some verbal suggestions. He was enthusiastic about the pictures of lakes, plant and animals. We are not sure wether he has really understood all the information, as, e.g., the location of the lakes.
- User U9 has chosen the historical itinerary; he was enthusiastic about the pictures of the underground trail and has declared he wanted to do it soon. Even in this case he needed some verbal help, and we are not sure he has completely understood all the information collected during navigation.

Table I synthesises the number of tasks realised by each user.

To summarise and provide a quantitative metric, we considered the task *completion rate - cr*, that represents one of main metrics used to evaluate the effectiveness of a Web site, that is, the accuracy and completeness with which users achieve specified goals.

$$cr = \frac{Number\ of\ tasks\ completed\ successfully}{Total\ number\ of\ tasks\ undertaken} * 100\%$$

This metric is recommended by the ISO/IEC 9126-4 Metrics [38]. In our test, each user had to complete three main tasks.

TABLE I. Only user U6 was not completely able to complete one of the three tasks.

User	Task 1	Task 2	Task 3
U1	Yes	Yes	Yes
U2	Yes	Yes	Yes
U3	Yes	Yes	Yes
U4	Yes	Yes	Yes
U5	Yes	Yes	Yes
U6	No	Yes	Yes
U7	Yes	Yes	Yes
U8	Yes	Yes	Yes
U9	Yes	Yes	Yes

Applying the results showed in Table I, the task completion rate is strongly positive.

$$cr = \frac{26}{27} * 100\% = 96.3\%$$

All the users, except a couple of them, had previous navigational experience on the Web, and were able to follow the itineraries and to use the mouse.

Figure 10 shows the percentages related to the chosen itineraries.

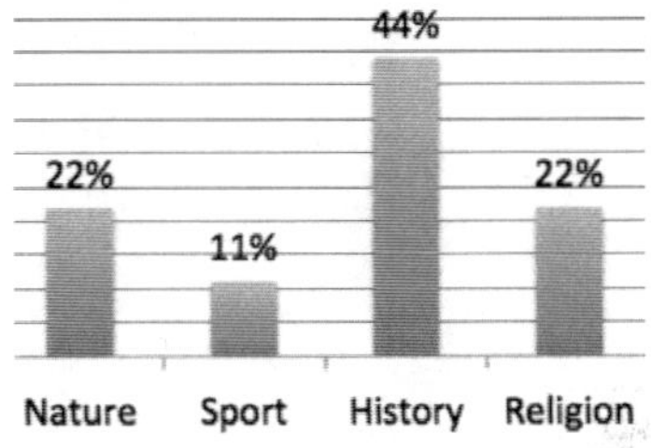

Figure 10. The typologies of chosen itineraries.

All the users were enthusiastic, and have really liked the site and all the itineraries. In particular, the elements they have preferred are: the pictures, an explanation on how to prepare the backpack, the proposed games. All of them have also asked questions about specific pictures, have discussed the site contents, and some of them really wanted to try the real itinerary right away. The use of the mouse facilitated the navigation. We also tried to run the tests using touch screens, but this seemed to complicate it.

As a limit, we have found that some users have shown some small difficulties on finding the location of some specific pictures, and on the sequences of some itineraries.

4) Assessment questionnaire: We have finally proposed an assessment questionnaire to the families. We have first met them and we have illustrated and shown the Web site. Some of them had already seen it previously and had given suggestions on its development. The families were asked to fulfil a questionnaire of 7 items, focused on four constructs, extracted and adapted from [39]: graphical layout (GL), structure and navigation (SN), language (LA), satisfaction (SA).

The levels of agreement were expressed by a 5-point Likert scale ("completely disagree", 5: "strongly agree").

Table II shows the 7 items.

TABLE II. The 7 items of the questionnaire

Construct	Item
01-GL	Accessible and easy graphical layout
02-LA	Accessible and easy language
03-SN	Accessibile and easy structure and navigation
04-SA	Easy to use
05-SA	Autonomous choices
06-SA	Satisfactory content
07-GS	Future use of the Web site

Figure 11 shows the results of the questionnaire, and for each statement, relatives' ratings as well as their mean (μ) and standard deviation (σ).

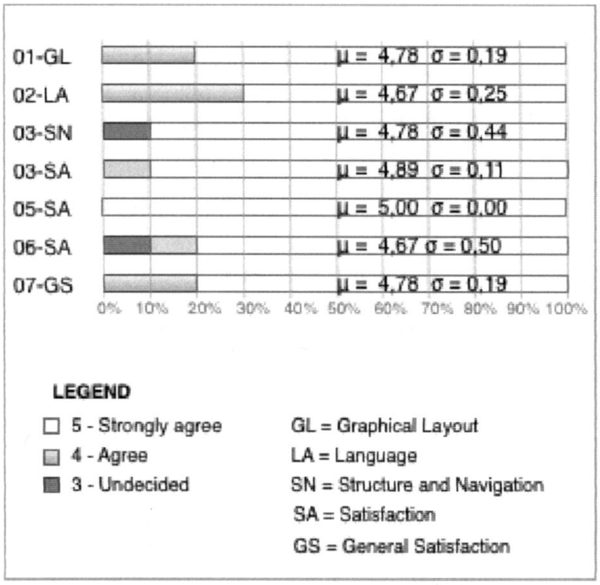

Figure 11. Relatives' assessment questionnaire results.

This project and the Web site have been really appreciated and the parents have claimed they will surely use it for future trips (μ=4.78, σ=0.19), since relatives were happy to have a site where their kids could "autonomously navigate". What emerged is that some of the itineraries were known, others were new. Parents really liked the section on the preparation of the backpack, and the indications on how the trip evolves in order to stimulate their kids orientation capabilities. The site has been appreciated in all the three main components: graphical layout (μ=4.78, σ=0.19); language (μ=4.67, σ=0.25); structure and navigation (μ=4.78, σ=0.44). It has been considered easy to use (μ=4.89, σ=0.11); in the opinion of the parents, the objective to sollecitate autonomous choices has been completely reached (μ=5), and content has been considered satisfactory (μ=4.78, σ=0.19).

C. Outcomes of the tests and of the assessment questionnaire for the ScopriMestre site

We have tested the site on a single user U, and we now illustrate the results.

1) User: We have tested the site with one non-verbal male user of 10 years old, with autism and a limited level of comprehension.

2) Computer skills: The user U had some basics skills on how to use computers/smartphones/tablets. In particular he had very good abilities on how to navigate on YouTube, but needed help to open specific pages (e.g., the main page of the ScopriMestre site).

3) Interests: The user U had some general interest on this touring activity.

In the development phase we have tested the site on the user, and noticed that only few changes were necessary in order to improve the presentation of the photo galleries. The rest of the site well suited the user.

4) Test results: We have started a formal test of the ScopriMestre site with the user U first with operator A and then with operator B.

a) Results with operator A: Operator A tested the site both on a computer and on a smartphone. User U needed some help during navigation, in particular he had to found the homepage open. He chose the sport itinerary all the times, by pointing it with his finger on the computer, or by independently choosing it with a smartphone. He intensively observed all the pictures, and videos, and tried to navigate also on the other itineraries.

b) Results with operator B: Operator B tested the site with U both on a computer and on a iPad. Even in this case user U needed some help to start the navigation from the main page, and he was then able to autonomously navigate inside the site using the iPad (with the computer needed some help). He again chose the sport itinerary and decided to first navigate on pictures and then to view the video. U also showed interest in the pictures related to the preparation of the backpack. He did not show much interest on the maps, because this requires some kind of concept abstraction level. The only exception was the map of Venice, since it is a real photo taken from an airplane. He also liked the pages on "what you can do" on the itinerary or "how you should behave". As a second itinerary U chose nature, then culture, then parks. Finally, operator B noticed that U preferred the video related to culture, which is an accelerated video, compared to the normal speed one.

To summarise the result, from the opinion of both operators A and B (since U is non-verbal and is not able to write), the user was able to complete the test up to the end. What also surprised both operators is that he was able to understand the symbolic pictures of the itineraries, at least the one used for nature and sport. User U was enthusiastic, and really liked the site and all the itineraries (in a scale of 5, 5 out of 5). With user U touch screens worked better than the computer with mouse.

As a limit, we have found that user U has shown some small difficulties on the use of the back command, and needed some initial help before understanding how it worked.

5) Assessment questionnaire: We have finally proposed the assessment questionnaire to the family of user U, after we have met the parents to illustrate the site. As for the case of the ScopriRieti site, this project and the ScopriMestre site have been really appreciated. What emerged is that all itineraries were known. Parents really liked the idea of letting the user choose a preferred itinerary. They also found useful being able to show the site to prepare their kid before starting the trip, given that users with ASD decrease the stress when they know in advance what they will be doing. Globally, the site has been really appreciated both for the contents and for its graphical design (in a scale of 5, 5 out of 5). What the parents suggested is to eventually add some sub-itineraries inside a single one, in order to have multiple options.

IX. Conclusion and Future Work

In this paper, we analysed how an accessible and usable website should be developed in order to be enjoyable by a user with ASD. We have also considered, how such a user can benefit of travelling around the world, and which tips it should follow in order to be able to face in serenity the trip. We have finally presented two interesting accessible and usable websites we have developed for users with ASD, the first for verbal, the second also for non-verbal users. With these sites, the users were able to plan and almost independently decide the trip they wanted to do. We have also shown the appreciation results it has received while experimented on a group of users with ASD. This work represented a pilot, prototypal project and the encouraging feedbacks allow us to plan future work:

- Extend the test on the ScopriMestre site to a wider set of users (results on a single users were very encouraging).
- Validate the usability and the accessibility of the websites on a *systematic analysis of data collected on a statistically significant sample* of users. This will require: (a) the creation of new websites that will use our conceptual model; they will contain as case study other cities; (b) the definition of the features of potential user profiles, in order to generating a taxonomic analysis of the experimentation results.
- Implement *a mobile app*, that will contain all the case studies and will become a concrete tool for the tourism; it should be an adaptive and ubiquitous app able to follow the users on their trips. We will collect all the information generated by the use of the app, like navigation paths, user profiles, user preferences, geo-localisations, etc.; the aim will be to analyse these data and define a reference model for an adaptive, and semantic app for the tourism of specific class of users.
- Implement the *social aspects*, in such a way users could vote their preferred pages, insert a personal comment or share their trips and their experience.

Acknowledgment

We want to thank centro Nemo, all its therapists and operators, the families, the users, and all the people that collaborated to this project. Moreover, we are grateful to dr Marco Lestuzzi, which made responsive the ScopriMestre and ScopriRieti websites.

References

[1] A. Dattolo, F. Luccio, and E. Pirone, "Webpage accessibility and usability for autistic users: a case study on a tourism website," in ACHI 2016: The Ninth International Conference on Advances in Computer-Human Interactions. Venice, Italy, 24-28 April 2016.

[2] "Web content accessibility guidelines (wcag) 2.0," 2008, https://www.w3.org/TR/WCAG20 [accessed: 2016-11-30].

[3] "W3 consortium," 2016, http://www.w3.org [accessed: 2016-11-30].

[4] M. C. Lisa Seeman, "Techniques for the the cognitive and learning disabilities accessibility task force (coga)," 2016, w3C, https://w3c.github.io/coga/techniques/ [accessed: 2016-09-15].

[5] "Act n. 4 of 9.01.2004," 2004, http://www.normattiva.it/uri-res/N2Ls?urn:nir:stato:legge:2004-01-09;4!vig= [accessed: 2016-11-30].

[6] American Psychiatric Association, Ed., The Diagnostic and Statistical Manual of Mental Disorders: DSM 5. BookpointUS, 2013.

[7] K. Still, R. Rehfeldt, R. Whelan, R. May, and S. Dymond, "Facilitating requesting skills using high-tech augmentative and alternative communication devices with individuals with autism spectrum disorders: A systematic review," Research in Autism Spectrum Disorders, vol. 8, 2014, pp. 1184–1199.

[8] A. Dattolo and F. Luccio, "Modelling volo, an augmentative and alternative communication application," in ACHI 2015: The Eighth International Conference on Advances in Computer-Human Interactions. Lisbon, Portugal, 22-27 February 2015.

[9] A. Bondy and L. Frost, "Mands across the water: A report on the application of the picture-exchange communication system in Peru," The Behavior Analyst, vol. 16, 1993, pp. 123–128.

[10] T. Grandin, "How does visual thinking work in the mind of a person with autism? A personal account," Philosophical Transactions of the Royal Society, vol. 364, no. 1522, May 2009, pp. 1437–1442.

[11] C. Sicile-Kira, "What is sensory processing disorder and how is it related to autism?" 2010, https://www.psychologytoday.com/blog/the-autism-advocate/201003/what-is-sensory-processing-disorder-and-how-is-it-related-autism [accessed: 2016-11-19].

[12] "World tourism organization," 2016, uNTWO, http://www2.unwto.org/ [accessed: 2016-11-25].

[13] "World tourism organization - publications and documentation," 2016, uNTWO, http://publications.unwto.org/ [accessed: 2016-11-15].

[14] "Accessibility is gaining ground," Smart Guide 2015 Travelling in Europe, 2015, p. 25, eNAT, http://www2.unwto.org/ [accessed: 2016-11-16].

[15] A. J. Baxter, T. Brugha, H. E. Erskine, R. W. Sheurer, T. Vos, and S. J. G., "The epidemiology and global burden of autism spectrum disorders," Psychological Medicine, vol. 45, no. 3, 2014, pp. 1601–613.

[16] H. Hamed, "Tourism and autism: An initiative study for how travel companies can plan tourism trips for autistic people," American Journal of Tourism Management, vol. 2, 2013, pp. 1–14.

[17] A. Goehner, "Six tips for traveling with an autistic child. time special.2012." 2012, http://content.time.com/time/specials/packages/article/0,28804,1893554_1893556,00.html [accessed: 2016-11-11].

[18] "Autism asd, travelling through the airport," http://www.dublinairport.com/gns/at-the-airport/autism-asd.aspx [accessed: 2016-11-20].

[19] "Royal air force museum, london, autism friendly trail," 2016, http://www.rafmuseum.org.uk/london/things-to-see-and-do/autism.aspx [accessed: 2016-11-29].

[20] "Metropolitan museum of art in new york, resources for visitors on the autism spectrum," 2016, http://www.metmuseum.org/events/programs/met-access/visitors-with-developmental-and-learning-disabilities/for-visitors-with-autism-spectrum-disorders [accessed: 2016-11-29].

[21] "Autistic globetrotting, autism travel made easy," 2016, http://www.autisticglobetrotting.com/ [accessed: 2016-11-30].

[22] "Convention on the rights of persons with disabilities," 2006, http://www.un.org/disabilities/convention/conventionfull.shtml [accessed: 2016-11-11].

[23] S. Abou-Zahra, "How people with disabilities use the web: Overview," 2012, https://www.w3.org/WAI/intro/people-use-web/ [accessed: 2016-11-15].

[24] "Web accessibility and usability working together," 2010, w3C, http://https://www.w3.org/WAI/intro/usable [accessed: 2016-11-25].

[25] J. Nielsen, "Usability 101: Introduction to usability," 2012, https://www.nngroup.com/articles/usability-101-introduction-to-usability/ [accessed: 2016-11-26].

[26] J. Nielsen and R. Budiu, Mobile Usability. New Riders, 2013.

[27] C. Putman and L. Chong, "Software and technologies for people with autism: what the users want?" in ASSETS 2008: 10th International ACM SIGACCESS Conference on Computers and Accessibility. Halifax, Nova Scotia, Canada, 13-15 October 2008, pp. 3–10.

[28] W. Chen, "Multitouch tabletop technology for people with autism spectrum disorder: A review of the literature," in DSAI 2012: 4th International Conference on Software Development for Enhancing Accessibility and Fighting Info-exclusion. Douro Region, Portugal, 19-22 July 2012.

[29] T. Britto and E. Pizzolato, "Towards web accessibility guidelines of interaction and interface design for people with autism spectrum disorder," in ACHI 2016: The Ninth International Conference on Advances in Computer-Human Interactions. Venice, Italy, 24-28 April 2016.

[30] M. Kamaruzamana, N. Rani, H. Nor, and M. Azaharia, "Developing user interface design application for children with autism," Procedia - Social and Behavioral Sciences, vol. 217, 2016, pp. 887 – 894.

[31] "An interview with jamie knight: autism and accessible web design," 2015, http://www.iheni.com/an-interview-with-jamie-knight-autism-and-accessible-web-design/ [accessed: 2016-11-02].

[32] M. G. Friedman and D. N. Bryen, "Web accessibility design recommendations for people with cognitive disabilities," Technology and Disability, vol. 19, no. 9, 2007, pp. 205–212.

[33] A. Darejeh and D. Singh, "A review on user interface design principles to increase software usability for users with less computer literacy," Journal of Computer Science, vol. 9, no. 11, 2013, pp. 1443–1450.

[34] "Cognitive accessibility user research," 2016, w3C, http://www.w3.org/TR/coga-user-research/ [accessed: 2016-11-26].

[35] E. Pirone, "Scopririeti," 2015, http://volo.uniud.it/scopririeti/ [accessed: 2016-11-30].

[36] ——, "Turismo e autismo: realizzazione di un sito web turistico per soggetti autistici," Master's thesis, Università Ca' Foscari, Venezia, 2015.

[37] A. Dattolo, F. Luccio, and E. Pirone, "Scoprimestre," 2016, http://volo.uniud.it/scoprimestre/ [accessed: 2016-11-30].

[38] N. Bevan, "Iso and industry standards for user centred design," 2000, http://usabilitynet.org/trump/documents/Usability_standards.ppt.pdf [accessed: 2016-11-30].

[39] T. Lavie and N. Tractinsky, "Assessing dimensions of perceived visual aesthetics of web sites," Int. Journal of Human-Computer Studies, vol. 60, no. 30, 2004, pp. 269–298.

Improved Knowledge Acquisition and Creation of Structured Knowledge for Systems Toxicology

Sam Ansari, Justyna Szostak, Marja Talikka, Julia Hoeng
Research & Development
Philip Morris Products SA (part of PMI)
Neuchatel, Switzerland
sam.ansari@pmi.com, justyna.szostak@pmi.com, marja.talikka@pmi.com, julia.hoeng@pmi.com

Juliane Fluck
Fraunhofer Institute for Algorithms and Scientific Computing
Sankt Augustin, Germany
juliane.fluck@scai.fraunhofer.de

***Abstract*—Ever-increasing scientific literature enhances our understanding on how toxicants impact biological systems. In order to utilize this information in the growing field of systems toxicology, the published data must be transformed into a structured format suitable for knowledge modelling, reasoning, and ultimately high throughput data analysis and interpretation. Consequently, there is an increasing demand from systems toxicologists to access such knowledge in a computable format, here biological network models. The BEL Information Extraction workFlow (BELIEF) automatically extracts biological entities and causal relationships from any text resource and converts them into a formalized language, the Biological Expression Language (BEL). BEL is a machine- and human-readable language that represents molecular relationships and events as semantic triples: subject–relationship–object. In addition to the automatic extraction through text mining, BELIEF also features a curation interface to verify and modify the proposed triples and benefits from BEL's human-readability. The curation interface facilitates this curation task by providing relevant information to ensure high curation accuracy and fast processing. The resulting BEL triples are then assembled to biological network models that represent specific biological processes for a given context, e.g., organism, tissue type, disease state. These biological network models can then be verified in a crowd-based approach utilizing a collaborative web-based platform before finally sharing them through a publicly available and specialized repository. In this strategy paper, we summarize over various solutions to challenges in the knowledge-based systems toxicological assessment.**

Keywords-component; text mining; BEL; knowledge management; network models; curation; systems toxicology.

I. INTRODUCTION TO SYSTEMS TOXICOLOGY

Systems toxicology supports the detailed understanding of the mechanisms by which biological systems respond to toxicants. This understanding can be used to assess the risk of chemicals, drugs or consumer products. In this work, we give a summary on critical developments within the biological, toxicological, as well as computational domain that has led to this knowledge-based systems toxicological assessment approach [1].

A. *Omics Profiling in Systems Toxicology, from Basic Research to Risk Assessment*

Towards the end of the second millennium a new field emerged in toxicological assessment when the high throughput analysis of the transcriptome called transcriptomics was first used to identify the cellular components and signaling pathways involved in toxicity response and in relationship to harm and disease [2][3]. This approach was called Omics profiling. Omics is the name of a group of biological fields, such as genomics for the discipline in genetics, proteomics for the study of proteins, metabolomics for the study of chemical processes involving metabolite (and the related field of lipidomics for lipids), transcriptomics, and others. This addition of Omics profiling to toxicological assessments opened new possibilities to better understand how different compounds cluster into similar mechanistic classes based on the molecular response profile they inflict on the test system. It has also enabled the discovery and validation of exposure-response biomarkers, as well as the classification and ranking of drug candidates [4]. Omics profiling also guides the development of new and more precise toxicological endpoints and targeted cellular assays [5] and can be valuable in the approximation of the lowest dose that results in the perturbation of the system, especially when data are sparse and when the toxicant affects more than a single pathway [6][7].

Initiatives, such as the ToxCast [8], Adverse Outcome Pathways (AOPs) [9], FutureTox [10], and the Comparative Omics profiling Database (CTD; http://ctdbase.org/) [11], share the common goal to turn the sole use of apical endpoints into predictive toxicology by gaining better understanding of toxicological mechanisms [12]. The

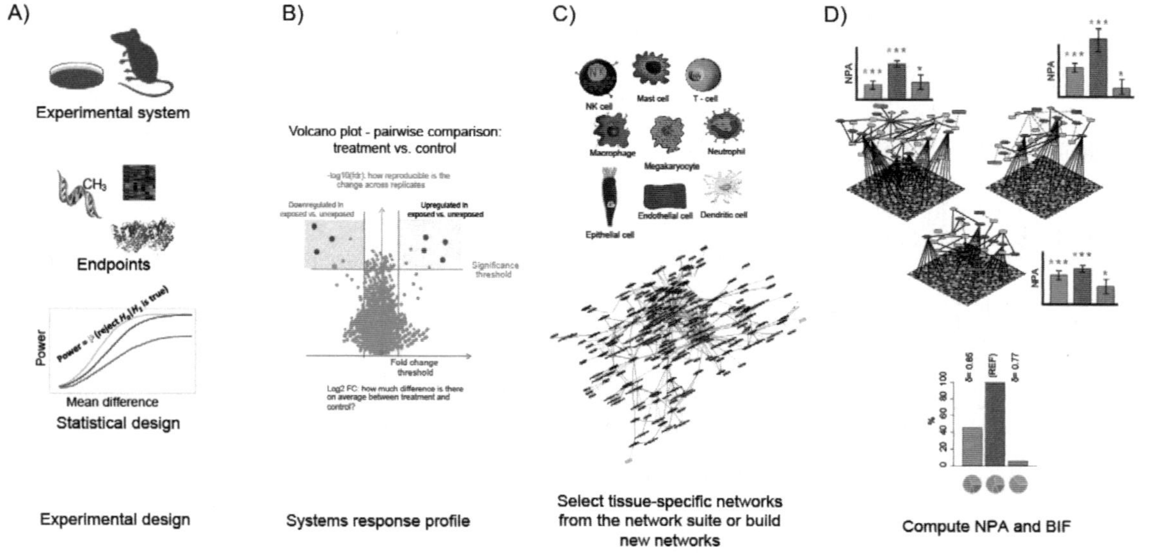

Figure 1. Overview of the Systems Toxicology Workflow. A) The workflow starts with a careful selection of the experimental system and robust statistical design based on the choice of endpoints to be measured. B) Common endpoints in systems toxicology are the high throughput Omics measurements. Transcriptomics data can be analysed to obtain the systems response profiles triggered by the exposure. C) The transcriptomics data is often analysed in the context of causal biological network models for mechanistic interpretation and the models have to be carefully chosen to align with the biological context that the experiment has been conducted in. D) Finally, sophisticated algorithms are used to compute the Network Perturbation Amplitude (NPA) for each network and the aggregated overall Biological Impact Factor (BIF).

ultimate goal of the toxicity testing in the 21st century is to use the insight gained from in vitro assays to predict adverse effects observed in laboratory animals and humans [13].

The comprehensive integration of classic toxicology approaches and Omics profiling forms the basis of systems toxicology. There is also strong emphasis on the development of computational platforms to enable quantitative analysis of molecular changes in response to a stressor and to accurately model the toxicological system [14].

B. Knowledge Modelling and Computational Analysis as Core Instruments in Systems Toxicology

There are a number of ways to extract meaningful signals from high throughput measurements involving a variety of suited software solutions on top of sophisticated laboratory instruments. At the same time the data generated by these high throughput methods creates a challenge in the analysis and interpretation with traditional data processing applications. Different pathway tools provide ways to analyze high throughput data, mainly differentially expressed genes to highlight the biological processes, in which the genes are known to function. These partially web-based tools that include the DAVID Bioinformatics Database [15] or the Gene Set Enrichment Analysis (GSEA) method [16] employ pathway repositories, such as Reactome [17], Biocarta, [18], and the Kyoto Encyclopedia of Genes and Genomes (KEGG) [19] that enable the mapping of the regulated genes into pathways. Fortunately, as the number of datasets increases, more literature that describes biological relationships is published. Using this knowledge, the creation of causal relationships from scientific evidences is becoming an efficient and popular methodology to analyze molecular data. Such models allow the interpretation of data in the context of directional graphs with signed interactions (edges) between biological entities (nodes) [20][21][22][23], providing the network perspective for the stressor response [24].

Causal biological network models are also the cornerstone of the workflow for impact assessment that we have developed over the past years [25][26]. The workflow starts with the design of appropriate experiments for data production. This includes the choice of exposure regimen, experimental test system, and the selection of measurements that will be made (see Figure 1). In addition to apical endpoints, transcriptomics profiling is almost always included to allow mechanistic interpretation of exposure effects. After conducting rigorous quality controls and statistical analyses, the gene expression changes are converted into a systems response profile that illustrates how the level of each molecular entity (here mRNA) is changed in response to the exposure. Finally, the transcriptomics data are analysed in the context of causal biological network models that summarize the *a priori*

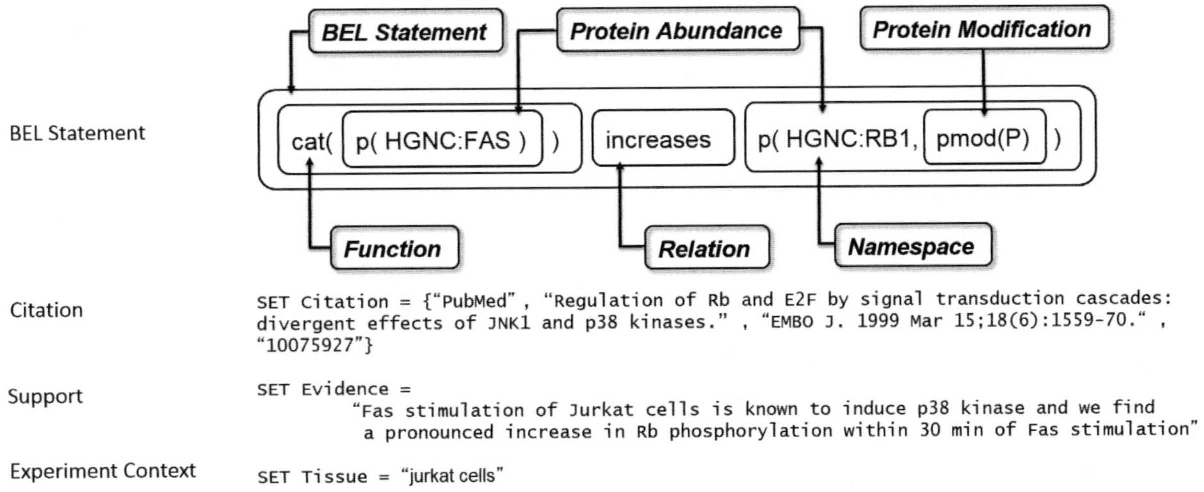

Figure 2. BEL Nanopub Overview. The three crucial elements of a BEL nanopub are the BEL Statement showing the knowledge statement in a triple and controlled terminology, as well as the citation information and actual evidence sentence. Experiment context is an optional field to simplify the triple assembly into biological network models. BEL nanopubs are coded in XML.

knowledge in a given context [25][26] (see Figure 1).

The network models, here causal network models, were built form scientific literature to reflect biological processes that are, e.g., assumed impacted in the lung and vascular system in response to cigarette smoke exposure [27][28]. The current suite of models consists of network families describing cell proliferation [29], cell fates [30], cell stress [31], pulmonary inflammation [32], tissue repair and vascular inflammation [33].

The causal network models are used in combination with transcriptomics data and computational algorithms that transform gene expression changes into Network Perturbation Amplitude (NPA) scores and the aggregated Biological Impact Factor (BIF) [34][35]. Such calculation requires the network model to contain a measurable layer reflecting gene regulations by some of the entities in the model backbone. The transcriptomics data are used to infer the activities of the backbone nodes based on the regulation of their target gene in the dataset.

These inferred changes in the backbone node activities are evaluated in the context of the overall network topology; the NPA score depicts the predicted effect that the exposure has on the biology described by the network model [26][34][35]. In some cases it is beneficial to get an overview of the overall biological impact that a stressor elicits on the test system. The BIF is an aggregation of the NPA scores stemming from perturbation of the individual biological processes included in the network model suite [35]. The advantage of computing a single holistic score is that it allows a high-level comparison of biological effects resulting from different exposures. Several use cases employing this approach have been published [36][37][38][39][40][41].

The following sections of this strategy paper will summarize the key modules in the implementation of this knowledge-based toxicological assessment.

II. Knowledge Acquisition for Knowledge Modelling

In this section, we describe the toolset and implementation strategy for knowledge acquisition and ultimately knowledge modelling in order to analyze and interpret systems toxicology data.

A. BEL and the BEL framework

Today, an overall accepted and widely spread exchange format for knowledge is through the unstructured text, natural language. Natural language contains many redundancies, uses varying vocabularies, introduces complication by using grammar and different sentence structures, as well as containing implications. All these factors make the unstructured text as such useless for knowledge management through computational tools. Our causal biological network models represent the aggregation of unstructured scientific knowledge formalized in the Biological Expression Language BEL [42]. BEL is a computer and human readable language specially designed to formalize biological relationships and allow the construction of network models to facilitate downstream computational analyses. While there are other conventions available that allow the formalization of unstructured biological knowledge, e.g., BioPAX [43] and SBML [44], BEL presents a considerable advantage that it is simple to read and edit by a biologist, because the formalization is close to natural language with simplification into a triple, here subject, predicate, and object as well as restriction of vocabulary via defined namespaces (see Figure 2). BEL conserves causal, e.g., increases and decreases, and non-causal correlative relationships, here positive correlation, negative correlation, and association. Based on Semantic-web technology, BEL uses a nanopublication model for publishing an assertion, together with attribution and provenance metadata [45] (Guidelines for Nanopublication http://nanopub.org/guidelines/working_draft/). A BEL nanopub is the smallest unit of information and represents a biological relationship with its provenance. Two elements are crucial, the BEL statement and the evidence, where evidence is the supporting text and citation information. Additionally, each BEL statement can be associated with experimental context information such as organism, organ, tissue, cell line, disease state and more. This context information is finally used to construct biological network models under specific experimental conditions.

In BEL subjects and objects are represented by a function of biological entities controlled by the BEL syntax and BEL terms that are managed in namespaces. A function can be the abundance of a particular biological entity, here chemical abundance *a()*, protein *p()*, genes *g()*, RNA *r()* or micro-RNA *mRNA()*. Biological process *bp()* or disease *path()* functions capture cellular parameters or processes in BEL. A detailed description of the BEL syntax and the use of BEL namespaces is described in detail on www.openbel.org.

BEL is accompanied by a set of tools packaged in the BEL framework (www.openbel.org). These tools allow the syntactic and semantic validation and compilation of single BEL triple into an assembled network model. The BEL framework also includes a knowledge assembly models managing software and a connector for Cytoscape network visualization software [46] in order to visualize and analyze the assembled networks in graph.

B. From Knowledge to BEL, the BEL Information Extraction workFlow (BELIEF)

An approach that is becoming more and more popular and that started back in the 80s is to either manually or automatically curate / parse and semantically annotate natural language word by word, sentence by sentence. The domain of text analytics with the help of linguistics was established and is increasingly developing tools and algorithms that better identify entities either via extended vocabularies or sophisticated statistical methods such as machine learning. At the same time, domain experts as well as curators and computational scientists focus on a way to define formats for a better and more applicable representation of knowledge. As it stands today, text mining either focuses on high recall and rather low precision, which is the case for most automated solutions, or on low recall but high precision, which is typically the case for manual curation.

The solution obviously lies in semi-automated knowledge extraction where linguistic tools identify relevant entities from natural language with a high recall and are manually curated for high precision.

Fluck et al. have addressed the challenge in efficiently extracting knowledge from the rich source of scientific articles by proposing a workflow combining both, the automated extraction method using text mining methodology and the manual curation of these results to ensure precision [47]. In the next two sections we will explain the challenges each methodology (manual versus automated) has and how in the third section both approaches can be combined into an efficient workflow, BELIEF.

1) Limitations of Manual Curation

Manual curation typically has the goal to bring unstructured text into structured data where various information sources are brought into one repository, here typically a database, and the data are annotated with controlled terminology. For a long time great effort has been made to build biological databases such as, e.g., UniProtKB/Swiss-Prot (www.uniprot.org/), MGI (www.informatics.jax.org/), HGNC (www.genenames.org/). The manual curation process is the dominant methodology for these efforts where a team of curators reviews literature and other knowledge sources for annotations given the specific context. These annotations are then stored in a structured format into databases [48][49][50]. However, one large source for variability even in these structured repositories is the variability from curator to curator defined by the experience the curator has and the effort the curator is taking. In fact, it was shown that expert curators present an accuracy of 90% for a specific task while the inter-curator agreement ranges from 77% to a minimum of 31% [51][52]. These results demonstrate the issues of this sophisticated yet very time consuming process that jeopardizes the quality and goal of these standardized repositories to some extent. Even when annotation guidelines are specified in order to create harmonization across different curators given a specific task, the personal variability is still high and impactful in the biological domain. Therefore, high-quality manual curation of the scientific literature is a very challenging and time-consuming effort and impacts the progress in the creation of these biological databases [53].

2) Limitations of Automated Curation

With the limitations of human curation, computational teams started focusing on automated curation processes to, in most cases, replace parts of the manual curation task [54]. Tools for named entity recognition (NER) for gene and protein name recognition are widely used within the database community. Typically tools such as Textpresso [55] and ProMiner [56] are employed to identify specific entity classes [57][58]. While the speed and output quantity in which automated annotations perform is impressive, the

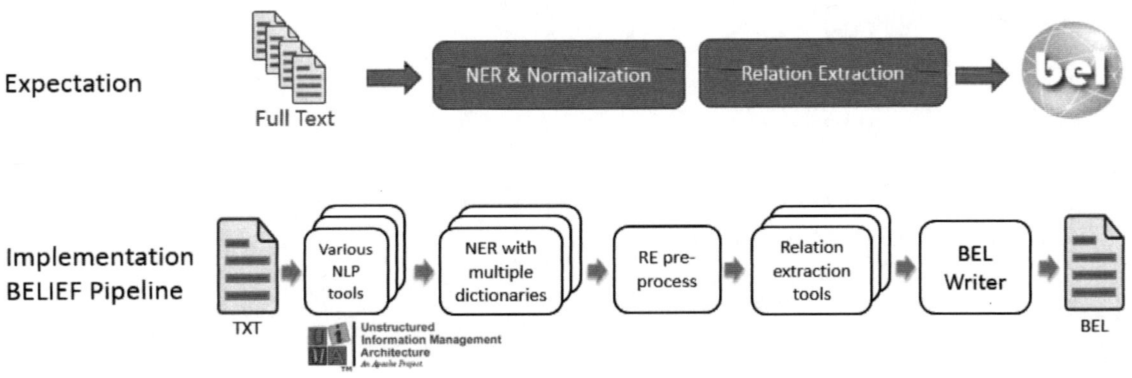

Figure 3. BELIEF Text Mining Pipeline Overview. The Expectation line shows a very high-level view on the functionality. In the row "Implementation BELIEF Pipeline" all relevant modules are shown. Various NLP tools are used for detecting and splitting sentences, identifying words etc. In the next step NER is used to detect relevant entities with given dictionaries, here namespaces. The relationships between these detected entities is captured in the next step and finally a BEL nanopub compliant output is generated.

results also have a greatly increased rate of errors compared to manual curation [53]. More specifically, it was shown that automated curation caused critical errors in assignments of particular functions that may affect as many as 30% of the proteins, and may even exceed 80% of individual protein families [48]. In a recent paper that modeled annotation errors in the Gene Ontology database, it was estimated that up to 49% of sequences functionally annotated by automatic sequence comparison methods could be mis-annotated [59]. Comparable errors were also observed in other analyses [48][60][61]. The introduction of errors in the public database could lead to severe error propagation that could make the data useless and even misleading when it comes to the interpretation of experimental data [48][62].

3) Solution and Performance of a Semi-automated Solution

Although text mining solutions did take over parts of the manual curation tasks, there was no approach shown before where a full knowledge statement was extracted and coded into a predefined syntax and become subject for manual curation. Looking at both approaches in annotating and extracting knowledge from unstructured text, the strength and weaknesses become obvious. The manual approach obviously results in much more reliable output at the cost of time, effort and harmonization / reproducibility. At the same time, automated annotation and extraction has a dramatically improved curation speed with full reproducibility but lacking precision. These strength and weaknesses are complementary and suggest a combined approach, here semi-automated approach. Especially in biological research and pathway modeling the identification of relationships between entities, e.g., protein-protein, drug-protein interactions or protein-disease relationships is crucial for mechanistic and network analyses. To be efficient, the automated curation process would have to be able to mimic the human ability to infer relations from the text. Text mining tools are currently not only able to detect and identify biological entities in the text, but they are also able to infer the relationships between these entities. The accuracy of text mining tools was estimated and demonstrated a high-performance of about 82-85% overall (Elsevier), 80 % for ProMiner for human and mouse gene/protein name recognition and about 50% for BioRat [63][64]. Altogether these evidences demonstrate that text mining is an efficient tool to curate unsolved amounts of data with a consistent quality for data detection and annotation.

In 2014 Fluck et al. released the BEL Information Extraction workFlow BELIEF (see Figure 3) [47]. The BELIEF infrastructure embeds an extraction information workflow combined with NER and relation extraction (RE) methods into a state of the art environment.

The combination of various linguistic tools into one workflow requires an extra effort in normalizing the results (see Figure 3).

BELIEF addresses the biological network model curation needs by identifying chemical, gene/protein, and biological process and disease terms in scientific articles. Additionally to that BELIEF identifies relationships through a combination of specialized ontologies and linguistics rules. On top of the BELIEF text mining pipeline sits the BELIEF Dashboard that provides users a manual curation interface for the automatically extracted BEL nanopubs (see Figure 4).

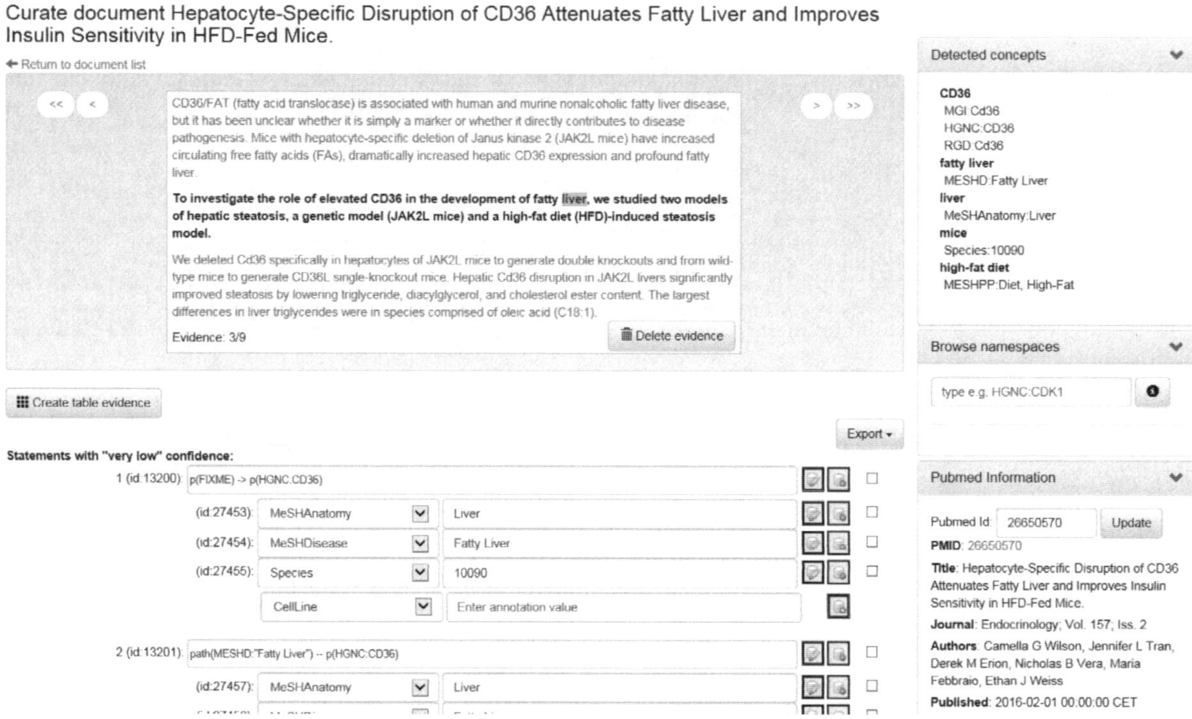

Figure 4. Overview of the BELIEF Dashboard. The main window contains two sections: 1. the evidence text with the next sentence and browsing buttons, and 2. the automatically generated BEL nanopub with options for modifications and export. The right banner contains supporting information such as the concepts together with the namespace sources that were detected, a namespace browser to assign new concepts that were not detected, as well as the citation information that automatically retrieved all required information based on the Pubmed ID provided.

The dashboard offers the possibility to the curators to visualize, edit, correct, and delete statements to ensure precision on the high recall output from the underlying text mining pipeline. In an assessment Fluck et al. not only showed a higher detection rate of this combined curation approach in BELIEF but also a much higher user acceptance rating on the simplification of the curation effort [47].

III. Knowledge Assembly and Verification

With the creation of BEL nanopubs extracted from and curated in BELIEF, causal network models can be assembled using the BEL framework tools. These network models are typically assembled from BEL nanopubs given a specific context. After assembling, these networks are further assessed either with experimental data or additional knowledge sources, e.g., databases or other scientific articles. However, this verification can also be carried out in a crowd-based approach. Boué et al. developed a web-based platform for a collaborative verification of these causal network models [27]. In their publication the authors show the outcome of a community challenge called Network Verification Challenge (NVC) by using their platform and verifying 50 biological network models relevant to lung biology and early COPD. Each participant was given the opportunity to confirm, reject, or modify the networks on a website (https://bionet.sbvimprover.com/) and to add mechanistic detail [65]. The challenge showed that even for a group of domain experts unfamiliar with BEL, the crowd performed well at representing scientific findings in BEL. In a similar setup Fluck and Rinaldi performed a BEL task in the BioCreative V challenge. The goal of the challenge was to address curation challenges presented in BEL with text mining solutions. The outcome was that even for computerized systems, BEL did not bring a challenge in adapting algorithms and addressing the challenge tasks well [66].

IV. Knowledge Sharing

As previously stated, the collaborative approach in verifying the representation of literature-based knowledge proved most useful. To ensure continuity in this review approach, network models must be shared and available to the public domain to allow the community use, review, and further provide feedback. In fact, one of the strongest motivators for participants going through the verification process was the use of these networks for their own research projects.

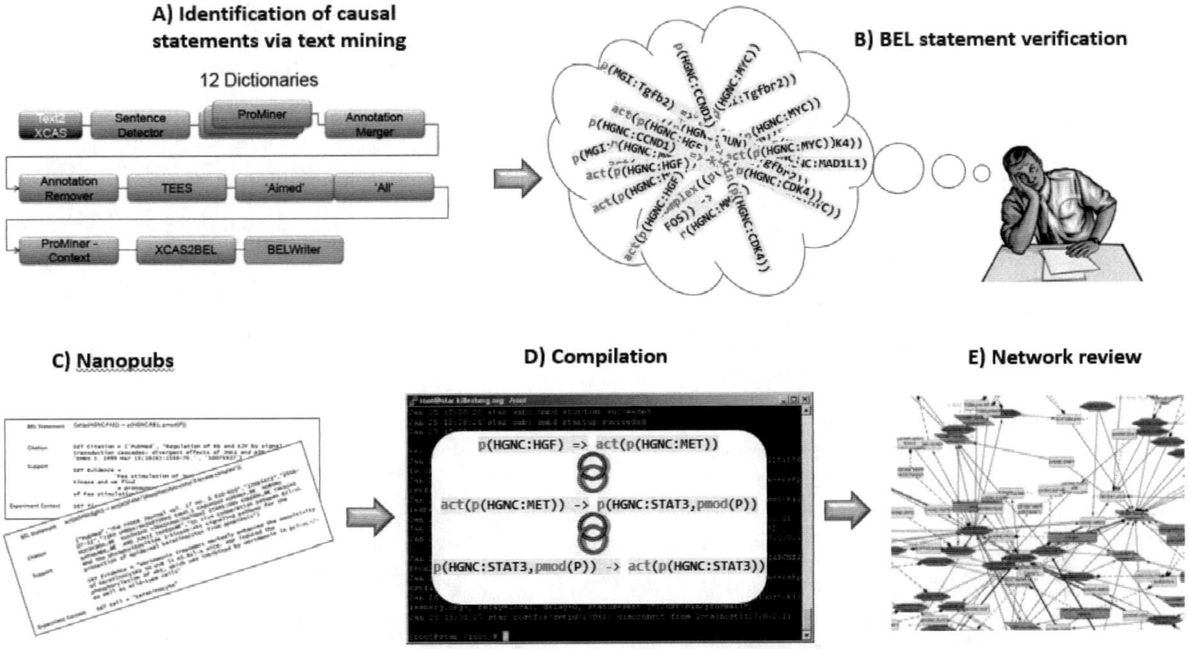

Figure 5. Process to build Causal Biological Network Models using BEL. A) The causal statements are identified in scientific literature and processed by text mining software within BELIEF. B) Domain expert biologists verify the statements using the curation dashboard in BELIEF. C) Semi-automated curation workflow gives rise to nanopubs that contain all essential information about the causal statements. D) The compilation process builds the isolated nanopubs into a coherent representation that can be E) visualized and verified using software such as Cytoscape.

However, the representation of network information in databases is not trivial. Boue et al. prepared an overview on network-based databases and their attributes [27]. In the same article, the authors present the Causal Biological Network database that is specialized for BEL triples and allows to query the data and find the right network model in the large number of available network models. Currently, the database contains biological network models that reflect causal signaling pathways across a wide range of biological processes, including cell fate, cell stress, cell proliferation, inflammation, tissue repair, and angiogenesis in the pulmonary and cardiovascular context. The database is openly accessed giving access to over 120 manually curated and well annotated biological network models. The database uses MongoDB that stores all network models and previous versions of each mode as JSON objects. With these objects users can query the database for genes, proteins, biological processes, small molecules, and keywords in the network descriptions in order to access the required network. On top of the database is a query and visualization layer that allows the users to browse the content and visualize the networks featuring filters for nodes and edges. A link to the supporting text in pubmed is available with each edge (http://causalbionet.com).

V. Future of Knowledge Modeling in Systems Toxicology

In the field of network toxicology, the current static models will eventually be replaced with dynamic models that can capture time and dose effects and provide better predictions on toxic outcomes [14]. This can be accomplished only by developing new and / or combining current modeling languages to handle differential equations that allow dynamic modelling in the context of a priori knowledge. There is also a need to invest in tools that can make conversions from one language to another (e.g., BEL to SBML) so that recorded knowledge is not syntax dependent and therefore limited in the toolset linked to the given syntax. There is still substantial work in enlarging the namespaces and controlled vocabulary to allow the curation of all species context. For instance, the zebrafish is a very attractive model system in modern toxicology research and it would be unfortunate if high throughput measurements from such species cannot be interpreted using causal biological network models. While the semi-automated curation workflow described here and in [67] is a major step towards efficient curation, it would further benefit from automated literature identification within a topic, and eventually the identified articles could be connected to the text mining tool, after which the exposed statements would be verified by curators. Such an approach would quickly pave the way to systems toxicologist's vision of a robust systems toxicology knowledgebase to keep up to date with the growing scientific knowledge. However, regardless how large, independent curation efforts cannot harness all the available information, leaving significant gaps in knowledge. One way to accomplish this is the education of researches about controlled vocabularies for expressing study results and enforcing a nanopub submission along with scientific manuscripts. In essence, a nanopub is the smallest unit of information and represents a biological relationship with its provenance derived from a publication. Ideally the entire content, including figures and tables with captions, could be represented in this format [68]. A community-driven approach the Concept Web Alliance has described guidelines for writing a nanopub, which has to consist of a statement, the origin of the statement, and the origin of the nanopub [45]. Even big datasets, often rather hidden in the supplemental parts, could be made more expedient when expressed in machine-readable formats with sufficient metadata on origin and context. Such approach has been tested in the assertion of differential gene expression in Huntington's disease [69], and the Open PHACTS Discovery Platform provides a guideline for precompetitive nanopub creation and outlines how nanostatements can be cited following their usage in a discovery project [70]. While nanopubs could be formalized in any modelling language, the aforementioned conversion tools would enable more efficient use of the growing toxicology knowledgebase.

VI. Conclusion

The emerging field of systems toxicology encourages new approaches in the processing of experimental data. Unlike many standard toxicological approaches, the amount of data generated for a single sample requires sophisticated computational approaches for the processing and computationally available *a priori* knowledge for the interpretation. In this work, we present a workflow that creates these computer-readable knowledge clusters, here biological network models (see Figure 1). The starting point is the identification of relevant knowledge sources. The workflow continues with computational approaches to create knowledge statements (here BEL Nanopubs) and their manual curation by experts to ensure correctness of the knowledge formalization. This leads to the creation of assembled network models that can be used with computational methods (here the network perturbation amplitude) to calculate an overall biological impact factor for toxicity assessment (see Figure 5). As knowledge changes and extends by time, a strategy must be put in place to ensure knowledge representation based on the current opinion in a specific biological field.

Acknowledgment

We acknowledge Stephanie Boué for the creation of various graphics used in the figures.

References

[1] S. Ansari, "Knowledge acquisition and application for product risk impact analyses in an industrial setup," Proc. Eighth International Conference on Information, Process, and Knowledge Management (EKNOW 2016), IARIA, 2016, Venice, Italy.

[2] E. F. Nuwaysir, M. Bittner, J. Trent, J. C. Barrett, and C. A. Afshari, "Microarrays and toxicology: The advent of toxicogenomics," Molecular Carcinogenesis, vol. 24, pp. 153-159, 1999.

[3] T. Storck, M. Von Brevern, C. Behrens, J. Scheel, and A. Bach, "Transcriptomics in predictive toxicology," Current Opinion in Drug Discovery & Development, vol. 5, pp. 90-97, 2002.

[4] C. A. Afshari, H. K. Hamadeh, and P. R. Bushel, "The evolution of bioinformatics in toxicology: advancing toxicogenomics," Toxicological Sciences, vol. 120, pp. 225-237, 2010.

[5] M. North and C. D. Vulpe, "Functional toxicogenomics: Mechanism-centered toxicology," International Journal of Molecular Sciences, vol. 11, pp. 4796-4813, 2010.

[6] I. Moffat, N. L. Chepelev, S. Labib, J. Bourdon-Lacombe, B. Kuo, J. K. Buick, et al, "Comparison of toxicogenomics and traditional approaches to inform mode of action and points of departure in human health risk assessment of benzo [a] pyrene in drinking water," Critical Reviews in Toxicology, vol. 45, pp. 1-43, 2015.

[7] R. S. Thomas, M. A. Philbert, S. S. Auerbach, B. A. Wetmore, M. J. Devito, I. Cote, et al, "Incorporating new technologies into toxicity testing and risk assessment: Moving from 21st century vision to a data-driven framework," Toxicological Sciences, vol. 136, pp. 4-18, 2013.

[8] A. M. Richard, R. S. Judson, K. A. Houck, C. M. Grulke, P. Volarath, I. Thillainadarajah, et al, "The toxcast chemical landscape: Paving the road to 21st century toxicology," Chemical Research in Toxicology, vol.. 29, pp. 1225-1251, 2016.

[9] D. L. Villeneuve, D. Crump, N. Garcia-Reyero, M. Hecker, T. H. Hutchinson, C. A. LaLone, et al, "Adverse outcome pathway development ii: Best practices," Toxicological Sciences, vol. 142, pp. 321-330, 2014.

[10] J. C. Rowlands, M. Sander, J. S. Bus, and F. O. Committee, "Futuretox: Building the road for 21st century toxicology and risk assessment practices," Toxicological Sciences, vol 137, pp. 269-277, 2013.

[11] A. P. Davis, C. J. Grondin, K. Lennon-Hopkins, C. Saraceni-Richards, D. Sciaky, B. L. King, et al, "The comparative toxicogenomics database's 10th year anniversary: Update 2015," Nucleic Acids Research, vol. 43, pp. 914-920, 2015.

[12] R. Benigni, "Predictive toxicology today: The transition from biological knowledge to practicable models," Expert Opinion on Drug Metabolism & Toxicology, vol. 12, pp. 989-992, 2016.

[13] D. Krewski, D. Acosta Jr, M. Andersen, H. Anderson, J. C. Bailar III, K. Boekelheide, et al, "Toxicity testing in the 21st century: A vision and a strategy," Journal of Toxicology and Environmental Health, Part B, vol. 13, pp. 51-138, 2010.

[14] S. J. Sturla, A. R. Boobis, R. E. FitzGerald, J. Hoeng, R. J. Kavlock, K. Schirmer, et al, "Systems toxicology: From basic research to risk assessment," Chemical Research in Toxicology, vol. 27, pp. 314-329, 2014.

[15] D. W. Huang, B. T. Sherman, and R. A. Lempicki, "Systematic and integrative analysis of large gene lists using DAVID Bioinformatics Resources," Nature Protocols, vol. 4, pp. 44-57, 2009.

[16] A. Subramanian, P. Tamayo, V. K. Mootha, S. Mukherjee, B. L. Ebert, M. A. Gillette, et al, "Gene set enrichment analysis: A knowledge-based approach for interpreting genome-wide expression profiles," Proceedings of the National Academy of Sciences of the United States of America, vol. 102, pp. 15545-15550, 2005.

[17] G. Joshi-Tope, M. Gillespie, I. Vastrik, P. D'Eustachio, E. Schmidt, B. de Bono, et al, "Reactome: A knowledgebase of biological pathways," Nucleic Acids Research, vol. 33, pp. 428-432, 2005.

[18] D. Nishimura, "Biocarta," Biotech Software & Internet Report: The Computer Software Journal for Scient, vol. 2, pp. 117-120, 2001.

[19] M. Kanehisa and S. Goto, "Kegg: Kyoto encyclopedia of genes and genomes," Nucleic Acids Research, vol. 28, pp. 27-30, 2000.

[20] L. Perfetto, L. Briganti, A. Calderone, A. C. Perpetuini, M. Iannuccelli, F. Langone, et al, "Signor: A database of causal relationships between biological entities," Nucleic Acids Research, vol. 44, pp. 548-554, 2015.

[21] L. Chindelevitch, D. Ziemek, A. Enayetallah, R. Randhawa, B. Sidders, C. Brockel, et al, "Causal reasoning on biological networks: Interpreting transcriptional changes," Bioinformatics, vol. 28, pp. 1114-1121, 2012.

[22] D. Djordjevic, A. Yang, A. Zadoorian, K. Rungrugeecharoen, and J. W. Ho, "How difficult is inference of mammalian causal gene regulatory networks?," PloS One, vol. 9, Nov. 2014, doi: 10.1371/journal.pone.0111661.

[23] A. Krämer, J. Green, J. Pollard, and S. Tugendreich, "Causal analysis approaches in ingenuity pathway analysis (ipa)," Bioinformatics, vol. 30, pp. 523-530, 2013.

[24] W. Zhang, "Network toxicology: A new science," Computational Ecology and Software, vol. 6, pp. 31-40, 2016.

[25] J. Hoeng, R. Deehan, D. Pratt, F. Martin, A. Sewer, T. M. Thomson, et al, "A network-based approach to quantifying the impact of biologically active substances," Drug Discovery Today, vol. 17, pp. 413-418, 2012.

[26] J. Hoeng, M. Talikka, F. Martin, S. Ansari, D. Drubin, A. Elamin, et al, "Toxicopanomics: Applications of genomics, transcriptomics, proteomics and lipidomics in predictive mechanistic toxicology," CRC Press, Boca Raton, USA, 2014.

[27] S. Boué, M. Talikka, J. W. Westra, W. Hayes, A. Di Fabio, J. Park, et al, "Causal biological network database: A comprehensive platform of causal biological network models focused on the pulmonary and vascular systems," Database, Apr. 2015, doi: 10.1093/database/bav030.

[28] S. Boué, M. Talikka, J. W. Westra, W. Hayes, A. Di Fabio, J. Park, et al, "Causal biological network database: a comprehensive platform of causal biological network models focused on the pulmonary and vascular systems," Database, vol. 2015, Apr. 2015, doi: 10.1093/database/bav030.

[29] J. W. Westra, W. K. Schlage, B. P. Frushour, S. Gebel, N. L. Catlett, W. Han, et al, "Construction of a computable cell proliferation network focused on non-diseased lung cells," BMC Systems Biology, vol. 5, Jul. 2011, doi: 10.1186/1752-0509-5-105.

[30] S. Gebel, R. B. Lichtner, B. Frushour, W. K. Schlage, V. Hoang, M. Talikka, et al, "Construction of a computable network model for DNA damage, autophagy, cell death, and senescence," Bioinformatics and Biology Insights, vol. 7, pp. 97-117, 2013.

[31] W. K. Schlage, J. W. Westra, S. Gebel, N. L. Catlett, C. Mathis, B. P. Frushour, et al, "A computable cellular stress network model for non-diseased pulmonary and cardiovascular tissue," BMC Systems Biology, vol. 5, Oct. 2011, doi: 10.1186/1752-0509-5-168.

[32] J. W. Westra, W. K. Schlage, A. Hengstermann, S. Gebel, C. Mathis, T. Thomson, et al, "A modular cell-type focused inflammatory process network model for non-diseased pulmonary tissue," Bioinformatics and Biology Insights, vol. 7, pp. 167-192, 2013.

[33] H. De León, S. Boué, W. K. Schlage, N. Boukharov, J. W. Westra, S. Gebel, et al, "A vascular biology network model focused on inflammatory processes to investigate atherogenesis and plaque instability," Journal of Translational Medicine, vol. 12, Jun. 2014, doi: 10.1186/1479-5876-12-185.

[34] F. Martin, A. Sewer, M. Talikka, Y. Xiang, J. Hoeng, and M. C. Peitsch, "Quantification of biological network perturbations for mechanistic insight and diagnostics using two-layer causal models," BMC Bioinformatics, vol. 15, Jul. 2014, doi: 10.1186/1471-2105-15-238.

[35] A. Sewer, F. Martin, W. K. Schlage, J. Hoeng, and M. C. Peitsch, "Quantifying the biological impact of active substances using causal network models," Computational Systems Toxicology, vol. 2015, pp. 223-256, doi: 10.1007/978-1-4939-2778-4_10.

[36] B. Phillips, E. Veljkovic, S. Boué, W. K. Schlage, G. Vuillaume, F. Martin, et al, "An 8-month systems toxicology inhalation/cessation study in apoe–/– mice to investigate cardiovascular and respiratory exposure effects of a candidate modified risk tobacco product, ths 2.2, compared with conventional cigarettes," Toxicological Sciences, vol. 151, pp. 426-464, 2015.

[37] A. R. Iskandar, I. Gonzalez-Suarez, S. Majeed, D. Marescotti, A. Sewer, Y. Xiang, et al, "A framework for in vitro systems toxicology assessment of e-liquids," Toxicology Mechanisms and Methods, vol. 26, pp. 389-413, 2016.

[38] S. Ansari, K. Baumer, S. Boué, S. Dijon, R. Dulize, K. Ekroos, et al, "Comprehensive systems biology analysis of a 7-month cigarette smoke inhalation study in c57bl/6 mice," Scientific Data, vol. 3, Jan. 2016, doi: 10.1038/sdata.2015.77.

[39] U. Kogel, I. G. Suarez, Y. Xiang, E. Dossin, P. Guy, C. Mathis, et al, "Biological impact of cigarette smoke compared to an aerosol produced from a prototypic

modified risk tobacco product on normal human bronchial epithelial cells," Toxicology In Vitro, vol. 29, pp. 2102-2115, 2015.

[40] A. R. Iskandar, Y. Xiang, S. Frentzel, M. Talikka, P. Leroy, D. Kuehn, et al, "Impact assessment of cigarette smoke exposure on organotypic bronchial epithelial tissue cultures: A comparison of mono-culture and co-culture model containing fibroblasts," Toxicological Sciences, vol. 147, pp. 207-221, 2015.

[41] B. Phillips, E. Veljkovic, M. J. Peck, A. Buettner, A. Elamin, E. Guedj, et al, "A 7-month cigarette smoke inhalation study in c57bl/6 mice demonstrates reduced lung inflammation and emphysema following smoking cessation or aerosol exposure from a prototypic modified risk tobacco product," Food and Chemical Toxicology, vol. 80, pp. 328-345, 2015.

[42] T. Slater, "Recent advances in modeling languages for pathway maps and computable biological networks," Drug Discovery Today, vol. 19, pp. 193-198, 2014.

[43] E. Demir, M. P. Cary, S. Paley, K. Fukuda, C. Lemer, I. Vastrik, et al, "The biopax community standard for pathway data sharing," Nat Biotech, vol. 28, pp. 935-942, 2010.

[44] M. Hucka, A. Finney, H. M. Sauro, H. Bolouri, J. C. Doyle, H. Kitano, et al, "The systems biology markup language (SBML): a medium for representation and exchange of biochemical network models," Bioinformatics, vol. 19, pp. 524-531, 2015.

[45] P. Groth, A. Gibson, and J. Velterop, "The anatomy of a nanopublication," Information Services & Use, vol. 30, pp. 51-56, 2010.

[46] P. Shannon, A. Markiel, O. Ozier, N. S. Baliga, J. T. Wang, D. Ramage, et al, "Cytoscape: A software environment for integrated models of biomolecular interaction networks," Genome Research, vol. 13, pp. 2498-2504, 2003.

[47] J. Fluck, S. Ansari, J. Szostak, J. Hoeng, M. Zimmermann, M. Hofmann-Apitius, and M. Peitsch, "Belief - a semiautomatic workflow for bel network creation," 6th International Symposium on Semantic Mining in Biomedicine (SMBM 2014), 2014, Aveiro, Portugal.

[48] A. M. Schnoes, S. D. Brown, I. Dodevski, and P. C. Babbitt, "Annotation error in public databases: Misannotation of molecular function in enzyme superfamilies," PLoS Computational Biology, vol. 5, Dec. 2009, doi: 10.1371/journal.pcbi.1000605.

[49] C. UniProt, "The universal protein resource (uniprot) 2009," Nucleic Acids Res, vol. 37, pp. 169-174, 2009.

[50] T. C. Wiegers, A. P. Davis, K. B. Cohen, L. Hirschman, and C. J. Mattingly, "Text mining and manual curation of chemical-gene-disease networks for the comparative toxicogenomics database (ctd)," BMC Bioinformatics, vol. 10, Oct. 2009, doi: 10.1186/1471-2105-10-326.

[51] C. N. Arighi, B. Carterette, K. B. Cohen, M. Krallinger, W. J. Wilbur, P. Fey, et al, "An overview of the biocreative 2012 workshop track iii: Interactive text mining task," Database, Jan. 2013, doi: 10.1093/database/bas056.

[52] J. L. Warner, P. Anick, and R. E. Drews, "Physician inter-annotator agreement in the quality oncology practice initiative manual abstraction task," Journal of Oncology Practice / American Society of Clinical Oncology, vol. 9, May 2013, doi: 10.1200/JOP.2013.000931.

[53] R. Winnenburg, T. Wachter, C. Plake, A. Doms, and M. Schroeder, "Facts from text: Can text mining help to scale-up high-quality manual curation of gene products with ontologies?," Briefings in Bioinformatics, vol. 9, pp. 466-478, 2008.

[54] Z. Lu and L. Hirschman, "Biocuration workflows and text mining: Overview of the biocreative 2012 workshop track ii," Database, Nov. 2012, doi: 10.1093/database/bas043.

[55] H.-M. Müller, E. E. Kenny, and P. W. Sternberg, "Textpresso: An ontology-based information retrieval and extraction system for biological literature," PLoS Biol, vol. 2, Nov. 2004, doi: 10.1371/journal.pbio.0020309.

[56] D. Hanisch, K. Fundel, H.-T. Mevissen, R. Zimmer, and J. Fluck, "Prominer: Rule-based protein and gene entity recognition," BMC Bioinformatics, vol. 6, May 2005, doi: 10.1186/1471-2105-6-S1-S14.

[57] K. Van Auken, P. Fey, T. Z. Berardini, R. Dodson, L. Cooper, D. Li, et al, "Text mining in the biocuration workflow: Applications for literature curation at wormbase, dictybase and tair," Database, Nov. 2012, doi: 10.1093/database/bas040.

[58] H. J. Drabkin, J. A. Blake, and M. G. I. Database, "Manual gene ontology annotation workflow at the mouse genome informatics database," Database, Oct. 2012, doi: 10.1093/database/bas045.

[59] C. E. Jones, A. L. Brown, and U. Baumann, "Estimating the annotation error rate of curated go database sequence annotations," BMC Bioinformatics, vol. 8, May 2007, doi: 10.1186/1471-2105-8-170.

[60] C. Andorf, D. Dobbs, and V. Honavar, "Exploring inconsistencies in genome-wide protein function annotations: A machine learning approach," BMC Bioinformatics, vol. 8, Aug. 2007, doi: 10.1186/1471-2105-8-284.

[61] D. Devos and A. Valencia, "Intrinsic errors in genome annotation," Trends in Genetics: TIG, vol. 17, pp. 429-431, 2001.

[62] P. D. Karp, "What we do not know about sequence analysis and sequence databases," Bioinformatics, vol. 14, pp. 753-754, 1998.

[63] A. A. Morgan, Z. Lu, X. Wang, A. M. Cohen, J. Fluck, P. Ruch, et al, "Overview of biocreative ii gene normalization," Genome Biology, vol. 9, Sep. 2008, doi: 10.1186/gb-2008-9-s2-s3.

[64] D. P. Corney, B. F. Buxton, W. B. Langdon, and D. T. Jones, "Biorat: Extracting biological information from full-length papers," Bioinformatics, vol. 20, pp. 3206-3213, 2004.

[65] A. A. Namasivayam, A. F. Morales, Á. M. F. Lacave, A. Tallam, B. Simovic, D. G. Alfaro, et al, "Community-reviewed biological network models for toxicology and drug discovery applications," Gene Regulation and Systems Biology, vol. 10, pp. 51-66, 2016.

[66] F. Rinaldi, T. R. Ellendorff, S. Madan, S. Clematide, A. van der Lek, T. Mevissen, et al, "Biocreative v track 4: A shared task for the extraction of causal network information using the biological expression language," Database, Jul. 2016, doi: 10.1093/database/baw067.

[67] J. Szostak, S. Ansari, S. Madan, J. Fluck, M. Talikka, A. Iskandar, et al, "Construction of biological networks from unstructured information based on a semi-automated

curation workflow," Database, Jun. 2015, doi: 10.1093/database/bav057.

[68] B. Mons, H. van Haagen, C. Chichester, J. T. den Dunnen, G. van Ommen, E. van Mulligen, et al, "The value of data," Nature Genetics, vol. 43, pp. 281-283, 2011.

[69] E. Mina, M. Thompson, R. Kaliyaperumal, J. Zhao, Z. Tatum, K. M. Hettne, et al, "Nanopublications for exposing experimental data in the life-sciences: A huntington's disease case study," Journal of Biomedical Semantics, vol. 6, Feb. 2015, doi: 10.1186/2041-1480-6-5.

[70] A. J. Williams, L. Harland, P. Groth, S. Pettifer, C. Chichester, E. L. Willighagen, et al, "Open phacts: Semantic interoperability for drug discovery," Drug Discovery Today, vol. 17, pp. 1188-1198, 2012.